Materials and Methods for Industrial Wastewater and Groundwater Treatment

Materials and Methods for Industrial Wastewater and Groundwater Treatment

S. K. Nataraj
Jain University
Bangalore, India

WILEY

Published by John Wiley & Sons, Inc., Hoboken, New Jersey.
Published simultaneously in Canada.

For general information on our other products and services or for technical support, please contact our Customer Care Department within the United States at (800) 762-2974, outside the United States at (317) 572-3993 or fax (317) 572-4002.

Wiley also publishes its books in a variety of electronic formats. Some content that appears in print may not be available in electronic formats. For more information about Wiley products, visit our web site at www.wiley.com.

Library of Congress Cataloging-in-Publication Data Applied for:
Hardback: 9781119901501

Cover Design: Wiley
Cover Image: © exdez/Getty Images

Set in 9.5/12.5pt STIXTwoText by Straive, Pondicherry, India

Dedicated to Family
Ashwini-Anvita-Vrishank

Parents: Veeramma-Sharanappa

Contents

Preface

The Sustainable Development Goals (SDGs) set to achieve by 2030 have been adopted by United Nations with the consent of membrane states has brought back water-related issues to forefront. Water quality and sanitation (Goal 6) has been one of the main focus under SDGs, which is closely intertwined with other goals such as poverty; health; sustainable cities and communities; responsible consumption and production, heath, poverty reduction, and ecosystems and sustainable consumption and production; life below water, and life on land (Goals 1, 3, 11, 12, 14, and 15). These goals recognize a vial links between water resources, accessibility, quality, and key socioeconomic factors associated with sustainable environmental and development.

On the other hand, increasing industrial manufacturing pursuits, economical activities, rapid urbanization and depleting freshwater sources due to which world is concerned on impending water scarcity. In the last two decades, freshwater consumption on a global scale has doubled that of population growth. This drastic water consumption pattern have reached across communities, societies, and larger geographical locations. This extended water consumption not restricted to limited sectors or region are unsustainable. Thus, the drastic increase in water consumption has disrupted the traditional water demand and supply patterns thereby arise of uncertain future. However, immediate measures may intervene in the escalating situation and may change for good in a long run. Nevertheless, as of today, water consumption is increasing and demand for freshwater driven in population growth and urbanization where the higher per capita water consumption in growing, domestic, and industrial and/or socio-economic activities across the sectors.

In addition to this, poor water resource management policies, governance and inefficient implementation/development practices, unorganized as well as unscientific agriculture activities, and lack of water recycling efforts are also likely to diminish both quality and quantity of water supplies across the globe. While industries continue to produce and discharge wastewater as well as ground water level expected to deplete, which leads developing countries to experience more acute and pervasive water insecurity. It is also expected that some developed countries will also face strains on their water resources and supply chain. If this continues, regions and countries with poor water resource management policies are unlikely to prepare for future situation in which addressing water-related challenges might face collective challenges, including poor economic growth, growing inequality, less economical activities, poverty, deteriorating health and sensitization infrastructure, greater risk of disease, and risk of internal political instability.

World witnesses addition of 2–2.5 billion people to current population count by 2050. This will add up to the stress on food production, power generation, and water supplies. The efficient management of these interconnected nexus between water, energy, and food mainly depends on the water managers. Per capita water consumption across the globe considerably varies, in developed regions it is estimated to an average value of ~200 L per person per day. However, on an average, water requirement for human basic needs is estimated ~50 L per person per day at a global level. However, there has been a substantial reduction in water consumption recorded due to promising innovations and technologies adopted in developed countries in sectors like agriculture, domestic usage, municipal consumption, power generation, and so on. However, at larger global level, present water usage patterns do not reflect the sustainability in the face of increasing population growth and climate change scares.

On the other hand, by now, major freshwater source of water has been exploited and water that has been consumed is laying in the form of wastewater in treatment plants, discharge ponds, disposal zones, or in the open environment. Even though there have been efforts to recycle reusable quality of water, however, efforts are limited. It is estimated that at global level, an average of ~15% of wastewater is being recycled immediately after their utilization. At the same time, advanced techniques are being rapidly developed to improve water security through establishing long-term infrastructure construction for both desalination and/or wastewater treatment targeting various sectors and capacities. These efforts are also being made to minimize the energy consumption and impact on environments. Therefore, shifts have been made toward design and development of energy efficient, economical, high-performing desalination, and wastewater treatment technologies using cleaner energy such as solar-assisted, biofuels, wind, and nuclear energy that can significantly address increasing demand for portable water and generation of reusable quality freshwater from wastewater sources. Therefore, seawater desalination and water recycling from waste sources are increasingly becoming major issues for water managers and priority policy issue for governments at the international level.

In addition to this, today world is better positioned to address emerging crisis than ever. Today we have vast information, the knowledge, innovation that is taking place, the technologies that are being developed, and the economic prosperity that can be a vital resources to create infrastructure to manage existing water resources as well as impending water crisis much more efficiently and effectively than ever. The enormous scientific research undertaken across the sectors at global level ensuring systematic understanding of water crisis and measures suitable to address them which is continuously adding to our information/knowledge, new findings, and tools.

In this context, present book provides detailed account of various policies, guidelines and norms framed for water management and usage from various agencies and government across the world. Sections of the chapters also discuss in detail the recent advances that are taking place at global level to design, develop, and implement the global water resource monitoring information system to obtain and distribute the information needed for efficient water management. Also, various regulatory norms followed for various contaminants and their presence in various sources have been discussed in detail to provide a wider spectrum of information. These sections include the specific norms and guidelines for dangerous conventional as well as emerging pollutants that potentially cause serious threat to treatment techniques. For this purpose, various reports, public documents, publications, case studies, monitoring data, and national and international guideline documents on both water quality and treatment guidelines using appropriate technologies have been discussed. On the technologies front, timely, comprehensive, and forward-looking detailed information on various advanced technologies that involve for desalination are mainly phase change technologies including thermal desalination (humidification/dehumidification, solar chimney, multistate flash distillation, multi-effect distillation, vapor compressed distillation, electrodialysis, reverse osmosis, nanofiltration, forward osmosis, electrodialysis, electrodialysis reversal and electro-osmosis, functional polymers, magnetic materials, and various functional nanomaterials, whereas wastewater treatment processes involve physical treatment techniques (screening, grit separation, floatation, filtration, and so on), chemical treatment methods (advanced oxidation/reduction, ion-exchange, photocatalysis), and biological treatment technique (aerobic and anaerobic, biological filtration) processes, physico-chemical techniques like granular-activated carbon (GAC) adsorption and absorption, advanced membrane-based separation processes like forward osmosis, capacitive deionization, advanced ion-exchange methods precipitation, coagulation, flocculation, multi-flash distillation processes, and so on, which have been discussed in detail.

Nonetheless, even in the presence of advanced technologies, water resources will continue to experience newer challenges which will directly affect the environmental sustainability at large. These challenges include (i) increasing water scarcity, (ii) vulnerability to pollution and contamination, (iii) degrading ecosystem, (iv) deforestation, (v) continuing urbanization, (vi) excessive use of agricultural chemicals, (vii) disturbance to biodiversity, (viii) intense droughts and shifts in precipitation patterns, (ix) rising sea levels, (x) looming climate change impacts, etc. Nevertheless, collective efforts and innovative strategies can provide futuristic solution to achieve water and environmental sustainability. Some of these may include (i) integrated water resource management, (ii) watershed management, (iii) conservation and restoration of ecosystems, (iv) adaptation to climate change (v) sustainable agriculture, (vi) promotion of green infrastructure, (vii) educational outreach, and so on. This book is a small contribution toward understanding some of the aspects made in the field of water treatment and wastewater recycling mentioned here. In addition to giving detailed accounts of recent advances in materials development and process designs aiming water treatment and wastewater treatment, this book hope to raise awareness among young and next generation regarding the crucial aspects of conserving natural water resources, safeguarding groundwater, minimizing water consumption, recycling used water, discharge of wastewater and environmental sustainability. Educating individuals, researchers, and managers in the water sectors on recent advances and innovations in materials and technologies in water sector empowers to make informed choices about water use, wastewater disposal, and overall sustainable living practices.

Finally, I would like to take this opportunity to thank Wiley & Team for giving me an opportunity to publish this book. Special thanks to Ms. Summers Scholl, Executive Editor, and Ms. Kubra Ameen, Managing Editor, at Wiley for coordinating throughout the drafting of this manuscript. I also thank my team of research scholars at Sustainable Materials and Processes Lab, Centre for Nano and Material Sciences, Jain University for their constant queries which contributed to the betterment of the manuscript. Also, my sincere thanks to my family, especially my wife Ashwini Nataraj and my children Anvita Nataraj and Vrishank Nataraj for sacrificing their time and allowing me to work on this book without disturbance. My parents (Veeramma, Sharanappa), in-laws (Yashodha, Mallikarjun), and brothers (Girish, Santosh, and Harish) for providing me with the right atmosphere and encouragements.

S. K. Nataraj
Kadakolla-Bengaluru

1

Fundamentals and Importance of Water Quality and Treatment

1.1 Introduction

Water is an essential natural resource and vital commodity, yet its importance and value are understated. Water is essential for life on earth and without access to safe and clean water, flora and fauna do not survive. On the other hand, contaminated and unsafe water leads to diseases and illness. Safe and adequate quantities of drinking water in humans and animals help regulate body temperature; safeguard body organs and tissues; and keep body tissues like eyes, nose, and mouth tissues in a moistened state. Water acts as a carrier of nutrients and oxygen to different parts of the body and cells. The adequate presence of water in living beings lubricates joints, water helps in keeping of kidneys and liver healthy by flushing out waste products regularly, and so on. Water, being a universal solvent and its ability to dissolve a wide range of vitamins, nutrients, substances, and minerals, acts as a medium through which all essentials are transported in the bodies of living beings and organisms. Water also plays an important role in regulating and facilitating work of biomolecules and macromolecules like enzymes and proteins in living organisms. Generally, an average human body is composed of about 60% water. Therefore, water acts as a multifunctional and multitasking facility essential to transport and regulate bodily fluids that help in food digestion, nutrient absorption, bodily fluid circulation, creation of saliva, transportation and distribution of nutrients, and maintenance and regulation of body temperature [1–3].

In the larger context, the health of large water bodies such as lakes, rivers, seas, and oceans can affect the weather of an area and climate of a region. Water has unique physico-chemical properties like heating and cooling more slowly than land masses which regulate maintaining atmospheric as well as climatic conditions of a region. For instance, due to the temperature regulatory

Materials and Methods for Industrial Wastewater and Groundwater Treatment, First Edition. S. K. Nataraj.

properties of water, the coastal regions maintain cooler temperatures in summer and warmer in winter. Thus, physico-chemical properties of water help in creating more moderate weather and climatic conditions maintaining narrower temperature ranges. On the other hand, changes in temperature due to climate change in many areas increase the water temperature which causes eutrophication and excessive algal growth, which directly deteriorates the drinking water quality. Further, recent decades have seen increased industrial activities, high water consumption, and environmental irregularities in treating and managing water resources that are now showing extreme weather events directly affecting the natural resources. Cumulative effects of pollution, extreme weather events, and ever-increasing water consumption make water more scarce, unpredictable, polluted, unsafe, inaccessible, or all of these. In addition to this, the contaminants and pollutants in the water cycle impact life and environment throughout, which, in turn, threatens biodiversity, sustainable development, and access to water and sanitation. Nonetheless, continued climate change is disrupting weather patterns, leading to unpredictable and extreme weather events, thereby reducing the limited access to water, aggravating water scarcity and contaminating all natural resources and supplies. Similarly, limited freshwater and drought situations can worsen the effluent concentration runoff, chemical composition, COD, BOD, and pH of surface water [4–6].

On the other hand, a large population worldwide draws groundwater for domestic as well as drinking purposes. Generally, ground water sources have been perceived to be safe for domestic and drinking purposes. However, the quality of drinking water extracted from groundwater sources is now being increasingly compromised by elevated concentrations of fertilizer components, sediments, minerals, nutrients, and emerging chemical substances due to increased pollution and extreme storm events. Also, the excessive use of synthetic substances like fertilizers for agricultural purposes enters soil sub-layers and ultimately reaches groundwater sources. This type of water pollution through human, industrial, and agricultural activities leaves a series of man-made toxic substances, chemicals, and/or by-products into groundwater sources. Therefore, this leads to inducing toxicity to the groundwater. Also, the presence of multiple substances creates a series of intermediates and unknown substances with hazardous consequences [7]. Among many side effects, water pollution is known to induce imbalance in the water cycle, which leads to the alteration of weather patterns that may substantially contribute further to global warming activities or severe environmental deterioration at large. Therefore, it is important to devise, design, and implement water quality monitoring, treatment, reuse, and management strategies to cater future water demand. This will also help in documenting patterns of water contamination, surveillance source monitoring, and nature of contamination, which can effectively guide future water resources, and their protection, safety, and management [8].

Modern manufacturing techniques use complex and hazardous raw materials to produce large quantities of commodities. Also, large quantities of freshwater are being drawn from natural resources, most of which end up as wastewater at the end of the product recovery. In this situation, advanced water quality monitoring, treatment, and management play a vital role in ensuring drinking water quality for their safety for human consumption. Similarly, maintaining surface water quality is essential for wildlife and marine life. Among many parameters, pH, temperature, conductivity, dissolved oxygen (DO), total dissolved solids (TDS), chemical oxygen demand (COD), biological oxygen demand (BOD), suspended sediment (SS), metals, nutrients, hydrocarbons, synthetic substances, and industrial chemicals are the main indicators that determine water quality. In a larger context, direct discharge of wastewater into the environment and retaining it in reservoirs in the long term leads to rapid deterioration of ecosystems and environment at large. Therefore, it is important to monitor and determine water quality to understand environmental impacts. These strategies also help in generating, evaluating, and forming policies on environmental water quality that aim to provide the safeguarding mechanisms to protect water bodies, natural water resources, and freshwater reservoirs in the environment against the adverse chemical and biological effects through multiple chemical and biological contamination arising from anthropogenic, agricultural, and industrial diffuse emissions at the point sources as well as their reservoirs [9–12].

In this direction, worldwide efforts are being made to safeguard water quality and environmental health. As water is essential for life on earth, it is obligatory for every state and national governments to take appropriate measures within their authority to formulate and implement norms and regulations to ensure sustainable water resource management. The United Nations (UN) has long been advocating regulating sustainable utilization of resources, in particular water as essential for everyone. World regulatory bodies have recognized the right of every human being to have access to enough water (50 and 100 L) for personal and domestic uses per person per day. In this regard, the UN, through its several programs and initiatives, has emphasized addressing the global water crisis caused mainly due to lack of quality water to meet industrial, agricultural, and human needs. Accordingly, the World Water Development Report (2021) on "Valuing Water" evaluates the status and challenges in valuing water across different sectors. This report also gives wider perspectives and identifies ways and means in which valuation can be promoted as a method and tool to help achieve sustainability [13].

More importantly, in 2015, under the umbrella of the UN, world leaders agreed to focus on 17 global goals concerning each nation and every human being. Accordingly, the United Nations Sustainable Development Group (UNSDG) a consortium of 36 UN departments, funds, programs, specialized agencies, and offices conglomerate to play a role in sustainable global development. Among 17 UNSDGs, Sustainable Development Goal-6 (SDG-6) is to "ensure availability and sustainable management of water and sanitation for all." This goal further aims to achieve universal and equitable access to safe and affordable drinking water for all by 2030. In the sanitation section, it aims to achieve access to adequate and equitable sanitation and hygiene for all by 2030 and end open defecation, paying special attention to the needs of women. In the larger context, UNSG's 6th targets cover both water cycle and sanitation in all aspects and their achievement is designed in regular intervals. Further, access to safe and/or healthy drinking water is defined as the major percentage of the population having access to affordable drinking water with improved and clean sources. On the other hand, with increased stress on water resources to sustain an ever-growing population, increased demand from the irrigation sector, industrial supplies, and other needs causing water is increasingly causing scarcity to this essential commodity [14].

Today, due to dire water scarcity, the world needs revolutionary and transformative methods to conserve existing water resources and treat wasted or used contaminated water sources. Even though major portions of natural streams untouched by human activities have maintained good quality, increased human and industrial activities have contaminated several sources of groundwater to a greater extent [15]. Groundwater contamination mainly occurs through a series of regular human activities. Recent times have witnessed increased industrial activities which resulted in the release of humongous amounts of contaminated water to unregulated reservoirs which led to the seeping of toxic as well as hazardous contaminants into groundwater streams. Among many contaminants, groundwater resources are mainly characterized by the presence of man-made products like synthetic substances or chemicals, oil contamination, gasoline, and so on. Major source of ground contamination is from the materials diffused through the land's surface to make soil contamination at first and further end up in the groundwater. Nevertheless, groundwater is mainly characterized by three types of contaminants generally categorized as synthetic chemicals, sediment, and fecal coliform. In addition to human activities, natural influences like soil types and their compositions also play a role in groundwater contamination. This makes the overall process complex as the different chemicals in soil react differently and may transform in a variety of forms in soil, thereby in groundwater upon reaching the reservoirs. Among these major categories, the significantly farming chemicals, storage tanks, septic waste, landfills, uncontrolled hazardous waste, and atmospheric pollutants contribute majorly [16–18].

On the other hand, groundwater contamination can result in lack of quality water, loss of good quality water supply, degradation of surface water system, poor drinking water quality, high cleanup costs, high costs for alternative water supplies, and risk to human and animal health upon consumption. This situation also forces the water management and policymakers to drastically reduce water withdrawals, and consumption and devising suitable water treatment technology for eliminating hazardous chemicals from contaminated groundwater. However, recent studies have concentrated on documenting detrimental effects and new classes of substances that are known as emerging pollutants (EPs). With this, there have been several efforts in recent times that enlisted several petroleum-based organic compounds, plastics, and fuels, and recommended a series of measures and guidelines to extract, use, and dispose after use. Also, the studies have emphasized the fundamental understanding of water treatment processes, training human resources to handle impending crises, and understanding the basics of wastewater treatment and its usage parameters in advanced water treatment techniques like advanced oxidation process, membrane-based separation processes, the activated sludge process, advanced activated sludge treatment process, advanced aeration techniques, biological nitrogen removal, and so on. Advanced water treatment technologies have been increasingly used in processes that are capable of drastically reducing or eliminating specific contaminants from groundwater or wastewater that are normally not achieved through conventional treatment techniques. In case of groundwater contamination, the level of impurities will be at a trace level that may require sensitive as well as advanced analytical tools [19, 20].

The quality of groundwater is mainly determined by its natural, human, and industrial influences. In the absence of human and industrial influences, groundwater quality would be determined by the nature of bedrock minerals, deposits of minerals, atmospheric influences like evapotranspiration, and the nature of dust and salt by wind. Further, the nature of leaching of mineral deposits, salts, organic matter, and soil nutrients due to the hydrological factors into groundwater can alter the physical, chemical, physico-chemical, and biological composition of water. On the other hand, typically contaminated water quality is estimated and determined following a comparative evaluation of physical, chemical, and biological characteristics of a water sample in alignment with water quality guidelines or global standards. Interestingly, modern drinking water safety and quality guidelines and standards are formulated to enable the enforcement of the provisions of clean, safe, and healthy water for human consumption that helps improve human health. These guidelines and standards are designed based on scientific rationale methodologies acceptable across the spectrum. These globally acceptable norms are guided by regulations and universal standards issued by global organizations like the World Health Organization (WHO) [21]. Even though many developed countries follow their own specific standards in addition to international norms to be applied in their sovereign country. For instance, European nations follow the European Drinking

Water Directive and the United States follow the United States Environmental Protection Agency's (EPA's) standards that have a series of norms and guidelines through their Safe Drinking Water Act [22]. Similarly, China formulated its own drinking water standard GB3838-2002 (Type II) issued by the Chinese Ministry of Environmental Protection in 2002. On the other hand, the Indian Standard was adopted by the Bureau of Indian Standards for determining the quality of drinking water for identification, collection, and distribution [23]. On the other hand, those countries without legislative or administrative norms and frameworks for safe water standards and the global standards issued by the WHO design and publish international guidelines to be followed in comparison to other countries [24].

1.2 Nature and Sources of Conventional Groundwater Contamination

Water has been considered an essential commodity around the globe. Water in its safe and clean state is used for domestic purposes such as washing, drinking, cooking, irrigation, industries for the purpose of cooling machines, and raw material solvents among many others. Water supports all aspects of life from drinking, cooking, washing, cleaning, and other human activities. Even though a major portion of the earth is covered with water, unfortunately, there is a limited amount of freshwater (<3%) that can be utilized for human consumption and domestic purposes. In this situation, increased pollution is inducing deep freshwater stress that is accessible for the daily life of millions of people. On the other hand, increased exposure of freshwater resources like wells, ponds, lakes, rivers, and streams to a series of emerging contaminants renders them unsafe, hazardous, and unusable. Among many trillions of liters of water available on earth, two-thirds is stored and isolated in glaciers and ice caps and merely less than 3% is freshwater available for consumption. Interestingly, water can be a renewable resource. However, overuse and lack of efficient management or contamination with a series of pollutants reduces the quality and quantity of freshwater resources. This led to water scarcity, and recent estimates suggest that more than 2 billion people around the world lack access to sufficient quantity of safe and clean water [25–27].

Nevertheless, groundwater is a major source of freshwater resources for the global population. Groundwater has been extracted throughout the world for domestic, industrial, and agricultural uses. However, nearly one-third of the world's population depends on groundwater for drinking and domestic purposes [28]. Most of the developing countries with large arid and semi-arid regions solely depend on sources of groundwater and limited surface water sources. Nevertheless, major arid and semi-arid regions only find a secured supply of groundwater for drinking and domestic purposes through sustainable means. However, ever-increasing urbanization, modernized agricultural practices, rigorous industrial activities, and looming climate change are posing significant threats to surface quality of surface as well as groundwater sources. Among many contaminants, toxic metals, organics, hydrocarbons, trace heavy metal contaminants, microplastics, pesticides, nanoparticles, and other emerging contaminants are posing serious threats to human health, ecological systems, and the environment at large, which directly affect the sustainable socio-economic development [29–32].

The recent past has seen increased use of synthetic substances and millions of chemical compounds. This has led to the contamination of several natural resources significantly, which has shown increased threat in the past three decades. Chemical contamination in groundwater has reached alarming levels due to prior mentioned reasons. While increasing contamination of groundwater poses a serious challenge to human populations, which requires greater attention from researchers, innovators, technologists, industrial partners, decision-makers, and policymakers to formulate strategies and programs to protect both the quality and quantity of water resources.

Groundwater contamination is defined as the intentional or accidental inclusion or addition of undesirable and unwanted substances to groundwater due to human, agricultural, and industrial activities. In a broader definition, the most commonly occurring contaminants in groundwater can be categorized into three types, namely chemical, sediment, and fecal coliform. Other than direct addition of substances, type of soil and landfills, and nature of waste dump yards play a role in groundwater contamination. Other vital source materials that enter into water resources that turn into a contaminant in groundwater can be subcategorized into farming chemicals (fertilizers, insecticides, pesticides, additives, and so on); septic waste from domestic households and industrial outlets; landfills originated from various sources; uncontrolled domestic, industrial, and agriculture hazardous waste; waste storage tanks and reservoirs; and atmospheric pollutants. Further, residues that originate during agricultural activities, spreading of nutrient slurry; diffusion of excessive fertilizers, insecticides, pesticides, fungicides, herbicides, and human and animal waste, or discharges on the land that can turn into pollutants. Release of such domestic and animal waste is generally characterized by high concentrations of substances such as nitrates, phosphates, organics, and pathogenic contamination such as viruses, bacteria, and fungi, which seep into underground water sources over a period of time. These situations create an aggravated situation when the soil facilitates the spreading and diffusion of excessive synthetic as well as natural through their favorable interaction with different types of soil. For instance, sandy soils facilitate the spreading and diffusion of pollutants much faster than clay soils do. This leads to diffusion

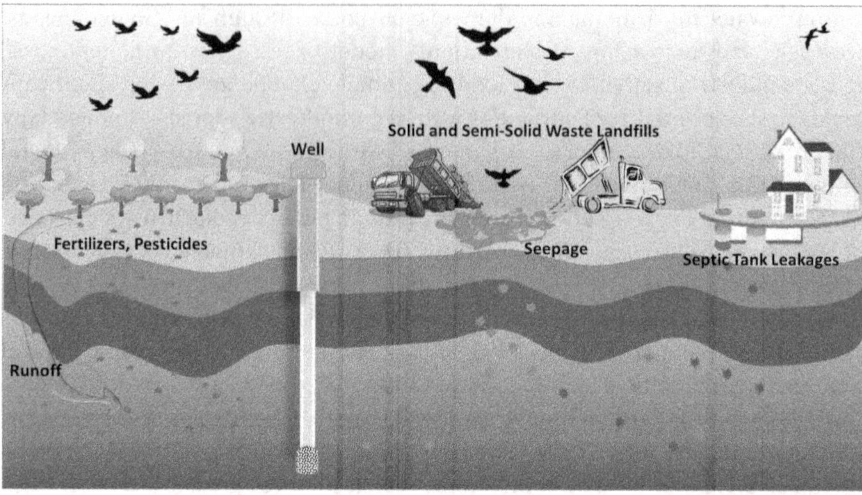

Figure 1.1 Schematic representation of different sources of groundwater contaminants originating from various sources like waste landfills, domestic wastewater, leakage from domestic septic tanks, diffusion of mineral deposits from earth's crust, and so on.

of fertilizers, chemical additives, viruses, bacteria, road salt, medications, and fuel. On the other hand, many of the pollutants in groundwater originated from geogenic origins as a result of dissolution of the natural mineral deposits over the period of time within the Earth's crust under the influence of excessive runoff or flood irrigation as shown schematically in Figure 1.1 [33–35].

On the other hand, hardness in water can be traced to percolation of limestone, chalk, or gypsum deposits that are largely composed of calcium, magnesium carbonates, bicarbonates, and sulfates. Among many low-solubility minerals, less to medium hardness can be characterized when the concentration of calcium carbonate is quantified near 150 mg/L. However, continued pollution due to rapid expansion of the global population led to uncontrolled and unregulated urbanization and industrialization, thereby inducing huge stress on agricultural production and the socio-economic programs. This is also posing serious challenges and contaminants have been known to induce negative impacts on natural resources due to their anthropogenic origin. Even though global policymakers and resource managers have devised a series of norms and regulations to tackle the issues associated with water pollution, the most severely affected population of resource pollution mainly belongs to those rapidly growing economies and middle-income countries. Therefore, it is important to document and design futuristic resource management guidelines considering the fate and consequences of natural resources in particular ground and surface contamination which directly affect living beings [36–39].

1.3 Emerging Pollutants

Modern definition of water pollutants and the nature of pollution is slowly and steadily shifting with the change in the nature of the contaminants and their properties. Futuristic pollutants have been now expanded with the inclusion of a broader list of substances, chemicals, and matters that are widely used in day-to-day life and have never been earlier considered a threat to humans or living beings or environment at large. Conventional pollutants or contaminants have been broadly categorized into physical, chemical, and biological contaminants. Conventional chemical and biological contaminants have been broadly identified with known side effects, toxicity, and end-use consequences. Among these biological contaminants such as pathogenic family, bacteria, and parasites have been known to induce serious health risks to humanity that date back to recorded civilization. Conventional pollutants have been well studied in the past for their origin, nature, properties, toxicity, and metabolic effects upon consumption by humans and living beings. However, a new set of findings reveals that the intermediary and transformative forms of conventional pollutants in water for a long period of time induce completely different properties and potential toxicity to water bodies and resources. This is causing a lot of anxiety and worries among researchers. The conventional pollutants transform into new intermediates with unknown physical, chemical, and physico-chemical properties. These unknown sets of offspring from the conventional pollutants have caused a serious threat to every possible water resource. Therefore, conventional pollutants and contaminants in their new outlook have raised a global concern [40].

Globally, today the most prevailing and prevalent problem water resources face in maintaining their quality, health, and usability is eutrophication. Eutrophication occurs as a result of high concentrations of nutrient loads such as phosphorus and nitrogen, which significantly impairs quality of water, thereby restricting direct use and health of water [4]. Natural eutrophication is a slow process that occurs as a result of accumulation of nutrients in lakes or other bodies of water. In this natural process, an increased concentration of nutrients feeds excessive algae growth which leads to unsightly scum on the surface of water, low penetration of light, and obstruction of air circulation, which decreases recreational value of water, which results in clogging of the water supply system. Eutrophication is increasingly considered a serious environmental threat in aquatic ecosystems since it often results in the degradation and deterioration of water resource quality and drastically decreases the DO in surface water resources. Once the concentration of dissolved water drastically decreases, eventually the aquatic system turns into a "*dead zone*" that is incapable of supporting aquatic life, which at large turns unusable. Major algal nutrients such as phosphorus and nitrogen originate mainly from leaked or discharged domestic sewage, which also increases microbial pollution, agricultural runoff, industrial effluents, and increased atmospheric pollution from fossil fuel burning, forest fires, and bush fires. This situation has worsened due to the increased human populations, excessive use and discharge of domestic wastewater, increased industrial activities with limited treatment measures, and agricultural runoffs due to flood irrigation that carries excessive fertilizers. Further, climate change and extreme weather patterns are threatening the natural water cycles and influencing the major alterations to the hydrological cycle [33, 41].

Increasingly, lakes and reservoirs have been exposed to increased amounts of synthetic and discarded materials. This makes them susceptible to the degradation and negative impacts of eutrophication as the complex dynamics of pollutants continuously change in relatively longer periods of time and many of these acts as substrates for relatively smaller pollutants. Among many nutrients, nitrogen concentrations give a broader indication for pollution of water in which an increase of nitrogen beyond 5 mg/L in water often indicates extent of pollution from domestic origin, particularly human and animal waste or fertilizer runoff from agricultural activities. Nevertheless, synthetic fertilizers and domestic sewage contribute to nutrient contaminants that majorly contribute to eutrophication. Overall, eutrophication results in uncontrolled algae growth, which prevents penetration of light in deep water, thereby causing deficiency of DO. The algal bloom also helps in acceleration of excessive water plant growth such as water hyacinth severely affecting reservoir-holding capacity, aquatic life, and overall ecosystem.

Another category of modern pollutants is termed persistent pollutants (PPs). The PPs, in particular persistent organic pollutants (POPs) are chemicals that endure or persist in the ecosystem or environment bioaccumulation through the food cycle and are known to pose a risk of causing adverse effects on health of living beings and the environment [42]. The POPs, in some cases, are also known as "forever chemicals" that the organic compounds remain in the environment and resist natural degradation through either biological, chemical, or photolytic processes. On the other hand, non-PPs (NPPs) are compounds that break down quickly in the environment. For instance, pesticides undergo natural degradation under the influence of external or environmental factors. Further, different classes of hazardous pesticides are termed cholinesterase-inhibiting pesticides that include organophosphates and carbamates. Nevertheless, both persistent pollutants and nonpersistent synthetic compounds and chemicals potentially tend to affect humans due to inherent toxicity immediately or within a few hours of contact. PPs tend to pose highly hazardous and poisonous effects over the long term of their presence in the ecosystem or environment. However, as per present estimation, nonpersistent chemicals like pesticides no longer pose a serious threat degeneration. Nonetheless, today PPs are known as a new class of substances and chemicals of global concern due to their potential toxicity, hazardous in their source, transportation, and use until their persistence in the environment. This is mainly due to their ability to biomagnify and bioaccumulate in ecosystems, and aquatic life which induce significant negative effects on human health and the environment. Today researchers are focusing more on PPs and POPs to study their potential toxicity as there is limited information and data available on their interaction and toxicological impacts on receptors in the environment [43].

On the other hand, new pollutants and EPs represent a new and emerging challenge to global water resource quality with potentially serious threats to humans and living beings' health, ecosystems, and environment at large. In water stress conditions, the need for good and safe quality water is essential and vital to sustain human well-being, rural and urban livelihoods, and a healthy environment for sustainable development. The major threat to sustaining good and safe quality water is a new class of EPs that can be understood in a general and broader definition as "any synthetic, man-made or naturally occurring chemical or any microorganism that is generally not monitored, regulated, and managed in the environment with potentially known, recorded, documented, or suspected adverse ecological and human health effects."

EPs originate from various anthropogenic sources like domestic households, agricultural activities, and industrial discharges and reach the environment in various forms and varieties as shown schematically in Figure 1.2. The EPs originated from various anthropogenic resources are distributed throughout aquatic ecosystems and environmental matrices on a daily basis. EPs are chemicals and compounds that have not been considered until recently and have been now classified or identified as

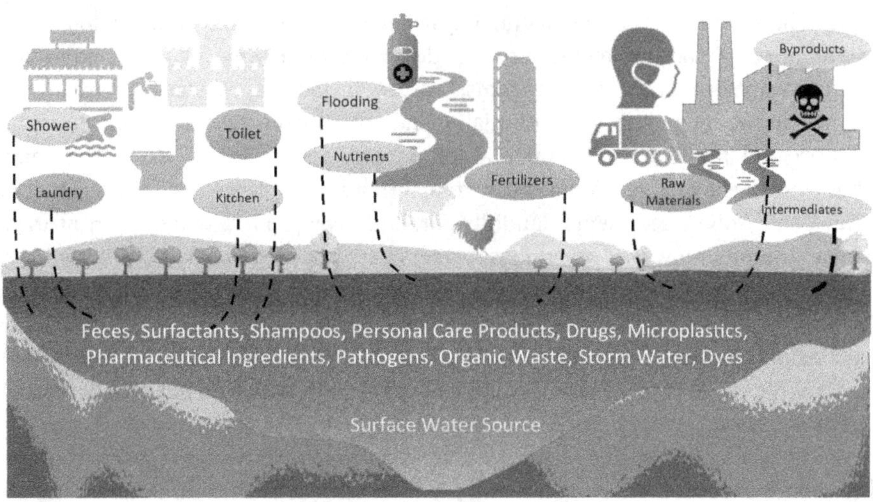

Figure 1.2 Schematic representation of origin of emerging pollutants, transportation of various man-made and synthetic substances through their excessive use which end up in lakes, reservoirs, and dams as a new class of pollutants.

dangerous to the environment, and human beings and living beings. By definition, these new classes of pollutants are labeled "emerging" due to their rising level of concern linked to an increasingly concerned list of substances that were ignored until now. EPs have been mainly identified as chemicals of a synthetic origin or man-made sources and may also be derived from a natural source and are increasingly categorized as public health risks that are yet to be established. EPs are generally traced back to their origins from household or domestic wastewater discharges containing surfactants that are enormously found in daily used soaps, shampoos, detergents, pharmaceutical ingredients, desecrated excessive drug molecules, pesticides, herbicides, personal care products, industrial discharges, agriculture runoffs, fertilizers, metals, industrial additives, solvents, industrial waste dumps, or landfills. Many of these man-made compounds, substances, and synthetic chemicals are used and discharged unregulated continuously into the aquatic resources and environment even in very low quantities on a daily basis. Among many drugs, pharmaceutical ingredients, pesticides, and organic compounds may cause chronic toxicity, disabilities, endocrine disruption, and disorders in humans and aquatic wildlife. Further, due to their long-term storage or holdings in waste reservoirs or landfills, EPs may develop bacterial pathogen resistance causing catastrophic consequences to the surrounding environment. Nevertheless, recent studies have shown increased interest and have made several attempts for the identification, detection, and analysis of EPs in both conventional concentrations and trace quantities in recent times. On the other hand, there has been great advancement in the development of analytical tools due to the continued synergic efforts by researchers, innovators, and technologists in improving conventional tools and refinement of specific techniques for detection of a wide array of emerging contaminants in aquatic ecosystems, natural resources, wastewater treatment plants, and so on both in terms of quality and quantity on emerging environmental concern and in various environmental components and biological tissues. It is noted that EPs may be mobile, continuously transformative, at the same time persistent in water, soil, solids, sediments, landfills, air, and ecological receptors even at trace concentrations. However, several researchers have engaged in advanced research to gather robust data on their origin, intermediates, transportation, fate, behavior in the environment, and impact on ecological sustainability and human health which are still lacking. Nevertheless, due to lack of data sets, case studies, and risk assessment reports, often risk of EPs cannot be affirmatively stated, which do not exist at present on the ecotoxicological significance [44, 45].

Another major emerging water quality concern is the impact of a series of excessive pharmaceutical and personal care products such as birth control pills, painkillers, surfactants, dyes, semi-volatile organic compounds, and antibiotics on aquatic ecosystems. Today very little is known about existence, effects, and risk of EPs in the long term on human health or ecosystems, although few of them have been known to exist in the list of conventional pollutants. For instance, some EPs like drug molecules or pesticides are believed to mimic natural hormones in humans and other living species. However, most of the newly found EPs have been traced natural sources of freshwater such as natural springs, seasonal streams, ponds, dams, rivers, rainfall, and boreholes. Today, there is limited access to freshwater resources available untouched by modern pollutants. But, the majority of population living in arid and semi-arid areas do not access clean and safe water due to the increased discharge of used water containing substances that are used in day-to-day life.

Most of the EPs of biological nature originate from the microbes and some of these biological EPs act as substrates with a conducive atmosphere to sustain and transform the other contaminants like farm pesticides, nutrients from fertilizers, raw sewage, and heavy metals in the water. Through agricultural activities, chemicals like nutrients from synthetic fertilizers, herbicides, and pesticides are washed away into runoff or flood streams, which seep into groundwater sources or join into surface water resources like streams and rivers that flow into ponds, lakes, and dams. Among many toxic substances, pesticides have, for example, severe adverse impacts on both human and environmental health. Some of the EPs like pesticides, herbicides, and insecticides may largely originate from agricultural activities confined to farm zones linked to river flow which carry them to a new class of toxic substances and chemicals downstream. Several reports have already recorded toxicity of pesticides are known to cause adverse human health complications such as cancer, behavioral disorder, mutation issues, reproductive health problems, and hormonal problems. Recent studies have also recorded the traces of pesticide residues in various lake and reservoir samples which are meant to supply freshwater to millions in urban and semi-urban populations. Therefore, presently most of the medium-to-large freshwater lakes and dams throughout the world are vulnerable to low-to-high concentrations of EP contamination. Nevertheless, the only way to prevent pollutant contamination in both groundwater and surface water is to regulate the excessive use of synthetic and man-made substances. Another way is to monitor and track the minimal use of agricultural synthetic nutrients in the form of fertilizers and pesticides. To prevent the entry of EPs, these substances and chemicals should not be applied or used close to water bodies [46–49].

1.4 Fundamentals, Properties, and Susceptibility of Water to Contamination

Water, in particular, showed a high affinity toward ionic compounds and other substances capable of forming through hydrogen bonds interacting with water molecules. Water molecule (H_2O) comprises one oxygen atom and two hydrogen

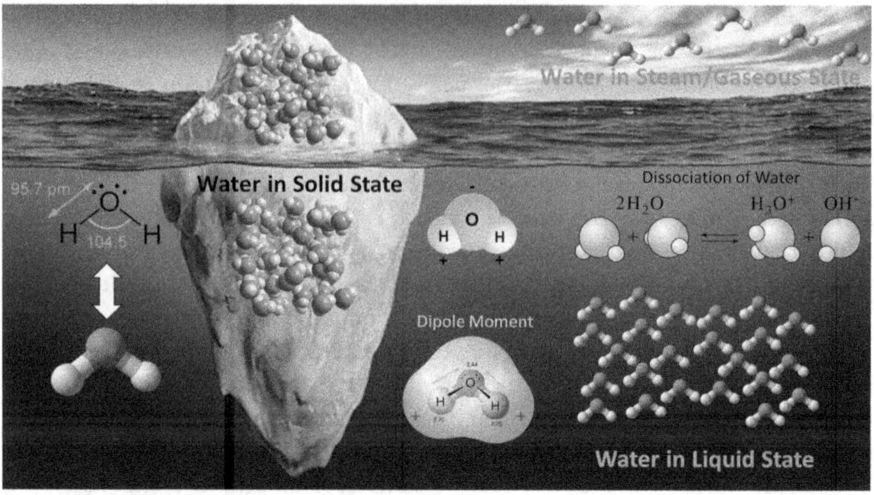

Figure 1.3 Illustration of water molecules in different states and their influence on physical, chemical, and physico-chemical properties of water in dissolution of different water-soluble substances or contaminants.

atoms in which an oxygen atom is connected with two hydrogen atoms via two covalent bonds as shown schematically in Figure 1.3. Water is a simple inorganic molecule comprising only three atoms. This tasteless, colorless, and odorless substance possesses unique properties which make it a very significant molecule vital to the existence of life on earth. However, the several significant physical, chemical, and physico-chemical properties make water a unique molecule. This exceptional importance of water molecules can be attributed mainly to their arrangement of atoms in the molecule and the bonds involved. The bond between oxygen and hydrogen is a polar covalent bond having a concentration of the electron density around the oxygen atom which is one of the most electronegative atoms. This specific arrangement in water molecules gives rise to a dipolar character in which the oxygen atom is the negative pole and hydrogen atoms are the positive poles. As a result, dipole character in water molecules enables atoms in water molecules to be connected among themselves by intermolecular forces known as "hydrogen bonds." On the other hand, hydrogen bonds are relatively strong in nature among intermolecular interactions in which a hydrogen atom from a certain molecule is connected to an electronegative atom such as nitrogen, fluorine, and oxygen of another molecule [50].

Even though hydrogen bonding can exist in many compounds such as NH_3 and HF; however, the main difference in hydrogen bonding in water molecule (H_2O) and other polar molecules is that each oxygen atom in water can form two hydrogen bonds, which is the same as the number of lone electron pairs on the oxygen atom. Due to this arrangement, water molecules are held or joined together through an extensive three-dimensional network in which each oxygen atom is approximately tetrahedral bonded to four hydrogen atoms, two by covalent bonds and two by hydrogen bonds [51]. On the other hand, oxygen and hydrogen in the water molecule possess electronegativity of 3.44 and 2.20, respectively. This broader difference in electronegativity polarizes each H—O bond, which influences the shifting of its electrons toward oxygen. These factors add as vectors to turn the overall water molecule polar. The dipoles in water make the overall molecule polar since it has a slight negative charge (oxygen) on one side and a slight positive charge (hydrogen) on the other, resulting in a net dipole. The solubility of water mainly arises from the polar nature of the water molecule, which helps other compounds and substances dissolve in water with ease. However, the dipole moment of a water molecule generally depends on its state of existence. In the solid-state form (ice), the dipole moment is $\approx$ 3.09 Debye (D), whereas in liquid water, it is $\approx$ 2.95 D, and in gaseous phase, it is $\approx$ 1.86 D, and these broader differences exist due to the difference in the nature of hydrogen-bonded environments. These extents of hydrogen bonding directly affect the solubility of a molecule, for instance sugars (sucrose) that have many polar oxygen–hydrogen (—OH) groups acquire overall higher polarity. Interestingly, the self-ionization of water, also known as auto-ionization of water, or auto-dissociation of water, is an ionization reaction in pure water or in an aqueous solution, in which a water molecule undergoes deprotonation by losing the nucleus of one of its hydrogen atoms to become a hydroxide ion, (OH^-). On the other hand, the hydrogen nucleus (H^+) immediately protonates neighboring water molecules to form hydronium (H_3O^+). It is an example of autoprotolysis and exemplifies the amphoteric nature of water, which makes water a good solvent to acquire many substances and chemicals within [52, 53].

Among many, five main properties of water, namely high specific heat, high polarity, low density as a solid, high heat of vaporization, and attraction to other polar molecules, can be used for a variety of applications. These properties yield water a name called a "universal solvent" as water can dissolve a larger number of substances than any other solvent or liquid found in nature. However, water cannot dissolve each and every substance. It is well documented that water only dissolves compounds or chemicals of polar nature. In this, the polarity of water molecules in which the hydrogen carries a positive charge, while the oxygen side carries a negative charge. Salt dissolves in water at the molecular level due to electrical charges as both water and salt are polar in nature with positive and negative charges on opposite sides in the molecule. The charge distribution on water molecules is vulnerable and gets attracted to a variety of other different molecules. The charge distribution on both hydrogen and oxygen in water molecules dissociates or separates ionic compounds into their negative and positive ions in which the positively charged species of an ionic compound is attracted toward the oxygen of water, while the negatively charged portion is attracted toward the hydrogen moiety of water. Even though water has been termed a universal solvent, water does not dissolve all substances or chemicals. For instance, hydroxides exhibit very low solubility in water and nonpolar substances such as organic compounds (oil, fats, and waxes) have low solubility in water due to the low solubility of these substances having oppositely charged globules when they are doped or mixed with the aqueous system [54].

With unique interactive chemical, physical, and physico-chemical properties, water becomes vital for all life and environmental processes on earth. Since age, water has been known to be an excellent solvent for many substances and has been widely used for day-to-day life.

1.5 Physical and Chemical Properties of Water

Water can exist in all three phases – solid, liquid, and vapor. When the water molecule transforms from one phase to another, distinctive variations in its physical properties occur. The molecular arrangement in the solid phase water (ice) leads to an increase in volume and at the same time decreases the overall density. However, expansion of the water molecule arranged in freeze condition makes ice float on top of liquid water. Among many other properties, water poses several other unique physical properties that are summarized in Table 1.1. Nevertheless, many of the water properties greatly depend on external influences such as temperature. Therefore, water property values have always been referred to for accuracy and dependency at different temperatures [51]. Among many physical, chemical, and physico-chemical properties, the table lists important parameters which influence the dissolution or carriage property of water such as (i) density, (ii) viscosity, (iii) surface tension, (iv) dipole moment, (v) specific heat, (vi) latent heat of fusion, (vii) latent heat of vaporization, and (viii) boiling point [55].

On the other hand, the colligative properties of water such as freezing point depression, boiling point elevation, vapor pressure lowering, and osmotic pressure are of central importance to various natural phenomena, processes, and driving forces for technological applications. Oceans contribute a total of 70% of earth's water and the human body comprises an average of 65% of water. Water conducts and regulates heat more easily in nature than any liquid except mercury. On the other hand, the specific heat is defined as the amount of energy needed to change the temperature of a substance, in this case water. Interestingly, water is characterized uniquely for its high specific heat, which helps in absorption of large amounts of heat energy before it begins to boil or get hot. However, upon cooling, water releases heat energy gradually. In a larger context, high specific

Table 1.1 Physical properties of water in different conditions and their significance.

Property	Temperature (°C)	Value	Effect and significance
Density	0	0.9999	Water molecules in solid (ice) are less dense than its liquid form. Therefore, ice floats at the surface of liquid water
	3.98	1.000 g/mL	
	25	0.9971 g/mL	
	100	0.9584	
Viscosity	0	I. 79 cP	Due to decreasing viscosity with temperature, fluidity increases, for instance, hot molasses flows much faster than cold molasses
	25	0.89 cP	
	100	0.28 cP	
Surface tension	0	75.6 dyn/cm	Ambient contamination results in a drastic decrease in the surface tension of natural or pure water.
	25	72.5 dyn/cm	
	100	58.9 dyn/cm	
Dipole moment		1.8546 D	Reflects the polarity of water molecule
Specific heat		1 cal/g/°C	Water can absorb a substantial amount of heat while its temperature rises only slightly
Latent heat of fusion	0	80 cal	
Latent heat of vaporization	100	540 cal	
Boiling point		100°C at 1 atm	

heat of water helps in controlling, regulating, and moderating of climate on earth. High specific heat of water also helps organisms regulate body temperature more effectively.

On the other hand, water pH is neutral in its pure form due to which water is neither basic nor acidic. However, when chemicals or substances are dissolved in water, it changes its pH. Interestingly, rainwater carries acidic pH (~5.6) as it carries naturally derived carbon dioxide and sulfur dioxide in its formation. Water can easily dissolve minerals and nutrients in groundwater flow, runoff, infiltration, and living organisms. Water has relatively high surface tension. Surface tension is defined as the property of the liquid surface that allows it to resist an external force as the cohesive forces in water dominate over adhesive forces in the water. Uniquely, water through its adhesive and elastic forces tends to aggregate in drops instead of spreading out over a surface as a thin film. This phenomenon results in water sticking to the sides of vertical solid surfaces or structures despite gravity that pulls downward. The surface tension of water is a very significant property in many industrial, biological, and environmental processes. Water activity is higher compared to other nonpolar liquids like oils with ~threefold, which is largely attributed to hydrogen-bonding capabilities of water and its dipolar interactions. The dynamic surface tension (DST) or interfacial tension (IT) is the value that refers to a particular surface age or interface age in dynamic mode. For instance, liquids such as surfactants (surface-active chemicals) can differ from the equilibrium value. In such a case, the formation of water drops which has a surface tension of ~72 mN/m (water at equilibrium RT conditions). However, freshly formed water surface can even show higher than 90 mN/m than under a dynamic condition with a relaxation process that occurs on a long-time scale ~1 ms. This variation in the surface tension of water in dynamic conditions is due to the adsorption effects of hydroxides or protons in water molecules. However, it is also important to know that a change in the pH does not significantly alter the DST. On the other hand, in a relatively longer time, it is unlikely that dipole orientation or hydrogen bonding properties of water molecules may not influence the altering DST of water [56, 57].

1.6 Types of Water and Contaminants in Water

Water molecules are extremely polar and formed as a result of hydrogen bonds. The main five physico-chemical properties like high polarity, high heat of vaporization, low density as a solid, high specific heat, and extreme polarity of water molecules attract other polar molecules. The water is amphoteric in nature, which implies that water acts as an acid and a base. Amphoteric nature of water molecules makes it both a proton acceptor and a proton donor. On the other hand, deionized water undergoes auto-protolysis in which ionization of water results in deprotonates in which protons are removed in order to form a hydroxide ion OH^-. At the same time, the hydrogen nucleus H^+ immediately accepts protons protonated with another water molecule to form a hydronium ion, H_3O^+. For example:

$$H_2O + NH_3 \Rightarrow OH^- + NH^{4+} \text{ (acid)(base)(conjugate base)(conjugate acid)}$$

$$H_2O + H_2S \Rightarrow H_3O^+ + HS^- \text{ (base)(acid)(conjugate acid)(conjugate base)}$$

In general, natural water in various sources, both underground and surface reservoirs, may contain low to high amounts of dissolved nutrients, minerals, or ions, such as calcium and magnesium. Based on the concentration and the extent of dissolved substances in water, water can be classified as (i) soft water and (ii) hard water. Soft water is generally more found in surface water resources that contain very low concentrations of ions, minerals, or nutrients. The main sources of soft water are rainwater, freshwater streams, and distilled water. On the other hand, hard water contains medium to high concentration salts of magnesium (Mg^{2+}) and calcium (Ca^{2+}) as chlorides, hydrogen carbonates, and sulfates. Soft water is largely free from salts of magnesium and calcium. The hard water can be further subdivided into two types, namely temporary hardness in which hardness is due to the presence of magnesium bicarbonate, $Mg(HCO_3)_2(aq)$ and calcium bicarbonate, $Ca(HCO_3)_2(aq)$. Temporary hardness can be easily removed by boiling. Simple heat treatment of temporary hardness through water can remove the dissolved salts in which boiling promotes the formation of carbonate precipitate from bicarbonate salts dissolved, which can be separated. Permanent hardness occurs due to the presence of sulfates, chlorides, and nitrate salts of calcium and magnesium. Permanent harness is rigid and cannot be removed either by heat treatment or conventional distillation processes.

Due to their forms and nature, water can be described with some important nomenclatures such as (i) Demineralized water which is free from all types of nutrients, minerals, salts, and only consists of H^+ and OH^- ions i, and (ii) Heavy water (D_2O) is made up of deuterium (an isotope of hydrogen) which is not suitable for drinking and general domestic use purposes because it lacks the ions required for proper body growth. However, heavy water has several commercial and industrial important applications, for instance, D_2O is used as a moderator in nuclear reactors. Heavy water cannot be used for drinking purposes because it slows down the metabolic process in the body [58].

In a solution, the general term "concentration" refers to the particular amount or standard quantity of solute in a solvent. Concentration may be expressed or represented in different ways depending on the state of solute, the amount of solute, and the amount of solvent and solution. The molar concentration or (molarity, M) of a solution is the number of moles of the solute species that is contained in 1 L of the solution (mol/L). Concentration of solute in water can also be represented in percent concentration that can be expressed as mass percent (w/w%, percentage of the mass of solute over the mass of solvent). However, solute concentration in solvent can simply be expressed in units of mass per volume in gram/liter (g/L) weight/liter (mg/L). Nonetheless, the solute that dissolves in water can be divided into two broad classes: (i) electrolytes and (ii) nonelectrolytes. An electrolyte is a substance that dissolves or dissociates in water to give rise to ions that make the solution an electrically conducting solution such as ionic salts, NaCl, KCl, $MgCl_2$, $MgSO_4$, and so on. A nonelectrolyte is a substance or chemical that dissolves in water to give a nonconducting end solution such as sucrose, table sugar, and so on.

Water facilitates or undergoes mainly three types of chemical reactions, namely (i) precipitation, (ii) oxidation–reduction, and (iii) acid/base reaction. In a precipitation process, a chemical reaction between aqueous and non-aqueous media where two ionic bonds combine and the insoluble salts that are formed. As a result, constituent species separate out as insoluble salt known as precipitate as a by-product and the reaction is called the precipitation reaction. For instance, a typical precipitation reaction that occurs in aqueous media is the formation of calcium carbonate solid when calcium solution is mixed with solution of carbonate as shown in the following reaction.

$$Ca^{2+}{}_{(aq)} + CO_3{}^{2-}{}_{(aq)} \rightarrow CaCO_{3(s)}$$

Oxidation–reduction reaction on the other hand is an electron transfer reaction between species or in a chemical reaction that the oxidation number of an atom, ion, or molecule changes by losing or gaining an electron. Redox reactions are common and significant to many of the natural functions of life and environment such as photosynthesis, corrosion or rusting combustion, and respiration. When the substance loses the electron, it is oxidized and is called a reducing agent. Whereas, when the substance accepts the electron, it is reduced and is called an oxidizing agent. Water in vapor form or moisture plays an important role in initiating oxidation reactions. For instance, iron pipe corrosion in which iron metal oxidizes and loses two electrons whereas hydrogen ions accept electrons and reduce to hydrogen gas.

$$Fe \rightarrow Fe^{2+} + 2e^-$$

$$2H^+ + 2e^- \rightarrow H_{2(g)}$$

Acid–base reactions occur when acids and bases undergo a reaction upon mixing. Among many definitions, acids can be termed as proton donors and bases can be termed as proton acceptors. The neutralization reaction in which a strong acid and a strong base react, in water it involves hydrogen (H^+) and hydroxide (OH^-) ions

$$H_3O^+ + OH^- \rightleftarrows 2H_2O.$$

whereas neutralization through the reaction of a weak acid and a weak base is more appropriately known to involve direct proton transfer from the acid to the base. Therefore, the neutralization reaction between an acid and a base result in an ionic compound and water. The ionic end compound produced as a result of a neutralization reaction is known as a salt and overall net equation for most acid–base reactions can be represented as:

$$H^+{}_{(aq)} + OH^-{}_{(aq)} \rightarrow H_2O_{(l)}$$

Nevertheless, the pH of a solution is a measure of hydronium ion (H_3O^+) concentration, which is considered a direct measure of acidity ($pH < 7$ acidic, $pH > 7$ basic).

Unless it is wastewater or contaminated water, the pH range in water is rarely below 4 or above 10. Therefore, pH of water is generally measured and analyzed for its nature and type of contamination and is essential for many chemical and biological processes in water.

1.7 Alkalinity in Groundwater and Surface Water

Alkalinity refers to the ability of water to neutralize acid in solution form. Today, water of alkaline nature is in high demand and water treatment technologies are continuously researched and developed to provide safe water for drinking purposes. The alkalinity occurs due to the presence of traces or impurities of potassium bicarbonate, sodium carbonate, or potassium carbonate in water. In many parts of the world, rainwater is characterized as alkaline in nature. The alkaline nature of water also occurs due to the buffering of acidity through the aerosols derived from soil that is rich in Ca and Mg. On the other hand, alkalinity is much greater than total hardness of the system which indicates that water need to be treated before using for drinking purposes through a softener. Stream water having relatively high alkalinities poses a greater possibility to neutralize acidic contaminants from wastewater or rainfall water which resists major alterations in pH. On the other hand, low alkalinity water is highly susceptible to changes in pH parameters.

Total buffer species in chemistry refers to the sum of acidic species form of the buffer and the conjugate base. The buffering species in water that induce alkalinity are mainly the base anions carbonate (CO_3^{2-}) and bicarbonate (HCO_3^-), even though hydroxide (OH^-), silicates, ammonium, phosphates, borates, sulfides, and organic ligands may also contribute to the alkalinity of water. Interestingly, alkalinity not only helps in regulating the pH of a water body but also manages the metal content. For instance, carbonate and bicarbonate ions in water can be used to effectively remove hazardous heavy metals, such as cadmium and lead via precipitation of metals out of aqueous solution. On the other hand, alkalinity in groundwater and natural surface water primarily occurs due to the dissolved carbonate minerals, and some contribution from CO_2 available in the soil and atmosphere above the water table. Nevertheless, mainly three carbonate species, namely CO_3^{2-}, H_2CO_3, and HCO_3^- contribute to total alkalinity of water. However, their relative concentrations and proportions depend on temperature and pH of water. Among all, dissolved bicarbonate (HCO_3^-) at neutral values of pH = ~7 is known to have a major contribution to alkaline water. However, at pH = ~9, the contribution from CO_3^{2-} and other anions plays a dominant role in their form and proportions [59, 60].

However, in surface water and streams with high acidic nature, dissolved CO_2 is dominant in the form of H_2CO_3. Carbon dioxide in water facilitates important reactions, which thereby controls the pH of natural water. Nevertheless, the realization of equilibrium between three carbonate species and H^+ and OH^-, which regulate pH is relatively rapid. However, equilibrium flowers and alters its reaction rate in stream water due to the variation in gaseous CO_2 in the atmosphere, which may not exist in equilibrium at all times. However, when surface water through soil enters streams, excessive free CO_2 gradually liberates and enters the atmosphere, which helps in equilibrating water through the outgassing effect of atmospheric CO_2 and its concentration balancing. Therefore, stream water at low-order has potential to have greater concentrations of dissolved CO_2 than high-order streams. Interestingly, alkalinity has an inverse relation with the total hardness in which lower alkalinity elevates total hardness of nitrate, chloride, or sulfate significantly. Also, low concentrations of levels of alkalinity with <150 mg/L are potentially more corrosive, which can also be termed as the sum total of all titratable bases down to pH = ~4.5–4. Nonetheless, the significant contribution to alkalinity of water is from the carbonate species and free H^+ or OH^-. The total hydrogen ion (H^+) which is absorbed by water containing mainly carbonate species can also be termed alkalinity which is represented as $HCO_3^-+2CO_3^{2+}+OH^--H^+$. In most natural water situations (pH 6–8), the OH^- and H^+ are negligible. Therefore, alkalinity is $HCO_3^-+2CO_3^{2+}$ and it is to be noted that CO_3^{2-} is multiplied by 2 because it can accept two protons and the reaction can be represented as [61]:

$$H_2CO_3 \rightarrow H^+ + HCO_3^- \rightarrow pKa1 = 6.35$$

$$HCO_3^- \rightarrow H^+ + CO_3^{2-} \rightarrow pKa2 = 10.33$$

From this, it can be inferred that below pH 4.5, every carbonate species available in water is in the form of H_2CO_3 and the alkalinity results in negative and at pH 7.5–8.3, all carbonates are in the form of HCO_3^-. Also, alkalinity can be equated to HCO_3^- above pH 11.5, all carbonates are CO_3^{2-}, and alkalinity overall can then be equated to $2CO_3^{2+} + OH^-$, even though concentration of alkalinity is represented in parts per million (ppm) and weight per volume (mg/L). However, by convention, alkalinity is not expressed in molarity.

1.8 Hardness of Ground and Surface Water

Hardness of water can be defined as concentration or contamination (in this negative connotation) of high amounts of dissolved magnesium and calcium. In general, hard water in the context of drinking water is not a health risk. However, it can be considered a nuisance due to the build-up of minerals (Ca and Mg) on plumbing fixtures or pipelines. On the other hand, hardness of water can be related to high amounts of calcium and magnesium. However, harnesses can also be referees to some time to iron content in the water. Therefore, hardness of water technically has been defined as the sum total of all polyvalent cations dissolved in water. Thus, overall concentration of hardness is estimated and quantified by adding up the concentrations of both calcium and magnesium and converting the sum of it to an equivalent concentration of calcium carbonate ($CaCO_3$) in milligrams per liter (mg/L) of water. The guidelines for different water management and regulating authorities targeting drinking water quality specifically define the hardness of water. Hard water used for domestic purposes requires extra soap and detergent for laundry and significantly contributes to scaling in boilers, pipelines, and industrial equipment. Even though general terms are used to describe hardness based on the poor detergent and/or soap performance; however, specific guidelines for classification of different types of hardness in waters are based on the equivalent concentration of calcium carbonate (Table 1.2) in which (i) soft water carries 0–60 mg/L, (ii) mild or medium harness is between 61 and 120 mg/L, (iii) hard water is categorized for calcium carbonate having concentration of 121–180 mg/L, and (iv) very hard water with minerals having calcium carbonate concentration equivalent or more than 180 mg/L [62].

Water is a universal solvent that can easily dissolve various types of minerals, nutrients, and salts both as impurities and saturated solutions. The optimum and safe range of hardness in drinking water is from 80 to 100 mg/L, greater than 200 mg/L harness is considered poor for direct drinking, and greater than 500 mg/L is categorized as unacceptable for domestic purposes. In many parts of the world, groundwater is estimated to be harder than surface water; in many regions, it is quantified to be greater than 1000 mg/L [63].

Among two types of hardness, carbonate temporary hardness can be easily removed from water by heat treatments, which results in precipitation. Upon heating, temporary hardness contaminants in water separate insoluble carbonates and most of the time precipitate deposits on container walls or form deposits in hot water

Table 1.2 Different types of hardness in water.

Type of hardness	Equivalent concentration of $CaCO_3$
Soft <60 mg/L	Soft <60 mg/L
Medium hard	60 mg/L–<120 mg/L
Hard	120–<180 mg/L
Very hard	180 mg/L or greater

heaters. Carbonates are the main constituents that form salt having lower solubility in water result in a condition refers to total hardness or alkalinity. Noncarbonated hardness referred to as permanent hardness is the total hardness in excess of the alkalinity. Therefore, in case alkalinity is equal to or greater than the total hardness, it means there is no noncarbonated hardness present in the water [64].

Water hardness in groundwater mainly occurs from natural weathering of limestone, calcium-bearing minerals, and sedimentary rocks. Interestingly, the major source of sweet or soft water is known to originate in regions or geographical areas having predominantly igneous rock formations. On the other hand, regions or geographical formations with sedimentary rocks are found to induce high concentrations of hardness in groundwater. Hardness in groundwater also occurs due to the leaching and seepage of synthetic chemicals, mining industry activities, and release of effluent or excessive application of fertilizers and lime to the soil in agricultural activities. In addition to chemical, physical, and physico-chemical properties, commercially, hard water poses a serious aesthetic concern, which is mainly due to unpleasant taste, poor lathering, scaling supply systems, blocking plumbing fixtures, and induced corrosion as a result of high concentration of Ca and Mg ions in water. Soft water, on the other hand, can also result in pipe corrosion and result in improved solubility of heavy metals such as zinc, lead, copper, cadmium, and excessive lime and nitrate where fertilizers are used for agricultural purposes. Typical groundwater contamination, chemical constituents, and their concentrations can be summarized as shown in Table 1.3 [65, 66].

Table 1.3 Nature of chemical constituents in natural groundwater extracted for drinking and domestic purposes.

Chemical constituent	Concentration (mg/L)
Bicarbonate Carbonate Calcium	150–200
Magnesium	25–50
Sodium	60 (irrigation) 20–120 (health)
Iron	Less than 3
Manganese	Less than 0.05
Chloride	250
Fluoride	0.7–1.2
Sulfate	300–400 (taste) 600–1000 (laxative action)

Permanent hardness in water has been removed using advanced water treatment methods such as ion exchange or oxidizing filters and reverse osmosis (RO). Conventional ion-exchange method uses an active exchangeable site on polymer or inorganic resins which can be used to exchange contaminants like calcium and magnesium ions from sodium or potassium ions in resins. However, the ion-exchange method has limitations in which the process increases sodium ion concentration in water.

1.9 Wastewater and Types of Wastewaters

Broadly and philosophically, wastewater can be defined as "a natural resource of water in the wrong place or with the company of wrong constituents." Wastewater contains waterborne waste materials of different natures and origins including human activities, domestic activities, communities, waste matter from industrial and agricultural activities, commercial establishments, and so on. In general, the main constituents and composition of wastewater are more than 90–99% water and the remaining 1–9% is waste or undesired constituents originating from various origins. The wastewater constituents other than water could be organic matter, inorganic compounds, biological constituents, microorganisms, plastic waste, oil contaminants, and so on. Generally, wastewater effluents or discharges are released either after partial treatment or untreated to a variety of environments, such as streams, lakes, ponds, estuaries, rivers, and oceans. On the other hand, to minimize the adverse effects of wastewater's direct release into the natural environment, artificial reservoirs are created to store before the treatment of wastewater. These types of reservoirs are artificial lakes created close to either wastewater treatment plants or industrial hubs, or residential areas, and so on. The main purpose of artificial reservoirs is to retain wastewater for easy management where contaminated water is collected. This helps in minimizing mixing of wastewater in rainwater, rivers, lakes, and other natural reservoirs, instead helping in the water treatment process [67, 68].

Wastewater can be mainly categorized into three types, namely (i) domestic wastewater, (ii) industrial wastewater, and (iii) storm sewage. Nevertheless, recent studies have also categorized runoff out of agricultural activities as one of the major wastewater categories. (i) Domestic wastewater or municipal sewage carries used water mainly from human activities involving houses, apartments, community, and municipal discharges. Due to its origin and nature domestic wastewater is also called sanitary wastewater or sewage. In particular, domestic wastewater originate from human living processes such as bathing, laundry, toilet flushing, dishwashing, etc. On the other hand, commercial domestic wastewater originates from commercial and nondomestic sources, such as taxidermy, beauty salon, furniture refinishing, household instrument cleaning, or auto-body or mechanical repair shops. (ii) Industrial wastewater on the other hand is the aqueous discard comprising industrial raw materials, traces of intermediates, by-products, residual products and substances that are of low-to-no commercial value in industrial manufacturing having been suspended or dissolved in outlet water. Most of the industrial wastewater is typically generated in the process of excessive use of water, during manufacturing process or during cleaning activities which takes place simultaneously or post product manufacturing process. The major sources of industrial wastewater are commodity chemical production units, food industry, electric power plants, battery manufacturing, mining industry, textile industry, dyeing industry, metal processing industry including steel, iron and aluminum processing units, dairy industry, pharmaceutical manufacturing, metal working, mines and quarries, coal-fired plants, nuclear industry, fossil-fuel power stations, oil and gas extraction, petroleum refining and petrochemicals, pulp and paper industry. From these industries, many manufacturing industries discharge medium to large quantities of wastewater with medium to high concentrations of contaminants like metals such as mercury, lead, chromium, cadmium, arsenic, selenium, and nitrogen compounds in various forms and types such as nitrates and nitrites. Other major constituents in wastewater include drug molecules, organic contaminants, oil globules, radioactive elements, ionic species, microplastics, pathogens, and so on. (iii) Storm wastewater runs as drain water that contaminates and carries surface water runoff from freshwater channels, rainwater, melting snow, and overflow from agricultural supplies and lawn water. These effluents convey these runoffs to water bodies such as lakes, rivers, and catch basins [69, 70].

On the other hand, agricultural wastewater comprises mainly residues of fertilizers, dissolved organic substances, chemicals from various operations, pesticides, nutrients, and debris from mechanical processing of harvests, food grains including peeling and trimming, and hydrodynamic forms due to impacts in washing and transporting. Agricultural wastes in water also include residues of weeds, sawdust, leaf litter, crop residues, forest waste, livestock waste, and so on. Many parts of the world, especially developing countries, practice stubble burning or combustion of agricultural solid wastes indiscriminately as an immediate solution. However, agricultural waste dumped is also practiced, which ends up in floods, rainwater runoff, and flood irrigation runoff. All these activities carry agricultural solid waste into public places, thereby resulting in the generation of large pollutants in water, air, and soil contamination. Further, fine traces of solid waste end up in water sources through various channels, thereby polluting the aquatic ecosystems.

1.10 Types and Nature of Contamination in Wastewater

With all these issues, it is important to extend discussion to larger contaminants available not only in groundwater but also in surface water resources and their origin from various artificial, human, and industrial sources. Most of the contaminants originated from synthetic or human activities have been known to pose serious health and environmental risks. Nonetheless, the most immediate and detrimental effects of both conventional and EPs in wastewater contribute severely to the environment. These contaminants in wastewater from various types of nature also contribute toward destruction of natural ecosystems, habitats, and the wildlife that lives in these habitats by exposing them to harmful chemicals, matter, and substances that would otherwise not be present in nature. Water quality and its analysis, characterization, management, and treatment strategies are mainly devised based on the type and nature of pollutants present in the wastewater. Therefore, the concentrations of organic, inorganic, and composite substances dissolved in wastewater and their concentrations need to be regularly monitored and documented prior to designing appropriate treatment techniques [71, 72].

On the other hand, advanced wastewater treatment infrastructures and facilities are designed and operated to speed up the natural process of separating contaminants and purifying water. With billions of population, thousands of industries, and millions of hectares of agricultural land activities use, discharge of used water is more complex and task even more humongous. Without appropriate wastewater treatment infrastructure in place, the amount of wastewater produced would cause catastrophe. This is especially true in the case of developing and widely industrialized countries in which utilization of freshwater is large and the treatment of used water is not taken so seriously. As of today, it is estimated that globally over 80% of all wastewater produced, disposed, and discharged into the environment is either partially treated or without treatment. Countries that are underdeveloped, developing, and heavily industrialized have water treatment facilities that are conventional in nature and can only handle a limited number of contaminants at a time. However, with limited facilities and infrastructure in place, they use various techniques and methods to treat water with one simple common goal, which is to purify the wastewater as much as possible and dispose of it into the environment to keep humans and the environment safe and thriving [73].

Conventional wastewater treatment facilities and infrastructures are designed to help purify the water by eliminating toxic, hazardous, and potentially life-threatening contaminants. However, new classes of EPs generally escape or bypass the conventional treatment techniques. Potentially, these are some concerns global agencies constantly monitor to eliminate their future threat. Partially treated, unclean, and unsafe water poses enormous health and environmental risks due to which around 1.7 million deaths occur annually of which >90% are in developing countries [74, 75]. This makes researchers and technologists frame novel water treatment techniques for effective removal of all contaminants from wastewater.

The main aim of establishing and maintaining wastewater treatment infrastructure and facilities is to protect humans and conserve the ecosystem from toxic and harmful contaminants found in wastewater. These infrastructures and facilities are managed in several stages and broadly can be classified into four major categories. Among these are monitoring, management, regulating, and advisory agents around the globe who constantly monitor and alter water quality issues from time to time in the larger interest of humans, wildlife, and environment. In this direction, an important paradigm shift is taking place at multiple levels to innovate and advance sustainable water treatment techniques toward a circular economy in which wastewater is considered an invaluable resource rather than a simple waste or liability.

Today sustainable water treatment techniques are designed to maximize resource recovery and value-added product recycling. Among many resource recoveries, clean water, energy, nutrients, fertilizers, minerals, and precious metals can be extracted from wastewater which can directly fulfil the Sustainable Development Goals (SDGs) set by the UN. Also, the World Bank (WB) launched one of the circular economy initiatives terming "Wastewater: From Waste to Resource" in 2018 to address the future wastewater challenge and propagate awareness among future generation, general public, policymakers, and decision-makers about the potential threat of wastewater and treat waste as a new resource. The global and national agencies have also initiated formulating and designing broader as well as specific guidelines to improve the planning, formulating, design, management, resource gathering, and financing of wastewater treatment. Although water recovery is the primary goal in most of the water treatment methods, however, modern and advanced initiative are expected provide resource recovery, and promotes much-needed measures to make realistic approaches in water recycling [76].

Before stepping into a detailed understanding of nature and types of contaminants, it is important to enlist the parameters for safe and clean water useful for drinking as well as domestic usage. According to global agencies, the definition of safe drinking water refers to water that does not represent or induce any significant risk to human's and living being's health over a lifetime of consumption, which also includes different sensitivities that may occur between life stages. Regulatory agencies around the globe have set maximum more or less similar permissible levels in a variety of water sources in their composition and properties to be safe for consumption. These include color, taste, temperature, acidity/alkalinity (pH), dissolved solids (specific conductance), DO, particulate matter (turbidity), hardness, and SS.

However, most of the wastewater or contaminated water is characterized by its various parameters in various forms and levels of presence. Major health concerns worldwide mainly arise from waterborne diseases. Among various affected, the greatest risk of waterborne disease is among infants and young children, the debilitated population, and the elderly who are vulnerable and living under unsanitary circumstances. Therefore, safe drinking water is essential for day-to-day life and conducts uninterrupted domestic purposes. Nevertheless, specific guidelines have been formulated and monitored on a regular basis to maintain the water quality. These guidelines are not only applicable for drinking water purification and wastewater treatment but also applicable for packaged water for human consumption. However, higher water quality may be desired for some special purposes, such as pharmaceutical industry applications, renal dialysis, cleaning of contact lenses, food processing and production applications, and so on. Nonetheless, the general guidelines may or may not be suitable for the larger protection or conservation of aquatic life or general industrial applications. Therefore, water quality is closely monitored across the spectrum. In general, types of water pollutants, contaminants, or parameters are broadly categorized into several large sections and specific sub-sections including physical, chemical, and biological contaminants. In further subsections, contaminants are classified into synthetic organic chemicals, inorganic substances, sediments, plant nutrients, pathogenic organisms, microplastics, oxygen-demanding wastes, radioactive substances, oil, heat, and so on. Further details are discussed in the following sections.

1.10.1 Physical Nature of Contamination

Physical properties of water are primary indicators of the extent of contamination and potentially the nature of contamination. Physical properties indicate the appearance of water and are characterized by turbidity, color, temperature, taste, and odor. Physical parameters play an important role in characterizing water quality and designing proper treatment techniques before deciding their end reuse. Further, physical characteristics can be analyzed to identify all impurities that are obstructive to the sense of smell, taste, sight, and one of the very significant physical characteristics, i.e. turbidity. On the other hand, odor in wastewater typically develops in various sources such as sewage and with time this parameter significantly develops a strong smell or odor. Wastewater at holding centers or reservoirs attains increased biological activities, which, in turn, elevates the temperature of the contaminated wastewater. Among all other parameters, turbidity is the measure of relative clarity of water. Turbidity is a physico-optical characteristic of water and it is measured as the amount of light that is scattered by suspended matter in the water when a light is incident through the water sample. The measurement of increased scattering directly indicates the extent of contamination in the water. Therefore, the higher the scattered light intensity, the higher the turbidity in water due to suspended matter and solid particles in wastewater. Turbidity is measured as the opaqueness of water or fluid, in general, in terms of nephelometric turbidity units (NTUs). The turbidity in surface water arises from suspended matter and substances such as silt, fine organic matter, colloidal contaminants, clay, plankton, and nanomaterial suspension. Nevertheless, relatively low-turbidity water can still be found as clear. On the other hand, high turbidity contamination looks cloudy or murky water. Turbidity larger than 5 NTUs is easily detected by holding a glass of water through bare-eye vision [77].

1.10.2 Chemical Contaminants

Chemical contaminants in water mainly occur due to dissolved organic compounds, and inorganic, elemental, and biological contaminants. These chemical contaminants may originate from natural occurrences or synthetic man-made sources. Some of the conventional pollutants that have strong interactive associations with water molecules through various interactions are salts; metals; nutrients; minerals including nitrogen, phosphate, and potassium; bleach, toxins produced by bacteria, drug molecules, personal care product residues, and other EPs. Chemical contamination majorly involves either dissociation of pollutants in water through various interactions and dissolved true solution forms such as humic substances, ionic constituents, surfactants, detergents, heavy metals, pesticides, and so on. With the set of contaminants, water can be analyzed for their chemical constituents in drinking water, domestic water supplies, and wastewater primarily includes the measurement of pH, conductivity, estimation of water hardness, alkalinity, and presence of anionic contamination such as sulfates, nitrates, chloride, fluorides, nitrate, and metal ions such as sodium, iron, manganese, copper, and zinc. Other sets of chemical contaminants include EPs such as toxic arsenic, lead, chromium, barium, selenium, cadmium, cyanides, mercury, and silver. The main contaminants largely found in trace amounts in natural water which may exceed those in wastewater sources are listed in Table 1.4 [78].

1.10.3 Biological Contaminants

Biological contaminants in water are living organisms that are harmful to humans and living beings upon consumption. Biological contaminants are broadly referred to as microbes or microbiological pollutants. Generally, biological, pathogenic, or microbial contaminants are classified into bacteria, viruses, protozoa, and

Table 1.4 Maximum permissible limits of different contaminants and parameters set by different regulatory authorities to be considered safe for consumption.

Contaminant or parameter	Maximum permissible limit (mg/L)
Turbidity	<0.1 turbidity units (TU)
Color	<3 color units
Odor	None
Taste	Pleasant
Aluminum	<0.05
Copper	<0.2
Iron	<0.05
Manganese	<0.01
Fluoride	<1.5
Total dissolved solids (TDS)	150–200
Zinc	<1.0
Hardness	80–120

parasites. Biological contaminants include bacteria, viruses, mites, animal dander, house dust, cat saliva, cockroaches, and pollens. Biological contaminants originate from various sources among which four main sources of biological contamination include pests, soil and water, cross-contamination, and various food raw materials; for instance, several types of biological contamination can be in food and food processing units or systems [79].

Even though some biological agents are essential in nature for life sustenance through various mechanisms, however, many uncontrolled biological contaminants pose serious public health risks. This may also be significant in modifying the physical and chemical characteristics of water. Therefore, water for drinking, cooking, and other human consumption activities should be free from disease-causing organisms such as bacteria, viruses, and other pathogens that are listed in Figure 1.4. Bacterial infection in the water causes harmful effects on surface water. Bacteria are typically single-celled microorganisms and are found largely in natural water sources as well as wastewater reservoirs. Mainly there are seven various types of bacteria that are largely found as harmful sources in drinking water such as *Escherichia Coli*, *Campylobacter Jejuni*, *Hepatitis A*, *Giardia Lamblia*, *Salmonella*, *Legionella Pneumophila*, and *Cryptosporidium*. On the other hand, viruses are a class of microorganisms that can be found in various surface water sources like cistern water and lake water. Generally, enteroviruses such as polioviruses, echoviruses, and coxsackie viruses make their way into our domestic water source through the fecal waste of infected animals and/or humans.

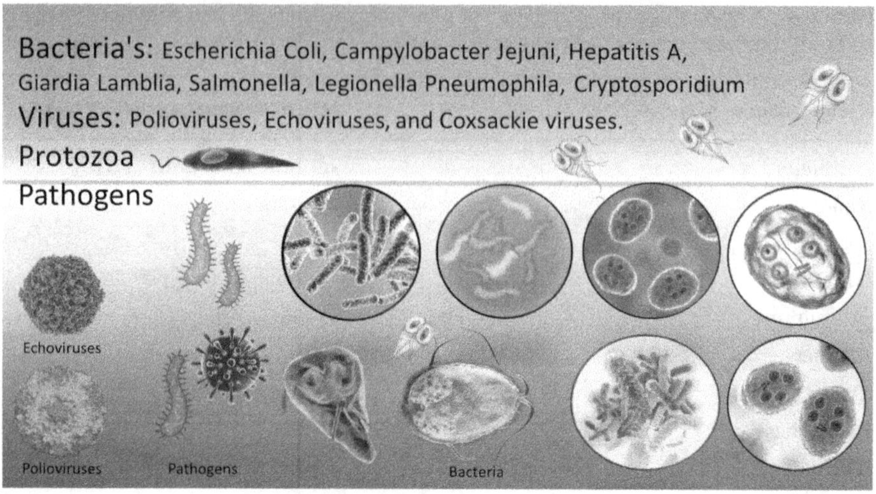

Figure 1.4 List of different bacterial contaminations commonly found in various sources of water such as bacteria, virus, protozoa, and other pathogens. Source: (a), (c), (d), (h) Dr_Microbe/Adobe Stock Photos; (f) crevis/Adobe Stock Photos

Another class of biological contaminants found in drinking water and wastewater are Protozoa that are microscopic unicellular eukaryotes having relatively complex internal structures but cause complex metabolic activities. Protozoal infection in water and consumption of contaminated water results in tissue damage in humans which leads to several diseases. In case of increased and chronic infections, the tissue damage further causes a weak immune response to the parasite and/or to host antigens as well as to changes in cytokine profiles.

1.10.4 Radioactive Contamination in Water

There are many heavy metals and trace elements that are most often categorized as associated with their environmental toxicity such as chromium, cadmium, lead, mercury, cobalt, silver, copper, iron, nickel, zinc, and tin, as well as the lighter elements such as arsenic, aluminum, and selenium. In some cases, trace elemental pollution is natural in origin. Among many, the commonly used radioactive elements are radium, uranium, polonium, thorium, etc. On the other hand, traces of radioactive elements are normally found in various water sources. However, the concentrations of radioactive pollutants, their forms, and compositions vary from sources of water and place to place depending mainly on the radio-chemical composition of the soil, rock strata, and wastewater. The extent of radioactive contamination is mainly considered in areas where water comes in contact with radioactive substances. Nonetheless, the major source of radioactive substances in drinking water could be from nuclear or atomic energy power sources and in the wastewater, it could be from the mining sources of radioactive materials. Also, seasonal variations also alter the extent of radioactive tracers in the water sources [80-83].

1.11 Conclusions

This chapter has overviewed the current scenarios and the technology opportunities to treat various types of water contaminants that exist in the environment. Detailed surveys of the contaminants and their level in various sources have shown the importance and the necessity of the water treatment system in the present world. Today, increasing water quality concerns and scarcity forecasts are forcing policymakers, water management agencies, and regulators involved in a series of educational initiatives to spread awareness. In this regard, stakeholder's participatory process has increased through activities such as workshops with stakeholders, multiple consultations, working on wastewater management projects, and so on. To assist such activities, there needs to be comprehensive information bank required to address data analysis and evaluation and give a detailed account of impending water scarcity-related issues. In this direction, the current chapter has made an effort to collect all scientific information from all sections of water sector. The chapter has given a list of important contaminants, their sources, and the extent of their existence in different ecological systems and potential treatment techniques. Also a list of EPs and their detailed account listed here give a base for further understanding and their potential treatment design and initiative in the near future. Nevertheless, this information greatly helps in raising awareness about water sources and the issue and promoting dialogue among various stakeholders in the water resource and treatment sectors including individuals, communities, local governing bodies, municipalities, governments, international organizations, the private sector, and so on. However, the challenges faced in the water treatment sector, in particular, groundwater treatment and wastewater handling are unique. Given the increasing interest in and importance of the issue, global agencies like the UN, WHO, and various other stakeholders are involved in dialogues to provide on-demand solutions to implement circular economy principles in groundwater management and wastewater treatment projects worldwide. On the other hand, EPs like microplastics, consumer and personal care products, nitrogen, and phosphates, in excessive amounts cause excessive plant growth, which releases toxins into the water. This leads to oxygen depletion and dead zone areas where fish and other aquatic life can no longer exist. Nevertheless, most of these contents found in extremely high concentrations in aquatic systems have posed a serious threat to fish and aquatic life, which requires immediate attention from the stakeholders. The detailed account in this chapter has provided a base for the next chapter to follow and discuss individual issues in each section.

References

1 Kontogeorgis, G.G., Holster, A., Kottaki, N. et al. (2022). Water structure, properties and some applications – a review. *Chem. Thermodyn. Therm. Anal.* 6: 100053.

2 Ball, P. (1999). *H2O. A Biography of Water*. Phoenix.

3 Ball, P. (2008). Water – an enduring mystery. *Nature* 452: 291–292.

4 Chislock, M.F., Doster, E., Zitomer, R.A., and Wilson, A.E. (2013). Eutrophication: causes, consequences, and controls in aquatic ecosystems. *Nat. Educ. Knowl.* 4 (4): 10.

5 Gray, N.F. (2008). *Drinking Water Quality: Problems and Solutions*, 2e. Cambridge: Cambridge University Press.

6 Spellman, F.R. (2017). *The Drinking Water Handbook*, 3e. Boca Raton: CRC Press.

7 Moore, J.W. (1986). Toxic substances in water. In: The Changing Environment. Springer Series on Environmental Management. Ed: Walker, L.R., Howarth, R.W., KapustkaL.A., Springer, New York, NY.

8 Rüdel, H., Körner, W., Letzel, T. et al. (2020). Persistent, mobile and toxic substances in the environment: a spotlight on current research and regulatory activities. *Environ. Sci. Eur.* 32: 5.

9 Hellawell, J.M. (1988). Toxic substances in rivers and streams. *Environ. Pollut.* 50 (1–2): 61–85.

10 Devi, R. and Dahiya, R.P. (2008). COD and BOD removal from domestic wastewater generated in decentralised sectors. *Bioresour. Technol.* 99 (2): 344–349. ISSN 0960-8524.

11 Li, J., Luo, G., He, L. et al. (2018). Analytical approaches for determining chemical oxygen demand in water bodies: a review. *Crit. Rev. Anal. Chem.* 48 (1): 47–65.

12 Aguilar-Torrejón, J.A., Balderas-Hernández, P., Roa-Morales, G. et al. (2023). Relationship, importance, and development of analytical techniques: COD, BOD, and, TOC in water – an overview through time. *SN Appl. Sci.* 5: 118.

13 The United Nations World Water Development Report (2021). Valuing Water, UN Water, UNESCO. 2 Mar 2021, Originally published, 21 Mar 2021.

14 The sustainable development goals report 2023: special edition (2023). Source UN DESA, Posted 10 Jul 2023 Originally published 10 Jul 2023. https://reliefweb.int/report/world/sustainable-development-goals-report-2023-special-edition (accessed 15 February 2024).

15 Naddaf, M. (2023). The world faces a water crisis – 4 powerful charts show how. *Nature* 615: 774–775.

16 Michalak, A.M., Xia, J., Brdjanovic, D. et al. (2023). The frontiers of water and sanitation. *Nat. Water* 1: 10–18.

17 Xu, R., Zeng, Z., Pan, M. et al. (2023). A global-scale framework for hydropower development incorporating strict environmental constraints. *Nat. Water* 1: 113–122.

18 Verbeke, R. (2023). FAIR and open data requires proper incentives and a shift in academic culture. *Nat. Water* 1: 7–9.

19 Bridgeman, J., Baker, J., Carliell-Marquet, C., and Carstea, E. (2013). Determination of changes in wastewater quality through a treatment works using fluorescence spectroscopy. *Environ. Technol.* 34 (23): 3069–3077.

20 Eckenfelder, W.W. (1980). *Principles of Water Quality Management*. CBI Publishing Company.

21 (2022). *WHO Report: Guidelines for Drinking-water Quality: Fourth Edition Incorporating the First and Second Addenda*. Geneva: World Health Organization Licence: CC BY-NC-SA 3.0 IGO.

22 US EPA. (1996). US-EPA Report: US EPA sets drinking water quality standards under the Safe Drinking Water Act (SDWA).

23 Jing, S., Ji, D., Lin, M. et al. (2017). Developing surface water quality standards in China. *Resour. Conserv. Recycl.* 117 (B): 294–303. ISSN 0921-3449.

24 Moore, J.W. (1989). Water quality guidelines and standards. In: *Balancing the Needs of Water Use. Springer Series on Environmental Management* (ed. L.R. Walker, R.W. Howarth, and L.A. Kapustka), 254. New York, NY: Springer.

25 Whelton, A.J., McMillan, L.K., Connell, M. et al. (2015). Residential tap water contamination following the freedom industries chemical spill: perceptions, Water Quality, and Health Impacts. *Environ. Sci. Technol.* 49 (2): 813–823.

26 Li, L., Haoran, Y., and Xiaocang, X. (2022). Effects of water pollution on human health and disease heterogeneity: a review, frontiers in environmental. *Science* 10: 880246.

27 Al-Taai, S.H.H. (2021). Water pollution its causes and effects. *IOP Conf. Ser.: Earth Environ. Sci* 790: 012026.

28 Li, P., Karunanidhi, D., Subramani, T. et al. (2021). Sources and consequences of groundwater contamination. *Arch. Environ. Contam. Toxicol.* 80: 1–10.

29 Johnbosco, C. (2020). Egbueri, groundwater quality assessment using pollution index of groundwater (PIG), ecological risk index (ERI) and hierarchical cluster analysis (HCA): a case study. *Groundw. Sustain. Dev.* 10: 100292.

30 Li, P. (2020). To make the water safer. *Expo. Health* 12: 337–342.

31 Li, P. and Wu, J. (2019). Sustainable living with risks: meeting the challenges. *Hum. Ecol. Risk. Assess.* 25: 1–10.

32 Hansen, B., Thorling, L., Schullehner, J. et al. (2017). Groundwater nitrate response to sustainable nitrogen management. *Sci. Rep.* 7: 8566.

33 He, X., Wu, J., and He, S. (2019). Hydrochemical characteristics and quality evaluation of groundwater in terms of health risks in Luohe aquifer in Wuqi

County of the Chinese Loess Plateau, northwest China. *Hum. Ecol. Risk. Assess.* 25: 32–51.

34 Li, P., Wu, J., Tian, R. et al. (2018). Geochemistry, hydraulic connectivity and quality appraisal of multilayered groundwater in the Hongdunzi coal mine, northwest China. *Mine Water Environ.* 37: 222–237.

35 Hayashi, A., Akimoto, K., Tomoda, T., and Kii, M. (2013). Global evaluation of the effects of agriculture and water management adaptations on the water-stressed population. *Mitig. Adapt. Strateg. Glob. Chang.* 18: 591–618.

36 Clement, M. and Meunie, A. (2010). Is inequality harmful for the environment? An empirical analysis applied to developing and transition countries. *Rev. Soc. Econ.* 68: 223–232.

37 Lam, T.P., Lee, T.-M., Chen, C.-Y., and Chang, J.-S. (2018). Strategies to control biological contaminants during microalgal cultivation in open ponds. *Bioresour. Technol.* 252: 180–187.

38 Adimalla, N., Qian, H., and Nandan, M.J. (2020). Groundwater chemistry integrating the pollution index of groundwater and evaluation of potential human health risk: a case study from hard rock terrain of south India. *Ecotoxicol. Environ. Saf.* 206: 111217. ISSN 0147-6513.

39 Wu, J., Zhang, Y., and Zhou, H. (2020). Groundwater chemistry and groundwater quality index incorporating health risk weighting in Dingbian County, Ordos basin of northwest China. *Geochemistry* 80: 125607.

40 Nataraj, S.K. (2022, Published March 31, 2022). *Emerging Pollutant Treatment in Wastewater*, 1e. CRC Press ISBN 9781032103242.

41 Wortmann, C., Helmers, M., Mallarino, A. et al. (2005). *Agricultural Phosphorus Management and Water Quality Protection in the Midwest*. Lincoln, NE: University of Nebraska.

42 Aravind Kumar, J., Krithiga, T., Sathish, S. et al. (2022). Persistent organic pollutants in water resources: fate, occurrence, characterization and risk analysis. *Sci. Total Environ.* 831: 154808. ISSN 0048-9697.

43 Winterlin, W., Walker, G., and Frank, H. (1968). Detection of cholinesterase-inhibiting pesticides following separation on thin-layer chromatograms. *J. Agric. Food Chem.* 16 (5): 808–812.

44 Kumar, R., Qureshi, M., Vishwakarma, D.K. et al. (2022). A review on emerging water contaminants and the application of sustainable removal technologies. *Case Stud. Chem. Environ. Eng.* 6: 100219. ISSN 2666-0164.

45 Tang, Y., Yin, M., Yang, W. et al. (2019). Emerging pollutants in water environment: occurrence, monitoring, fate, and risk assessment. *Water Environ. Res.* 91: 984–991.

46 Dhankhar, N. and Kumar, J. (2023). Impact of increasing pesticides and fertilizers on human health: a review. *Mater. Today Proc.* Available online 26 April 2023.https://doi.org/10.1016/j.matpr.2023.03.766.

47 Egbuna, C., Amadi, C.N., Patrick-Iwuanyanwu, K.C. et al. (2021). Emerging pollutants in Nigeria: a systematic review. *Environ. Toxicol. Pharmacol.* 85: 103638. ISSN 1382-6689.

48 Marcu, D., Keyser, S., Petrik, L. et al. (2023). Contaminants of emerging concern (CECs) and male reproductive health: challenging the future with a double-edged sword. *Toxics* 11: 330.

49 Sutton, P., Giudice, L.C., and Woodruff, T.J. (2010). Reproductive environmental health. *Curr. Opin. Obstet. Gynecol.* 22 (6): 517–524.

50 Chwastowski, J., Ciesielska, K., Ciesielski, W. et al. (2020). Structure and physicochemical properties of water treated under nitrogen with low-temperature glow plasma. *Water* 12: 1314.

51 Manahan, S.E. (2000). *Fundamentals of Environmental Chemistry*, 2e. CRC Press https://doi.org/10.1201/9781420056716.

52 Clough, S.A., Beers, Y., Klein, G.P., and Rothman, L.S. (1973). Dipole moment of water from Stark measurements of H2O, HDO, and D2O. *J. Chem. Phys.* 59 (5): 2254–2259. Bibcode:1973JChPh.59.2254C. doi:10.1063/1.1680328.

53 Gubskaya, A.V. and Kusalik, P.G. (2002). The total molecular dipole moment for liquid water. *J. Chem. Phys.* 117 (11): 5290–5302. Bibcode:2002JChPh.117.5290G. doi:10.1063/1.1501122.

54 Kendall, J., Davidson, A.W., and Adler, H. (1921). The prediction of solubility in polar solutions. *J. Am. Chem. Soc.* 43 (7): 1481–1502.

55 Smith, J.D., Cappa, C.D., Wilson, K.R. et al. (2005). Unified description of temperature-dependent hydrogen bond rearrangements in liquid water. *Proc. Natl. Acad. Sci. USA* 102 (40): 14171–14174.

56 Qazi, M.J., Schlegel, S.J., Backus, E.H.G. et al. (2020). Dynamic surface tension of surfactants in the presence of high salt concentrations. *Langmuir* 36 (27): 7956–7964.

57 Franses, E.I. and Basaran, O.A. (1996). Chien-Hsiang Chang, techniques to measure dynamic surface tension. *Curr. Opin. Colloid Interface Sci.* 1 (2): 296–303. ISSN 1359-0294.

58 Manahan, S.E. *Environmental Chemistry*. CRC Press Taylor & Francis Group 6000 Broken Sound Parkway NW, Suite 300 Boca Raton, FL 33487-2742 © 2010 by Taylor and Francis Group, LLC CRC Press is an imprint of Taylor & Francis Group.

59 Julian, K. (2018). Trick, Shaun Reeder, contaminated groundwater sampling and quality control of water analyses. In: *Environmental Geochemistry*, 1e (ed. B. De Vivo, H.E. Belkin, and A. Lima), 29–57. Elsevier.

60 Al-Hashimi, O., Hashim, K., Loffill, E. et al. (2021). A comprehensive review for groundwater contamination and remediation: occurrence. *Migration Adsorpt. Model. Mol.* 26 (19): 5913–4941.

61 Zeebe, R.E. and Wolf-Gladrow, D.A. (2009). Carbon Dioxide, Dissolved (Ocean). In: *Encyclopedia of Paleoclimatology and Ancient Environments*, Encyclopedia of Earth Sciences Series (ed. V. Gornitz). Dordrecht: Springer.

62 Sarath Prasanth, S.V., Magesh, N.S., Jitheshlal, K.V. et al. (2012). Evaluation of groundwater quality and its suitability for drinking and agricultural use in the coastal stretch of Alappuzha District, Kerala, India. *Appl. Water Sci.* 2: 165–175.

63 Le, T.M.K., Schreithofer, N., and Dahl, O. (2020). Dissolution test protocol for estimating water quality changes in minerals processing plants operating with closed water circulation. *Fortschr. Mineral.* 10: 653.

64 Solt, G. (2017). *Plant Engineer's Handbook, 2001, Water and Effluents in Carbonate Hardness Carbonate hardness, caused primarily by calcium bicarbonate; From: Industrial Water Treatment Process Technology*. Elsevier.

65 Büker, O., Stolt, K., Kroner, C. et al. (2021). Investigations on the influence of total water hardness and pH value on the measurement accuracy of domestic cold water meters. *Water* 13: 2701.

66 Ahn, M.K., Chilakala, R., Han, C., and Thenepalli, T. (2018). Removal of hardness from water samples by a carbonation process with a closed pressure reactor. *Water* 10: 54.

67 Kesari, K.K., Soni, R., Jamal, Q.M.S. et al. (2021). Wastewater treatment and reuse: a review of its applications and health implications. *Water Air Soil Pollut.* 232: 208.

68 Nepogodin, A.M. et al. (2021). Types of wastewater treatment technologies for facilities with a capacity up to 1000 m^3/day. *IOP Conf. Ser.: Earth Environ. Sci* 848: 012124.

69 Kaur, K., Reddy, S., Barathe, P. et al. (2022). Microplastic-associated pathogens and antimicrobial resistance in environment. *Chemosphere* 291 (2): 133005. ISSN 0045-6535.

70 Tavelli, R., Callens, M., Grootaert, C., and Abdallah, M.F. (2022). Andreja Rajkovic, Foodborne pathogens in the plastisphere: can microplastics in the food chain threaten microbial food safety? *Trends Food Sci. Technol.* 129: 1–10. ISSN 0924-2244.

71 Sharma, S. and Bhattacharya, A. (2017). Drinking water contamination and treatment techniques. *Appl Water Sci* 7: 1043–1067.

72 Nathanson, J.A. (2023). Environmental infrastructure. Encyclopedia Britannica.

73 UNESCO (2017). World water development report, wastewater: the untapped resource. https://reliefweb.int/report/world/2017-un-world-water-development-report-wastewater-untapped-resource (accessed 10 March 2020).

74 UN Water Report: World Water Assessment Programme (UNESCO WWAP). https://en.unesco.org/wwap (accessed 11 October 2022).

75 UN Environmental Programme Report: Better sewage treatment critical for human health and ecosystems. https://www.unenvironment.org/news-and-stories/story/better-sewage-treatment-critical-human-health-and-ecosystems (accessed 10 March 2020).

76 UN-SDGs. https://www.unep.org/explore-topics/sustainable-development-goals/why-do-sustainable-development-goals-matter/goal-6.

77 Hussain, M.A. (2016). Food contamination: major challenges of the future. *Foods* 5 (2): 21-22.

78 Shah, A., Arjunan, A., Baroutaji, A., and Zakharova, J. (2023). A review of physicochemical and biological contaminants in drinking water and their impacts on human health. *Water Sci. Eng.* 16 (4): 333–344.

79 Zulkifli, S.N., Rahim, H.A., and Lau, W.J. (2018). Detection of contaminants in water supply: a review on state-of-the-art monitoring technologies and their applications. *Sensors Actuators B Chem.* 255: 2657–2689.

80 Abada, E., Al-Fifi, Z., Al-Rajab, A.J. et al. (2019). Molecular identification of biological contaminants in different drinking water resources of the Jazan region, Saudi Arabia. *J. Water Health* 17 (4): 622–632.

81 Ma, R., Zheng, C., and Liu, C. (2018). Groundwater impacts of radioactive wastes and associated environmental modeling assessment. In: *Encyclopedia of Sustainability Science and Technology* (ed. R. Meyers). New York, NY: Springer.

82 Davis, M.L. (2010). *Water and Wastewater Engineering*, 1e. USA: The McGraw-Hill.

83 Davis, M.L. and Cornwell, D.A. (2012). *Introduction to Environmental Engineering*, 5e. USA: McGraw-Hill Education.

2

Ground, Surface Water, and Wastewater Quality Management and Treatment Regulations

2.1 Introduction to Water Conservation and Management

Sustainable development requires effective recycling and reuse of wastewater or contaminated water originating from various sources to provide alternatives along with existing water supplies. This approach can also help in addressing issues like water scarcity, and increase water security, including sustainability aspects in water resource and supply management. Water is essential and the reuse of recycled water can be planned or unplanned. Planned water use can be termed as providing a clearly defined allocation of water entitlements for consumptive use that includes both domestic as well as industrial applications. Water is allocated for environmental sustainability, development implementation, social programs, and agricultural operations. Unplanned water use and reuse can be termed situations in which a source of water significantly originates from previously used water. For instance, unplanned water reuse is necessitated when communities draw or obtain from supply sources from rivers that receive treated contaminated or wastewater discharges from communities upstream [1, 2].

On the other hand, water resources management (WRM) is the well-designed process of planning, developing, implementing, and managing water resources both in terms of water quantity and quality sustainability across all water sources and end uses. Water management includes the establishment of institutional organization, creation of infrastructure, providing incentives to stakeholders, and training and instituting information systems that support and guide water management protocols. Overall, water management methods can be mainly classified into four categories: (i) Conservation, (ii) Allocation, (iii) Retrofit program, and (iv) Behavioral practices. The water conservation methods largely help in recharging groundwater sources through various routes such as by

Materials and Methods for Industrial Wastewater and Groundwater Treatment,
First Edition. S. K. Nataraj.

reducing consumption and drawing good-quality water from alternative sources; for instance, sources such as groundwater recharge, reuse of Greywater, rainwater harvesting, and recycling contaminated water or wastewater. Rainwater harvesting has been an effective method of collecting rainwater that directly receives rainfall on the surface or rooftop rainwater harvesting, which can be stored as artificial groundwater recharging. Among many approaches, rooftop rainwater collection or harvesting has been a popular method that can be stored in tanks, directed into bore wells, or diverted to artificial or well-organized recharge systems [3–6].

On the other hand, groundwater recharge is a natural hydrologic process in which water moves downward or seeps through the soil from the surface to groundwater reservoirs. Groundwater recharge can occur both naturally and artificially through various methods. In the natural groundwater recharge process, groundwater is largely recharged through natural phenomena like rainwater diffusion, snow melt, and, to some extent, via sinking or seeping of surface water from reservoirs, rivers, and lakes. On the other hand, artificial groundwater rejuvenation or recharge is a method in which the purification of surface water is used to artificially recharge the groundwater. In this process, artificial groundwater rejuvenates through which the infiltration of surface water is carried out into shallow aquifers to improve the quantity and quality of groundwater. Further, greywater is also used to recharge and rejuvenate the groundwater. Greywater is defined as the domestic wastewater from nontoilet plumbing systems such as washing machines, hand basins, baths, and showers. With appropriate water management steps in these activities, greywater can be easily recycled and safely reused for purposes like gardens due to their lower contamination levels. The recycling and reuse of greywater can be easily achieved following standard water treatment techniques. As the total greywater accounts for nearly 35–40% of total domestic water consumption, recycled water from greywater sources can be effectively utilized for conservation purposes. However, greywater source systems must be noticeably labeled in order to avoid mix-ups with the drinking water supply systems [7–10].

On the other hand, recycling effluent or blackwater from sewage treatment plants (STPs) can also be effectively utilized for water recycling and reuse purposes. It is well established that recycling greywater is much easier than the blackwater system. Blackwater has severely contaminated wastewater originating from toilets, while greywater is less contaminated from other washing sources due to which blackwater is characterized by its high levels of contamination of disease-carrying or causing bacteria. Due to these characteristics, blackwater sources are complex to treat and require well-organized community-level wastewater treatment plants with regular monitoring and management protocols in place. Therefore, to avoid direct discharge of blackwater into the environment, STPs are used to recycle and reuse water from blackwater [11, 12].

In another method, water allocation is a simple method used as a systematic process to manage water conservation. In this process, water is distributed systematically to meet the various requirements of a community for domestic purposes. To successfully implement the effective allocations of water, it is important to consider certain factors such as the source of water, quality of water, methods for extraction, and effective and efficient supply system [13, 14].

Retrofit water conservation is a water-saving or conserving retrofitting measure that is used in existing buildings including mainly the following three aspects of infrastructures. These measures include (i) the use of water-saving equipment, tools, and appliances, (ii) the treatment of contaminated domestic wastewater and reuse of recycled water, and (iii) the collection, cleaning, and use of rainwater. All these activities are greatly influenced by the existence or the replacement of prevailing plumbing equipment with tools that use less water. In general, most of the retrofit arrangements are permanent, most of them are one-time conservation processes that can be implemented with little or no added cost over their lifetimes. However, the effective retrofit program also involves the use of awareness education programs to make water conservation efforts reach end users, the general public, and water management stakeholders [15–17].

On the other hand, behavioral practices are the social awareness measure that involves modifying water use practices and habits to realize more effective and efficient use of water. These awareness programs and measures thus reduce overall water misuse and consumption. However, water use behavior among the population can be implemented without altering the existing infrastructure. However, it requires improved understanding and awareness of water conservation methods in addition to currently used equipment and tools. Further, advancement in hardware development and efficient and effective equipment can still be adopted to replace outdated systems for better water conservation measures. Therefore, behavioral practices largely involve water conservation procedures such as repairing leaky faucets, replacing faulty faucets, awareness activities, and educational programs both at school and college levels to promote sustainable WRM.

2.2 Importance of Groundwater Management and Protection

Groundwater is one of the vast, widely used, and vital resources for drinking water in a large part of the world, commercial use, industrial purposes, and agricultural use. Effective practices in groundwater conservation, management, and use are critical to maintain sustainable water supplies intact and safe for future generations. These efforts help individuals, communities, and organizations achieve greater consistency in activities and tasks, which include the production of commercially important consumer products and community services. Systematic monitoring and management efforts increase efficiency in manufacturing processes, minimize waste, and improve resource quality [18, 19].

In all these efforts, international organizations, monitoring agencies, and governments with the help and collaboration of institutes such as the World Health Organization (WHO) and Environmental Protection Agency (EPA) and local water authorities have formulated frameworks for groundwater and surface water management. However, these specifications vary from region to region based on the availability of groundwater, quality of resources, and intended end use of water such as drinking purposes, agricultural activities, industrial production, or disposal of wastewater into the aquifers or aquatic system or landfills or environment at large. Water quality is particularly defined by keeping their quantitative specification of contamination levels in the water sources specifically termed as maximum contamination levels (MCLs) or maximum permissible level (MPL) for each dissolved, dispersed, and mixed chemical, substances, and/or pathogen, which affects both groundwater quality and surface water in order to meet and satisfy the minimum requirement to retain safe and clean groundwater quality for the larger protection of public health [20–23].

Water reuse is directly associated with the process of water recycling or water reclamation through which reclaimed water is further assessed for its quality prior to reuse. Recycled water from a variety of sources such as domestic discharges, agriculture and irrigation runoffs, contaminated surface water intended for potable water supplies, groundwater replenishment or recharge, industrial discharges, and environmental restoration. Often, communities, industries, and various other sectors seek to improve and optimize their overall water supply and use. However, planned recycled water reuse often refers to water systems designed with the intended goal of beneficial reuse. For instance, planned reuse of recycled water includes industrial process water, potable water supplies, agricultural and landscape irrigation, and groundwater supply management.

On the other hand, there are four main types of groundwater and surface water monitoring arrangements in place, namely (i) ambient monitoring, (ii) source monitoring, (iii) enforcement monitoring, and (iv) research monitoring. Ambient monitoring includes the collection of water quality data from a specific aquifer as an essential part of contamination detection and evaluation of changes in existing water quality. On the other hand, source water monitoring refers to the identification, sampling, and analysis of water sources or watersheds, or reservoirs for a variety of water quality parameters. The source water monitoring process occurs in specifically identified locations with physical site visits in which personnel or equipment are employed to collect water samples for analysis. Further, to make the overall process systematic and sustainable, water analysis is repeatedly carried out in regular intervals and in defined frequency at permanent points in which data processing and prognosis are implemented to establish trends to support further actions. These actions are focused on continued analysis, interception, rectification, and remediation of adverse anthropogenic impacts on the aquatic ecosystem and environment. Therefore, water quality monitoring is a periodically recurring task, which allows data interpretation, processing, and documentation. These exercises help in robust decision-making, policy framing, and learning effective implementation. In every step, continued research and periodic evaluation of new methods and technologies is essential to address emerging water contaminants, limitations of technologies, and associated water conservation issues [24–27].

2.3 Source Pollutants in Groundwater and Wastewater

In general, based on their sources, groundwater contaminants or pollutants are generally categorized into two types: (i) point source and (ii) nonpoint source. Point source contamination or pollution occurs when water contaminants originate from a single source, for instance, chemicals from a single manufacturing unit or factory. Point source contaminants are all dry-weather pollutants, which enter water sources through systematic pipes or channels. However, point source pollutants mainly originate from industrial facilities and domestic and municipal wastewater treatment plants. On the other hand, nonpoint sources occur when contaminants or pollution originate from multiple sources affecting the entire body of water, for example, domestic wastewater and industrial wastewater treatment plants. Further, groundwater sources are mainly affected by nonpoint sources such as agricultural runoff and seeping, storm drainage, construction sites, landfills, and other land distributions that account for nonsource pollution [28, 29].

Nevertheless, sources of water contamination and list of contaminants are vast. Also, the nature of contamination and origins range from nonhazardous to extremely hazardous. However, major concern with groundwater and wastewater contamination mainly comes from their quantitative analysis of contaminated water. Therefore, this section critically overviews the major pollutants that generally affect groundwater quality and are excessively found in wastewater.

Among many widely found contaminants in water, decomposed material from natural flora and fauna, food waste, dead plant, and animal tissue contribute tremendously and consume substantial atmospheric oxygen during their decomposition process. On the other hand, natural organic matter (NOM) that originates from natural sources is a mixture of significantly complex organic compounds and is widely found in all surface waters and groundwater sources. Therefore, the extent and level of decayed organic matter in water is measured as biochemical oxygen demand (BOD) or chemical oxygen demand (COD). Chemical or biological oxygen demand is a measure of the amount of oxidizable contaminants or organic substances in a surface or groundwater source that can drastically lower the DO concentrations. Even though it is reported that NOM has no direct recorded impact on health, it is estimated to affect the efficacy of drinking water source treatment processes from groundwater and consequently the safety of drinking water. Organic matter initially enters the water sources in the particulate form and larger particulates almost immediately settle down and colloidal range particles suspend for a long period of time. These suspended colloidal organic matter and settled particulate matter in the long run experience continuous decomposition and decaying. This leads to the formation of fine organic compounds utilizing accessible oxygen in the surroundings. Also, sturdy organic matter, which either settles or remains suspended helps in leaching the soluble substances into the water. Therefore, based on origin, occurrence, chemical composition, and morphologic characteristic, organic matter in water can be divided into four categories, namely (i) bio-organic matter, (ii) dissolved organic matter, (iii) colloid organic matter, and (iv) aggregate organic matter [30–34].

On the other hand, oxygen-demanding materials can also originate from discharges of food-processing plants, domestic and municipal wastewater treatment plants, breweries, pulp and paper mills, distillery wastewater, and so on, which compose large amounts of organic pollutants. This is due to the occurrence of organic materials, which decompose in the water source and can deplete the water quality with the consumption of dissolved oxygen (DO).

In addition to NOMs, inorganic sediments and suspended solids (SS) may also contribute to the increased COD in which most of the inorganic sediment material washed into a stream and larger water reservoirs as a result of construction activities, agricultural activities, land cultivation, demolition, and mining operations. These new classes of sediment pollutants suspended in water sources in the long term interfere with fish spawning due to which they can cover gravel beds and block light penetration, thereby making food harder to find. The overall activities of inorganic sediments and suspensions also directly damage gill structures, smothering aquatic insects, and fishes, thereby drastically depleting the DO in the water bodies, creating anerobic conditions, further creating unhealthy conditions and unpleasant odors [35–38].

Interestingly, several nutrients in excess can also have a negative impact on water quality in groundwater. Natural nutrients such as nitrogen and phosphorus are estimated to promote accelerated eutrophication or hasten the biological aging of streams, lakes, artificial reservoirs, and estuaries. Nitrogen and phosphorus are found in abundance in residential, domestic, or municipal wastewater and agricultural runoff. They are largely associated with greywater sources, plant debris, animal wastes, or fertilizers through agricultural activities.

The temperature may not be directly classified as a pollutant causing or source of pollution in water sources. However, a rise in the contaminated water temperature may accelerate the decomposition or unwanted reaction among pollutants. Therefore, temperature or heat may be classified as an indirect water pollutant that increases the temperature of municipal or industrial effluents randomly or drastically. Also, heated wastewater indulges in alterations of the water stream or surface water bank vegetation that elevate the stream temperatures due to solar radiation. This also immediately increases the discharge temperature drastically, thereby affecting the ecology of a stream, lake, or water resource. Even though the localized rise in the temperature of water sources can have beneficial effects in maintaining aquatic life, however, increase in the temperatures of effluents lowers the solubility of oxygen in water sources. Overall, this phenomenon reduces the amount of DO in the water and drastically reduces the availability of oxygen-dependent species. Temperature also increases the metabolic pattern and metabolic rate of aquatic organisms, further decreasing the amount of available or DO as respiration increases in aquatic life [33].

Synthetic substances, chemicals, and products including personal care products, plastics, fertilizers, pesticides, and herbicides, pharmaceutical chemicals such as antibiotics, cosmetics, detergents, toxic chemicals, heavy metals, and biological contaminants in water can adversely affect aquatic ecosystems, thereby making the water resource unusable and unsafe for human consumption. These contaminants may originate from various sources such as industrial effluents, municipal wastewater, urban runoff, or agricultural activities and so on.

On the other hand, pathogenic microorganisms are complex and hazardous pollutants that directly affect the groundwater as well as surface water quality and human health upon consumption. In particular, pathogen contamination of groundwater and surface water bodies is a major water safety concern throughout the world. Waterborne diseases such as diarrhea and gastrointestinal illness are mainly caused by various microorganisms such as bacteria, viruses, protozoa, algal, and fungi infection upon the consumption of contaminated water. Major pathogenic pollution occurs from the fecal coliform bacteria, *Escherichia coli*, which generally lives in the intestinal tract of warm-blooded animals. *Vibrio cholera* is another pathogenic bacteria that causes cholera and *Shigella* and similarly, *Salmonella* causes dysentery *Giardia lamblia*, *Cryptosporidium parvum*, *Entamoeba histolytica* cause diseases like Giardiasis, Cryptosporidiosis, and Amoebiasis. Viruses like Coronavirus and Hepatitis A virus (HAV) cause Hepatitis A. Poliovirus causes Poliomyelitis and algae like *Desmodesmus armatus* cause desmodesmus infection. There are several fungi such as *Aspergillus* that most frequently cause lung infections [39–43].

Nevertheless, organic, inorganic, and pathogenic species that originate from various sources in water bodies can cause diseases. In particular, the presence of these pathogenic organisms in water could alter and induce characteristic changes from the original microbial floral community in natural as well as contaminated water bodies. In another series of pollutants, various oil contaminations can result from leakage or accidents of oil from huge domestic sources as well as commercial oil transportation and storage tankers filled with crude oil. This contamination leads to inclusion of several petroleum compounds into water bodies. Similarly, acids and bases from mining activities also alter water quality in several surface water sources, which kills the aquatic organisms. Mining activities also leave sulfur-laden wastewater that causes oxidized forms to yield sulfuric acid when they come in contact with air. All these materials and activities discussed can be categorized as oxygen-demanding materials that are of significant importance [44].

2.4 Oxygen-Demanding Materials

DO refers to the measure of oxygen dissolved in the water available for living aquatic organisms in both surface and groundwater sources. In surface water, the amount of DO in a stream or lake or river or reservoir provides a quality of water in the form of free and non-compound oxygen present in water. However, bonded oxygen in water within a dissolved compound does not account for the DO levels in the water. Therefore, DO is a crucial and very significant parameter considered in assessing water quality. Due to its influence on the living atmosphere, it helps organisms within a water body. Naturally, oxygen enters water bodies from the surrounding atmosphere or air through diffusion, due to continuous aeration, or as a by-product in the photosynthesis process. Therefore, high concentrations of DO in water resources have always been considered healthy and safe for life to exist and consistently support various aquatic lives [45].

Oxygen-demanding materials can be broadly defined as materials that can be oxidized in the receiving water body in which drastic or gradual consumption of molecular oxygen leads to water quality deterioration. Oxygen-demanding materials are generally biodegradable organic compounds. However, these may also include several inorganic compounds. Nevertheless, there have been several studies, which highlight the critical level of DO that is crucial to sustaining sustainable life in the water body. For instance, freshwater fish brook trout (*Salvelinus fontinalis*) may require a minimum of 7.5 mg/L of DO, while crab life aquatic species require a minimum of 3 mg/L to survive in freshwater sources. In groundwater or surface water, oxygen-demanding materials mainly originate from domestic sewage or from the landfills and solid waste dump yards as primary sources in which human waste, domestic discharges, and food residue contribute majorly. Other major oxygen-demanding materials that drastically deplete the DO in groundwater and surface water originate from NOM such as crop residue, dead leaves, and animal droppings, which enter into water from nonpoint sources [46–48].

2.4.1 Biological Oxygen Demand

On the other hand, BOD is the most commonly used parameter for measuring the quantitative analysis of organic oxygen-demanding materials in water sources. The BOD is defined as a measure of the quantity of oxygen in water used by aerobic microorganisms, which is also a measure of the amount of molecular oxygen required for life to exist in water, which gradually decreases in contaminated water via oxidation of organic matter at certain temperatures over a period of time. The BOD is commonly represented in milligrams of oxygen consumed per liter (mg/L) of contaminated water in a prescribed period of time at 20°C, which directly indicates the degree of organic pollution in ground as well as surface water. Even though the natural plant and animal residues end up as oxygen-demanding materials, growth of decayed plants and animals may be unnaturally accelerated when overly abundant nutrients and sunlight interact over a period of time. On the other hand, domestic and municipal discharges carry pet and animal wastes from sidewalks and streets, leaves, grass clippings, nutrients from lawn fertilizers, and paper from residential areas, which upon decaying increase oxygen demand in water bodies. Moreover, oxygen consumed in the process of animal and plant source decomposition deprives the organisms of life. There are very few organisms that survive on low DO such as aerobic bacteria; most of them live on dead algae and other dead microorganisms, which are part of the decomposition cycle. Nevertheless, a small percentage of organisms like algae and other organisms produce significant amounts of oxygen through photosynthesis. Even though various waste components increasingly pose a threat to groundwater and surface water, there is a continuous competition between the consumption of oxygen from decaying waste constituents and production of oxygen through photosynthesis. However, the rate of oxygen consumption and availability depends on various factors like pH, temperature, the presence of friendly microorganisms, and the type of inorganic and organic material in the water [49].

Generally, the biochemical oxygen demand test known as BOD_5 is used to analyze the measure of BOD, which refers to the amount of DO consumed by microbes to break down organic constituents and dissolved material present in water sources at a certain temperature over a period of time. BOD is analyzed in water sources for incubating samples over five days at 20°C. For the same reason, it is called BOD_5. This test provides an amount of oxygen required for the biological oxidation of organic matter in a water sample at 20°C over a period of five days' incubation. During these five days of incubation, the sample of water is appropriately diluted; for instance, highly contaminated wastewater with freshwater and an inoculum of microorganisms is introduced. Here, initial oxygen concentration is analyzed and the sample is stored in darkness to avoid generation of any possible photosynthetic oxygen at 20°C for five days. The oxygen concentration over a period of five days is measured at regular intervals and at the end of the fifth day to conclude the result. The measurement before and after reading is taken into account by adjusting the dilution factor, which gives the end BOD_5 end result. The concentration of oxygen is further quantified following the traditional Winkler titration method or advanced BOD apparatus equipped with a DO sensor and a manometer. The BOD_5 method is the most widely used method for assessment of water quality [50, 51].

2.4.2 Chemical Oxygen Demand

Similar to the BOD, the COD is an important indicator of water quality, which is measured as the amount of DO that is required to oxidize chemical organic materials, like petroleum in water sources. Generally, COD is used to analyze the short-term influence and impact of wastewater constituents that enter freshwater or groundwater, which will have on the oxygen levels of the water source. Therefore, COD is a measure of the oxygen required to oxidize the inorganic material and organic nutrients, such as nitrate and ammonia in water. The COD measurement gives the electron donating capacity of all the organic compounds in the sample, soluble or particulate, biodegradable or nonbiodegradable. In general, the COD of wastewater is greater than the BOD and constituents and compounds can be oxidized chemically more than that oxidized biologically. Generally, COD is analyzed using a strong oxidizing agent like potassium dichromate, $Cr_2O_7^{2-}$ to oxidize the organic constituents and matter in solution converting it into carbon dioxide and water under acidic conditions. This is carried out by digesting COD-contaminated water for approximately 2 hours at 150°C. Further, the amount of oxygen needed to neutralize the demand is calculated from the quantity of chemical oxidant consumed [52, 53].

2.4.3 BOD from Nitrogenous and Carbonaceous Contamination

Carbonaceous-BOD ($CBOD_5$) is largely represented as oxygen demand associated with the biodegradation of organic constituents in absence of nitrogen in wastewater sources and this is generally used for designing aerobic wastewater treatment. Therefore, the amount of DO utilized by microorganisms to be able to oxidize organic matter in the absence of nitrification is called carbonaceous BOD

(CBOD). Nevertheless, additional amounts of DO can be depleted during the nitrification process from the sample, which is the conversion of ammonia (NH_3) to nitrate (NO_3^-) through the nitrogenous bacteria called nitrogenous BOD (NBOD). There are different forms of nitrogen in urine and proteins having amino acids, urea, ammonia, uric acids, and nitrates largely act as nutrients, which help in algae and aquatic plant growth. The nitrogenous waste in domestic, municipal, and industrial wastewater is consumed by autotrophic bacteria. These bacteria consume a significant amount of oxygen and are used as an energy source to convert ammonia into nitrates. The BOD_5 test also can be used to measure the nitrogenous and CBOD [54–56].

2.4.4 Phosphorus- and Nitrogen-Induced BOD in Water

Phosphorus and nitrogen are important nutrients present in natural water sources, which effectively maintain water quality. However, an excessive amount of nutrients helps in growth of biological contaminants like algae that affect water quality deterioration since these nutrients act as essential food for biological growth [57, 58].

2.4.5 Chemistry of Nitrogen in Various Water Sources

Nitrogen in nature and aquatic environment sources occurs in five important forms, namely organic nitrogen, nitrite, nitrate, ammonia, and dissolved nitrogen gas. Among these, ammonia widely occurs as one of the intermediate compounds during biological metabolism along with organic nitrogen and is mainly considered an indicator of deterioration of water quality and can be quantified for increased water pollution. Moreover, aerobic decomposition of organic nitrogen and ammonia in contaminated, domestic, municipal, and industrial wastewater eventually leads to production of nitrite (NO_2^-) and eventually nitrate (NO_3^-). Thus, high nitrate concentration increases the organic nitrogen pollution, which further catalyzes the dissolved organic matter to completely oxidize. Similarly, use of excessive nitrogenous fertilizers induces high concentrations of nitrate in groundwater. This follows the process in which direct land use of organic fertilizer leads to consumption of available oxygen in the soils, which allows oxidation of the organic nitrogen in the fertilizer [59, 60].

Generally, the presence of organic nitrogen and ammonia and their level in water can be used as indicators of pollution in water (see Figure 2.1). Nitrogen in both organic nitrogen and ammonia is often combined and measured as Kjeldahl nitrogen. Therefore, total Kjeldahl nitrogen (TKN) is a measure or the sum of nitrogen bound in organic substances, nitrogen in ammonia (NH_3-N), and in ammonium (NH_4^+-N) in the chemical analysis of water, soil, or wastewater, such

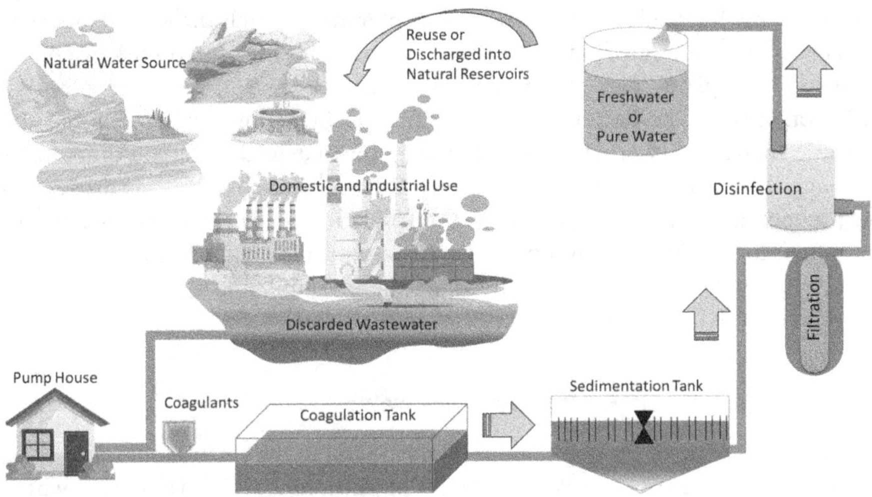

Figure 2.1 General schematic of water life cycle in a typical sourcing, usage, discharge, and recycling process.

as municipal domestic wastewater and industrial wastewater treatment plants [61]. Nitrate and nitrite can be qualitatively and quantitatively analyzed separately using various spectrophotometric methods. Even though nitrogen is present in abundance in the atmosphere, however, generally plants cannot utilize atmospheric nitrogen directly. Instead, fixed nitrogen in a compound is expected to be used by plants. However, to get nitrogen in the fixed form, nitrogen needs to go through a cycle, which involves the following steps: (i) Nitrogen fixation, (ii) Nitrification, (iii) Denitrification, (iv) Ammonification, and (v) Assimilation. Importantly, nitrogen fixation proceeds with atmospheric nitrogen being converted into ammonia through a pair of bacterial enzymes known as nitrogenase present in some species of bacteria such as cyanobacteria. Even though ammonia forms directly as an end product in this reaction. However, it quickly ionizes to form ammonium (NH_4^+). On the other hand, nitrification proceeds with two steps in which ammonia is converted to nitrites (NO_2^-) and thereby to nitrates (NO_3^-). Two different genera of bacteria, namely *Nitrosomonas* and *Nitrobacter* present in the soil oxidize the ammonia into inorganic forms of nitrogen. Further, in the denitrification process, reducing nitrate and nitrite leads to highly oxidized forms of nitrogen, which will be consumed by various groups of organisms converting into gaseous nitrogen [62].

In the assimilation step, living organisms consume NO_3^- and NH_4^+ (ammonium), which is produced through nitrogen fixation and nitrification. This form of nitrogen is largely consumed by plants through their roots and assimilated into plant nucleic acids and protein, which enter into animals through plant tissue consumption. Ammonification process, on the other hand, occurs when an animal and plant die or animals excrete waste. Further, decomposing microorganisms like bacteria and fungi initiate breakdown of proteins in the organic matter. This process releases ammonia that further dissolves in water through the soil. Simultaneously, ammonia further combines with a hydrogen ion to produce ammonium. In the end, denitrification leads to the breakdown of nitrates into metabolized oxygen using microorganisms such as bacteria. This step re-releases nitrogen gas back into the atmosphere, completing the overall nitrogen cycle in the environment.

2.4.6 Phosphorus in Environment and Water

Phosphorus is one of the abundant minerals found in the earth's crust and rocks, primarily in sedimentary rocks in the form of phosphate (PO_4^{3-}) minerals, which are further consumed by plants and animals as depicted in Figure 2.2. With time and climatic changes, weathering rocks erode into the soil, which further leads to the dissolution of phosphate minerals in water sources like streams and rivers to the ocean. Phosphorus in nature is a crucial chemical as it is essential to form DNA. On the other hand, phosphorus in water bodies occurs largely in the form of organic phosphate and inorganic orthophosphate or polyphosphates (APHA, 2012) [63]. Phosphorus is crucial for plant life and is abundantly available in water, which speeds up eutrophication in the environment leading to a drastic

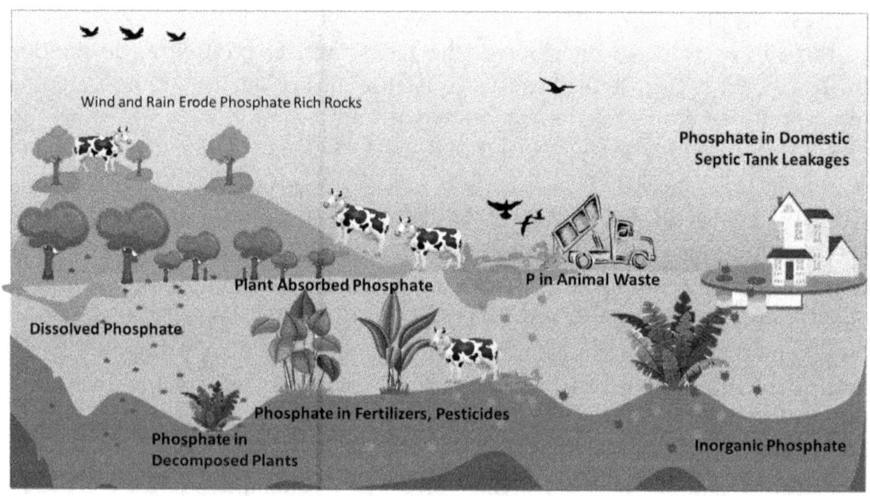

Figure 2.2 Different sources of phosphate and phosphorus in various natural sources in the environment and water.

decrease in DO in water sources due to the increased presence of organic nutrients and minerals [64, 65].

On the other hand, phosphorus and derivative nutrients in wastewater mainly occur from sources like human excreta, phosphorus-containing household detergents, industrial waste sources, and domestic and municipal effluents. It is well recorded that precipitation runoff contributes less to phosphorus loads in wastewater upon mixing with sewer municipal waste sources. Phosphorus is generally analyzed and quantified as total phosphorus in all forms or dissolved phosphorus. Nevertheless, dissolved orthophosphate (PO_3^-) in water is a crucial indicator of water pollution as it is rapidly taken up by bacteria. Therefore, PO_3^- does not exist in high concentrations in freshwater sources. The phosphorus cycle involves various forms of phosphorus at different steps, which can be easily measured using spectrophotometric techniques. The phosphorus cycle can be defined as the biogeochemical cycle that describes the movement of phosphorus in various sources like the lithosphere of solid earth, the hydrosphere that is water sources, and finally biosphere such as living organisms. The detailed phosphorus cycle can be briefly summarized in four following steps, namely [57, 66, 67]:

1) Weathering and rain lead to the washing away of phosphates from rocks over a period of time leading to the dissolution and distribution of phosphates into soil and water.
2) In the second step, plants consume phosphate from the soil, and further animals take up these plants. With this process, the phosphate is naturally incorporated into organic molecules (DNA). In the end, when plants and animals die, decay, and decompose, phosphate returns to the soil.
3) Upon decomposition, plants and animals feed bacteria, which break down the organic matter. This leads to phosphate changing its form to an inorganic form of phosphorus via the mineralization process.
4) Mineralization leads phosphorus back into the soil, which, in turn, transfers inorganic phosphorus into waterways and then oceans. Finally, mineralized phosphorus in various water sources is integrated into sediments and rock, and the cycle restarts and continues.

Nevertheless, excessive use of modern chemicals such as fertilizers artificially induces high concentrations of phosphorus in the natural cycle. This cycle is largely contaminated by the artificially induced fertilizers of phosphorus and animal manure, which has high concentration of phosphorus. As soil cannot absorb beyond critical concentration of phosphorus from artificial sources, it eventually washes away with water runoffs. This is the main reason for the increase in the phosphorus concentration in the water bodies, which leads to eutrophication of freshwater sources [68–72].

2.5 Eutrophication of Surface Water Sources

Eutrophication is characterized by growth of excessive plant and algal colonies in water sources and reservoirs due to the increased availability of nutrients like phosphorus and nitrogen. In eutrophication, lakes and reservoirs obtain nutrients like phosphorus and nitrogen and sediments from various surroundings and become more abundant, which turns into nutrients to serve as foods for algae and fish. Phosphorus and nitrogen enter into the lake through various natural phenomena like storm runoff, and other waste sources, which promote rapid growth of algae and microorganisms on the surface of the lake. Further, when the algae and other microorganisms die, they drop and sediment to the bottom of the lake and turn into a source of carbon for decomposing bacteria. On the other hand, aerobic bacteria, which happen in the presence of oxygen for living use almost all available DO in the process during the decomposition of organic matter, which further leads to the depletion of the DO to cause the bottom of the lake to turn water resources to anerobic. As the concentration of phosphorus and nitrogen increases, more and more algae die, and so as depletion of DO, which leads to further decomposition and increases anerobic characteristics of water sources. Further, aerobic biological activities and increasing frequency in which a water body becomes overly enriched with nutrients, leading to plentiful growth of simple plant life over the years are called eutrophication. Natural eutrophication process is a slow process and may take thousands of years to induce irreversible damage to aquatic ecosystems. However, the eutrophication process can be shortened to years as a result of excessive use of chemicals that provide nutrients such as detergents, fertilizers, and pesticides [73, 74].

2.6 Water Resources Management

There are limited sources of water present on planet earth and there is a humongous amount of used and wasted water all around. Technically, water cannot be created. Therefore, recycling wasted water is as good as creating a new water resource for potential reuse in various applications. Water is extensively used and wasted in day-to-day life including municipalities and industries resulting in generation of millions of liters of wastewater. Industrial processes largely use freshwater sources and at the end of the industrial process, over 90% of the water turns into waste. However, most of the wasted water could be recycled to reusable quality. For this, these recycled sources of water need to be adequately treated to meet "fit-for-purpose specifications" for the desired end reuse. More importantly, specific wastewater treatment requirements require meticulous assessment and analysis of particular source wastewater to assert the quality needed after the recycling process to ensure water safety, public health, environmental protection, and specific user needs. For instance, reclaimed water for agricultural applications would need to be of sufficient quality to prevent the entry of toxic and hazardous constituents to enter the food chain via plants and soils, which also maintains food safety [75].

On the other hand, groundwater sources mainly contain natural origin chemicals and constituents influenced by the nature of soil, depth of soil, constituents of soil such as minerals and subsurface geological nature and their formations through which groundwater is in contact. Even though a major part of groundwater may be untouched by direct contact with industrial and domestic wastewater, long-term exposure to contaminated soil and accidental leaching leads to groundwater contamination. Nevertheless, advanced and systematic water management regulations and policies in place help in developing efficient water management practices for safe, healthy, and sustainable water access in the country. Therefore, water management aims to control, monitor, and distribute water resources to minimize wastage and to maximize effective beneficial various end uses. Efficient water management at a larger scale is taken care of by building dams, reservoirs, and effective distribution systems through river channels, which reduces the risk of danger due to flooding. On the other hand, advanced water management measures improve the effective use of limited water resource supplies for domestic, industrial, as well as agricultural purposes.

Interestingly, water management sometimes also involves changing water storage, extraction, distribution, and usage practices such as groundwater withdrawal rates and distribution for different purposes. More importantly, suitable utilization of water in every household too can effectively minimize the consumption of precious water resources. All the water management and minimization efforts also need additional education and training programs. This topic could include topics of water sourcing, usage, treatment, and finally, conservation, namely: (i) Collection, (ii) Screening and Straining, (iii) Chemical Addition, (iv) Coagulation, and Flocculation, (v) Sedimentation and Clarification, (vi) Filtration, (vii) Disinfection, (viii) Storage, (ix) and, finally, Distribution [76, 77].

In addition to this, the identification of effective water reuse strategies is also required to increase the economics involved in water treatment technologies. Significantly, water treatments are designed to maximize water reuse and strategies for the reuse of recycled water are generally avoided for human consumption and food preparation applications. However, appropriately recycled water from waste sources can be directly used for many outdoor applications such as (i) Irrigation for agriculture, (ii) Watering landscapes such as golf courses, parks, and rights-of-ways, (iii) Municipal water supply for nondrinking purposes, (iv) Industrial process used in factories, power plants, mills, refineries, and automobile washing purposes, (v) Indoor applications like toilet flushing, (vi) Surface cleaning, dust control of roads, construction sites, and other motorways, (vii) Construction activities and use in concrete mixing processes, (viii) Creating artificial lakes and refilling inland reservoirs and coastal aquifers, (ix) Environmental restoration, and so on [78].

2.7 Water Reuse Regulations

Increasing climate change patterns push world leaders and policymakers to frame guidelines to counter aftereffects and outcomes of climate changes, thereby environmental degradation. Initially, many nations followed conventional approaches to tackle issues associated with water, sanitation, and health sectors. However, modern pollutants have forced the world to adopt environmental policies that are holistic in nature and cater to all water conservation and management protocols with a futuristic overview. More recently, water conservation, environmental policies, and management have been formulated as a broader project with the direct involvement of all stakeholders including policymakers, administrators, communities, individuals, nongovernment organizations, and other corporate sectors that are involved in the extraction and distribution of resources. Thus, a tendency is emerging to take responsibility for environmental protection and management with larger legislative and public support and interest converging broadly. In this direction, there have been several changes in the trend observed worldwide in which major responsibilities are distributed among governmental and nongovernmental organizations in every aspect of economic, social, and technological changes. This shift of power and resource flow strictly from government or nation-state to a more inclusive nongovernmental, community- or civil society-controlled, institution-managed, private sector or skilled individual-administered governance that recognizes the contributions of resources at various levels [79].

2.7.1 History of Drinking Water Regulations

History evidently reveals that ancient civilizations had well-organized water extraction, storage, distribution, and management systems. Our ancestors were apparent and well versed in recognizing the importance of water resources, water quantity maintenance for drinking, and other purposes. However, our ancestors may not have faced the high level of contamination in the surface or groundwater and never might have worried about the quality of drinking water. At least this information has not been well documented. However, it is evident that most of the traditional water management plans and strategies have been adopted in arid regions to harness rainwater through harvesting and conservation procedures.

This led to the series of experiments and subsequent development of techniques, which are later considered conventional methods over centuries. Further, centuries of experience led to the development of a series of traditional water management structures to catch, hold, and store rainwater in arid regions.

On the other hand, different strategies had been followed in areas like hilly topography to manage surface water aiming to harness for human consumption. Nevertheless, historical mentions have long recorded aesthetic issues associated with water sources in various forms and characteristics such as taste, odor, unpleasant appearance, and so on. These records and historical documentation have particularly mentioned aesthetic issues with drinking water. It is therefore estimated that water quality awareness gradually developed among the general public over the years and people recognize their senses alone were not accurate judges of the quality of water in various forms and sources. This led to the further systematic development of water treatment methods initially focused on improving the aesthetic characteristics of drinking water. Nevertheless, earlier efforts to improve the odor and taste of drinking water have been recorded as early as 4000 BCE. Ancient scriptures, mainly Greek and Sanskrit, have several recommendations and mentions on water treatment methods followed for ages such as the use of charcoal, boiling, filtering through sand beds, exposure to sunlight, and straining. On the other hand, the use of bronze and copper utensils has been in practice in almost every household application. Ancient Sanskrit writings mentioned using silver- and copper-based utensils to induce anti-bacterial and anti-pathogenic characteristics during the storage and drinking of water. Further, water that is characterized by its visible cloudiness and now referred to as turbidity was one more important basic parameter that pushed our ancestors to develop the earliest water purification techniques. As the turbidity or cloudiness in contaminated water comes from the dispersed particulate pollutants inducing objectionable taste and appearance, such turbidity was mostly treated with charcoal and sand bed. However, ancient Egyptians reportedly utilized chemical alum as early as 1500 BCE. Further, advanced treatment techniques have rigorously developed with the various industrial revolutions that were taking place in Europe, the United States, and other ancient civilizations. One such effort was recorded in the early 1700s in which filtration methods were developed as an effective means of removing particulate pollutants from water. Even though there were limited means to measure the degree of clarity, by the early 1800s, slow multi-layered sand filtration was popular and being extensively and regularly used throughout Europe [80–83].

Further, in the mid and late 1800s, scientists experienced a greater need and understanding of the natural sources and potential contaminants in their occurrence. These efforts were first focused on drinking water quality assessment for contaminants. As discussed, odor or smell and taste have been an earlier consideration. However, over the years, contaminants in drinking water were particularly focused on those that were invisible to bare eyes. These efforts immediately caught the attention when a series of epidemics hit the larger population throughout the world, which directly impacted the understanding of water quality. This further led to the identification of waterborne diseases like cholera, which was first discovered by "germ theory," and which was explained as the cause of waterborne disease. Pasteur explained microscopic organisms or microbes as causes of serious waterborne diseases like cholera. Further, in the late 19th and early 20th centuries, drinking water quality parameters were mainly focused on understanding new classes of disease-causing microbes like pathogens [84, 85].

Another major parameter turbidity was mainly considered as another source of water contamination, which may originate from particulate dispersions that may originate from dust of inorganic origin, fecal matter that could harbor pathogens, and so on. These findings have led to modern water treatment, in particular, drinking water treatment techniques in the early 1900s. These first modern efforts targeted removing or reducing turbidity, and microbial contaminants, which were the main causes of concern for dysentery, typhoid, and cholera epidemics around the globe. Even though multiple filtration techniques including sand bed, charcoal, and cloth filtration were fairly effective in treating turbidity to permissible levels, drinking water disinfection was further treated with chemical techniques like chlorine. These initial efforts (early 1900s) played a significant role in identifying, detecting, and reducing the causes of contamination, which led to a drastic decrease in the number of waterborne disease outbreaks as shown in Figure 2.3 [86].

Nevertheless, the first effort made in 1908 of using chlorine disinfection primarily was to disinfect drinking water for municipal supply. This successful adoption of chlorination technique further paved ways to other chemical and physico-chemical methods of water treatment such as ozone disinfection, adsorption, and ion exchange processes. Even though efforts started in the early 1900s to implement the new water treatment techniques, the United States and Europe began to employ these modern water treatment techniques a decade later (1914) in the form of federal regulation targeting bacteriological contamination in drinking water supplies for municipal purposes. Further, the United States Public Health Service formulated a set of standards and regularly revised them in 1925, 1946, and 1962 including several new contaminants with specific regulatory advice on 28 modern chemical substances. Later, a specific safe drinking water act was passed in 1974 to address the most comprehensive federal drinking water standards, which were then widely adopted by all 50 states with minor changes as the Public Health Service standards in their jurisdiction. By the late 1960s, industrial and agricultural advances led to the innovation and development of thousands of new man-made chemicals and

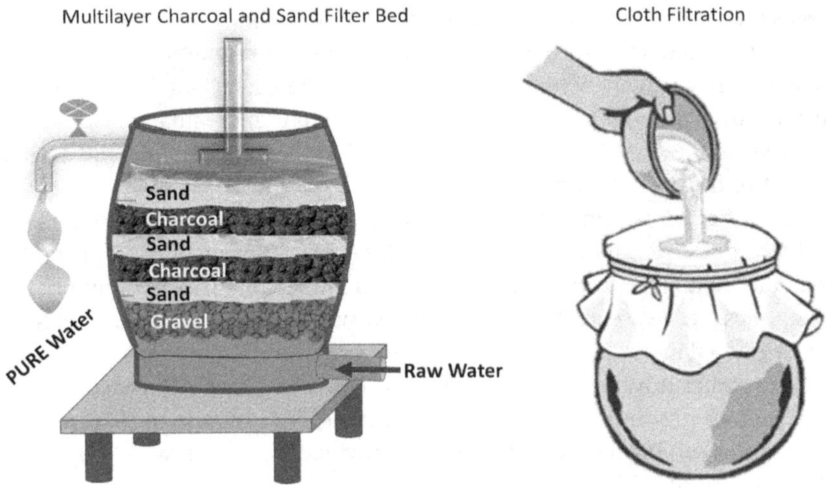

Figure 2.3 Historically known multilayer charcoal/sand filter bed and cloth filtration unit used popularly for drinking water purification.

substances. Many of these substances and chemicals were finding their way into water sources and supplies through manufacturing and product factory discharges, farm field runoff, municipal runoff, and leaking underground storage, disposal tanks, and so on that showed increasingly negative impacts on the water quality, environment, and, at large, public health [87–90].

Even though physical water treatment techniques such as aeration, flocculation, and activated carbon adsorption existed at the time, these techniques were underutilized due to the lack of awareness and need. On the other hand, earlier versions of these techniques were also ineffective in removing modern pollutants like organic molecules. At the same time, the rising health concerns in many parts of the world spurred the federal government, and local governing bodies to conduct several studies and formulate regulations and guidelines on the quality of drinking water supply. Among many such initiatives, the most significant survey on water systems was conducted by the Public Health Service in 1969, which showed only 60% of water supplies met all Public Health Service standards. On the other hand, over half of existing water treatment facilities showed major deficiencies in meeting standard clarification, disinfection, or experienced stress in the distribution system (Pipes and Fittings) and combinations of these deficiencies [91].

Interestingly, a study conducted in 1972 identified traces of 36 chemicals in treated water collected from various water treatment plants that drew samples from American water sources such as the Mississippi River in Louisiana. As a consequence, these extended studies led to the formulation of new legislative proposals for a federal regulation of safe drinking water law that were tabled and debated in Congress in 1973. This further initiated the awareness among general public and policy makers, which eventually led to the passage of various federal health and environmental laws. Among these, one important regulatory legislation was passed as the Safe Drinking Water Act of 1974. This was amended with significant vetting in 1986 and 1996, which has been administered till today by the US EPA's Office of Ground Water and Drinking Water and its partners. This created awareness and both industrial and domestic water managers constantly implemented a number of water systems for the treatment of various contaminated water sources. In between, EPA conducted surveys from 1976 to 1995, which showed a gradual increase in the use of treated water at different levels from small to medium community water supply systems. This trend also influenced further awareness of water sourcing and treatment patterns. For instance, in 1976, merely 33% of systems serving less than 100 people provided treatment, which, by 1995, rose to 69% [92–96].

Even though modern water treatment approaches started in the early 1900s, which only considered creating safer drinking water out of natural water sources like lakes, rivers, and reservoirs from which mainly municipal and domestic drinking water supplies were drawn, later became susceptible to pollution. The mid-20th century saw an industrial revolution in manufacturing, which created larger issues in the form of pollution. This led to the creation of larger water treatment systems to cater to the larger customer base. Today, drinking water filtration, ozonation, and chlorination are still in use as effective treatment techniques for domestic and municipal water supplies from harmful pathogens, microbes, and larger organic contamination. Even though advanced and additional multi-step water disinfection techniques have been developed over the years, conventional disinfection remains effective for microbial contamination treatment. Most advanced water treatment was adopted at a large scale in the early 1970s and 1980s in the form of reverse osmosis (RO) membrane filtration treatment techniques in hybrid combination with ozonation. At the same time, European regulatory and monitoring agencies adopted regulations like the Safe Drinking Water Act (SDWA) in 1974 and subsequent amendments in 1986 and 1996 set national standards to protect public drinking water and various sources of water and human-made contaminants in them. More significantly, SDWA standards over the years have also included guidelines for maximum level of contamination limits for chemical, microbiological, and other contaminants in drinking water. Further, the European SDWA also issued guidelines and criteria for removing contaminants from various water sources. Nonetheless, EPA also issues guidance or a health advisory from time to time instead of regulations and additional rules, which protect water sources against exposure to pathogens, which cause waterborne diseases. These innovative approaches mainly came into practice after observing the limitation of long-practiced chlorination techniques, which were ineffective against some pathogens. Similarly, several hybrid combinations of techniques have later come into practice with the discovery of more chlorine-resistant pathogens in drinking water that can cause illnesses like Legionnaires' disease, hepatitis, gastroenteritis, and cryptosporidiosis. Thus, following hybrid approaches, several modern chemical contaminants in drinking have also been effectively identified and treated. Further, a survey conducted by EPA in 1995 has shown nearly 64% of the population across the United States followed chlorine-disinfection procedures seriously for both groundwater and surface water sources before using them for household as well as direct consumption [97, 98].

Nevertheless, there have been several rules, regulations, and amendments adopted throughout the world for efficient groundwater and surface water treatment like filtration and disinfection. In all cases, performance criteria of each water sourcing, distribution, and treatment systems have been constantly amended to address ever-emerging issues associated with groundwater, surface water, and drinking water sources. In this direction, EPA has established several criteria to assess groundwater sources and their influential factors in which these guidelines help in identifying sources of contamination.

Even though a typical drinking water treatment plant used today still involves one or the other conventional disinfection step either in chlorination or ozonation form, advanced approaches involving membrane processes have lately gained acceleration. However, newer combinations of treatment approaches like RO and granular activated carbon (GAC) and UV lamps have also gained momentum in the 2000s in modern drinking water plants as shown in Figure 2.4 [99].

During all these developments, US EPA and European agencies have constantly monitored and formulated the regulations and guidelines over the years. Further, the US Food and Drug Administration has also entered the monitoring process to

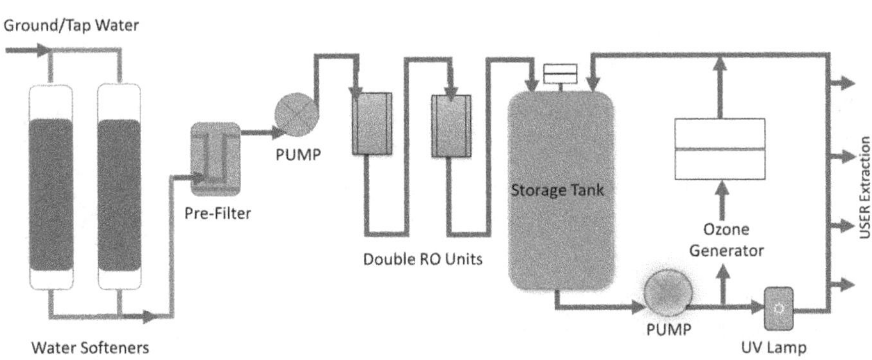

Figure 2.4 Advanced water treatment techniques used in domestic water treatment processes in which carbon pre-filters or UV post-filters are generally used for disinfection.

regulate commercial water treatment and bottled drinking water markets. Some of the important regulations adopted by the US and European countries throughout the 20th century, which were then largely adopted by the rest of the world have been listed in Table 2.1.

2.7.2 Modern Water Regulations

Generally, a wastewater treatment, reuse, and management policy is a well-organized and deliberate action plan to regulate and guide decisions to achieve rational outcomes in the water sector. This also provides a focused thrust and futuristic work plan for the activities toward achievement of the desired results. Modern waste and water management are relevant and timely strategies assist the policy makers, technologists and managers in operationalizing regulations and policies. Generally, Nations or States formulate and maintain primary regulatory framework

Table 2.1 Details of selected legislatures and policy framework enlisted by EPA and regulations framed and enacted regarding drinking water by the United States between 1974 and 2013.

Regulation/legislature or guidelines	Year
Safe Drinking Water Act (SDWA)	1974
Interim primary drinking water standards	1975
National primary drinking water standards	1985
SDWA amendments	1986
Surface water treatment rule (SWTR)	1989
Total coliform rule (TCR)	1989
Chemical contaminant rules phase I	1989
Lead and copper regulations	1990
Chemical contaminant rules phase II	1992
Chemical contaminant rules phase IIB	1993
Chemical contaminant rules phase V	1994
National Primary Drinking Water Regulations (NPDWRs)	1996
SDWA amendments	1996
Information collection rule	1996
Interim enhanced SWTR	1998
Stage 1 – disinfectants and disinfection by-products (D-DBP) regulation	1998
Drinking water contaminant candidate list	1998
Unregulated contaminant monitoring regulations	1999

Table 2.1 (Continued)

Regulation/legislature or guidelines	Year
Radionuclides rule	2000
Lead and copper rule – action levels	2000
Filter backwash recycling rule	2001
Long-term 1 enhanced SWTR	2002
Unregulated contaminant monitoring regulations	2002
Drinking water contaminant candidate list 2	2005
Long-term 2 enhanced SWTR	2006
Stage 2 D-DBP rule	2006
Ground water rule	2006
Aircraft Drinking Water Rule (ADWR)	2009
Drinking water contaminant candidate list 3	2009
Revised Total Coliform Rule (RTCR)	2013

and regulatory authority in identifying, allocating, developing, and distributing water resources. However, some states for years and many in the recent past have established programs to particularly address treatment, reclamation, and reuse into existing programs. Developed nations like the United States adopt drinking water programs at state and local government levels to implement regulations under the Safe Drinking Water Act and the Clean Water Act to protect and conserve the quality of drinking water sources and water bodies like rivers and lakes. Therefore, these two programs together provide a foundation from which local governing bodies and states at large can enable, regulate, and monitor contaminated groundwater and surface water treatment and reuse as they deem appropriate [100, 101].

Among many regulatory practices, there are several important methodologies being adopted to minimize the water consumption and production of effluent at the end of their use both at domestic as well as industrial levels, namely (i) The first measure among many to minimize consumption assumption comes from the efficient use and management of water through best practices such as use of advanced processes, techniques, and control of end-process operations, (ii) The effective use of process integration techniques in different domestic and industrial processes based on pinch analysis, (iii) Focusing on the process optimization in which water is extensively used based on mass/energy balances by applying advanced strategies aiming at effluent reduction for instance multi-objective optimization (MOO) based on mathematical modelling, (iv) Further integration of process optimization and product manufacturing planning techniques in concurrence with real time monitoring and control of product quality and minimization of water losses, and (v) More importantly, increased demand for water can be intelligently managed with the assistance of advanced artificial intelligence or machine learning or Internet of Things (IoT)-assisted, controlled, and monitoring devices. This will overall enhance the efficacy of water and futuristic decision-making in favor of water and environmental protection.

Nonetheless, wastewater treatment and reuse standards in different countries are either adopted or inspired by the WHO standards or other international standards without altering them to appropriate local conditions. However, localized broader as well as specific studies are essential to understand the effectiveness or suitability of rules for specific countries, regions, or states. Many of the developing nations do not have a time-tested policy on reuse of recycled or treated water recycling and reuse. However, majority of them borrow them from either WHO or US EPA or European Union (EU) standards. However, with recent climate change awareness and increased attention to environmental concerns, many nations around the globe are reviewing existing policies, regulations, by-laws, and wastewater-monitoring protocols as necessary steps to ensure consistency and future needs. Nevertheless, some of the Nations and States that have already adopted and implemented wastewater treatment and reuse policies have also initiated time-to-time reviews and updates to suit the local, regional, and case-specific requirements to ensure consistency in the national framework. These modified regulations may result in a relaxation of the specific guidelines and thus augment the quantities and qualities of reclaimed or recycled water. However, this may push for more stringent regulations and standards to protect overall public health without compromising supplies to match demand. In either case, extensive studies are being conducted and commissioned to ensure safe and effective implementation of wastewater treatment and reuse guidelines, which will enhance confidence in reusability and credibility of recycled water as a valuable resource. Nonetheless, it is essential to adopt international practices, and robust guidelines in addition to locally prevailing and regional sociocultural, epidemiological, and environmental conditions.

Therefore, a policy is a deliberate but systematically framed plan of action to guide specific decisions to achieve rational end outcomes and provide the desired results. However, timely application of relevant strategies greatly helps implement and operationalize policy. In this direction, the first international concerted effort was initiated by the United Nations through the Conference on Environment and Development in 1989, which adopted the resolution on conserving the environment through efficient waste management strategies. Further, Agenda 21, which is a comprehensive plan of action adopted globally in consent with several national and local organizations, stated that governments should "promote waste minimization and prevention as the principal objective in their respective national waste management programs. This agenda also aims to develop and implement national plans for waste management and prioritize waste recycling and reuse." Agenda 21 adopted as a result of the Rio Declaration, at the United Nations Conference on Environment and Development (UNCED), Brazil on Environment and Development follows the principles for the Sustainable Management of Forests with the consent of more than 178 Governments majorly aims to reduce human impact on every sector of environment [102].

Nevertheless, modern methodologies and regulations across the globe have included a comprehensive list of synthetic as well as conventional contaminants based on their origin and nature of toxicity. These contaminants have also been classified into various categories such as chemical, biological, and emerging pollutants. More significantly, each type of contaminant has been identified and specific regulations have been formulated to address the contaminants under legal framework. For instance, EPA sets legal limits on occurrence and residues of over 90 contaminants in drinking water into different categories with specific legal frameworks to address as listed in Table 2.2. Thus, many of the legal frameworks protect water sources and issue guidelines in larger contexts to prevent contamination of the sources beyond natural minerals. Therefore, these legal limits for a contaminant reflect the extent of protection water sources, environment and human health can achieve using the best possible available technologies. In addition to this, present trends show the framing and implementation of new regulatory guidelines from various agencies like WHO,

Table 2.2 The list of rules and regulations specifically formulated to regulate specific types of contaminants and their origin.

Contaminant type	Regulations
Chemical contaminants	• Arsenic rule • Chemical contaminant rules • Lead and copper rule • Radionuclides rule • Variance and exemptions rule
Microbial contaminants	• Aircraft drinking water rule • Ground water rule • Stage 1 and stage 2 disinfectant/disinfection by-products rule • Surface water treatment rules • Total coliform rule and revised total coliform rule
Right-to-know rules	• Consumer confidence report rule • Public notification rule

US EPA, SDWA, and many countries in collaboration and inspired by each other. Also, these agencies' regulations also set water quality monitoring, water-testing schedules, developing novel protocols and methods that water sourcing, distribution and waste management systems must follow. Nonetheless, regulatory authorities also give individual states and local governments the opportunity to set, modify existing guidelines and/or enforce their own drinking water standards if the existing standards are required based on emerging trends.

More significantly, it is important to identify and regulate the indigenized solutions to address the local, region-specific, and contamination-specific. Therefore, water use, wastewater generation, and waste discharges are subject to regional, national, and international standards. For instance, legislation in member countries of the EU formulates and follows the directives of the European Commission (EC), which is the executive body of the EU. For Integrated Pollution Prevention and Control (IPPC) Directive 96/61/EC (Directive 2008/1/EC codified version). This directive has been formulated to address industrial production processes that account for a considerable share of the overall pollution in Europe, which mainly targets emissions of greenhouse gases, acidifying substances, solid waste generation and disposal, and wastewater emissions. Therefore, the EU adopted the IPPC Directive (Directive 96/61/EC) in 1996, a set of common rules for permitting and controlling industrial installations in member countries. Also, to address emerging issues and emerging pollutants in wastewater, the IPPC Directive has recently been codified as Directive 2008/1/EC. On the other hand, in the United States, a number of federal regulations and statutes have been passed by Congress and later signed into law by the President, making them part of the national mission as they are central to the Office of Water's mission. Ultimately, all these efforts culminate into a national legal framework through Presidential Executive Orders (EOs), which play a central role in a number of Office of Water activities. Therefore, EOs are then legally binding orders that direct the US EPA and other federal water monitoring agencies in their execution of Congressionally established statutes, laws, and policies. Similarly, several other countries around the globe such as India, China, Southeast Asian countries, Australia, Latin America, and Africa have adopted their own legislations to address regional and local environmental issues in particular water management [103–105].

Meanwhile, water treatment technologies and water recycling efforts gained considerable attention during the 1990s, which further attracted increased acceleration in the water management sector. These initiatives have led to further develop methodologies for minimizing water consumption and gradually reducing the wastewater generation. However, efforts have suffered setbacks due to unavailability of suitable technologies and limited awareness. Nevertheless, early 2000s have seen several studies, which led to the education and creation of awareness dealing with partial or comprehensive water management topics.

However, exponential industrial and population growth led to a drastic change in water consumption patterns both in domestic as well as industrial sectors. Nonetheless, commissioning local studies is essential to address immediate needs of local WRM as they may facilitate a relaxation of the guidelines, thereby augmenting the quantities of reclaimed and recycled water without compromising public safety and health, and/or may recommend more stringent or modified standards to protect local industrial and community health. In either case, such specialized commissioned studies are deemed necessary to ensure effective, efficient, and safe implementation of wastewater reuse guidelines, which will increase confidence in the acceptance of recycled or reclaimed water as a valuable new resource.

An overall international, national, or regional wastewater treatment policy determines and governs the framework for each activity in the waste sector depending on the nature of water use and contaminant generation. Nevertheless, the increasing threat of emerging pollutants and persistent substances demands clear, concise, and consistent policies that are necessary requirements for the wastewater management and treatment industries. This is expected to necessitate well-established and supported wastewater management policies that are crucial in wastewater management. According to many international agencies like the United Nations Environment Programme (UNEP), one crucial element common to most waste policies is a waste hierarchy. This waste management hierarchy is a stepwise approach to waste segregation, prevention, recycling, reuse, and management in the order of environmental priority for different waste management options as illustrated in the waste pyramid in Figure 2.5 [106, 107].

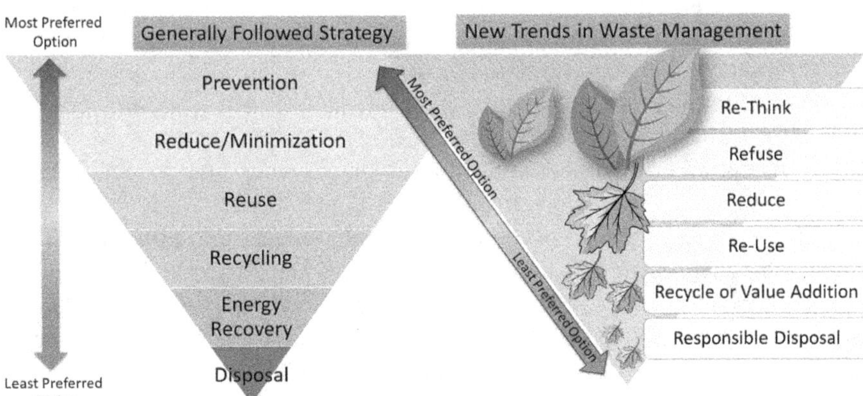

Figure 2.5 Hierarchical representation of traditional and modern approaches in both solid and liquid waste management and reduction strategies.

2.8 Water Treatment

Water treatment methods, technologies, or protocols are the systematic process that makes water more safe, clean, and suitable for specific end use. Traditional water treatments were restricted only to common water disinfection methods including chlorination, ozonation, and lately ultraviolet radiated techniques. Among these major disease outbreaks, attention toward biological disinfection material may also clog pipes in cooling systems. Chlorination destroys harmful parasites, bacteria, and other organisms, which turns contaminated water to be useful for household applications like drinking or cooking, industrial supply, agricultural irrigation, firefighting, water recreation, river flow maintenance, or simply for resource recharge or safe discharge into the environment. Largely, reuse strategies generally depend on the degree of treatment and the treatment method followed, which determines the quality of recycled water. Nevertheless, this section mainly focuses on the listing of water treatment plants that are employed to produce safe drinking water and protocols adopted in wastewater treatment plants that treat water for safe discharge or future reuse [108, 109].

2.8.1 Drinking Water Treatment

Drinking water treatment method generally followed in large-scale plant operations can be broadly classified into: (i) Disinfection plants that are used for the production of high-quality safe and clean water source from various sources like groundwater, surface water, and so on to ensure that product water does not contain any biological contaminants like pathogens, (ii) Filtration plants are used to treat less harmful contaminants from surface water, and (iii) Water softening plants are used to treat hardness in the groundwater. A typical water treatment and filtration plant is shown in Figure 2.6, which is designed mainly to disinfect bacteria and remove various physico-chemical characteristic contamination like color, odor, turbidity and other contaminants.

In this step, water treatment and filtration plant involves the following steps: (i) Screening: in this step, larger particulate contaminants are removed from entering next step, which larger contaminant may interrupt the treatment process in the subsequent steps, (ii) Rapid mixing: In this step, external chemicals are added and rapidly dispersed through the contaminated water, (iii) Flocculation: In this step, chemicals like alum or aluminum sulfate is added to the contaminated water to neutralize the fine particulate or colloidal contaminants through electrostatic interaction, which then brings fine colloidal particle to turn them to form large particles called flocs that can readily be settled out under gravitational force influence, (iv) Sedimentation: Flocculation initiated in the previous step leads to sedimentation as floc settles the suspended or colloidal contaminants to

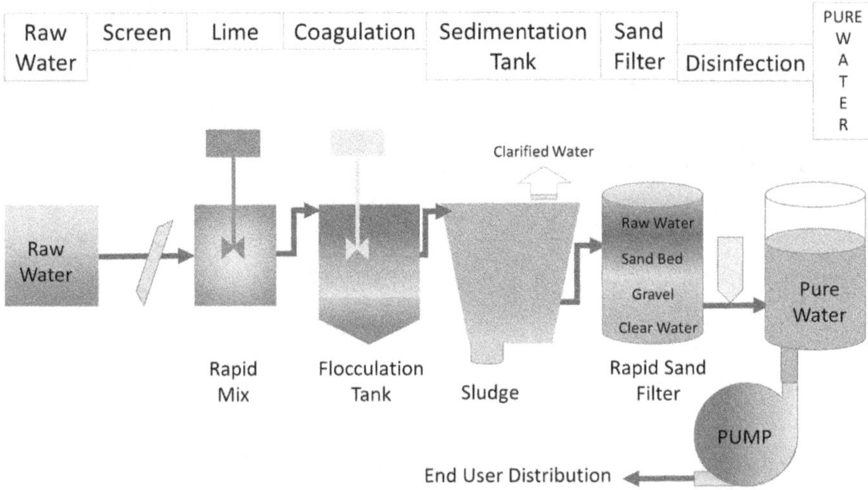

Figure 2.6 Schematics of different steps involved in water treatment/filtration plants used for production of drinking water from various contaminated sources.

the bottom of the influent water due to its weight, (v) Filtration: Sedimentation is the pre-final step of the treatment plan, which follows filtration step separating sediment from the supernatant product water from sand, gravel, and charcoal. However, filtration of purified supernatant can be effectively separated from fine residual particles such as dust, parasites, bacteria, viruses, and chemicals using suitable pore size filters, and (vi) Disinfection: This final step involves the addition of disinfection chemicals in order to eliminate residual pathogenic contaminants before approving for drinking or household applications [110, 111].

2.8.2 Wastewater Treatment

Municipal, industrial, and emerging wastewater treatment plants are designed with multiple steps involving three major categories: (i) primary treatment, (ii) secondary treatment, and (iii) advanced treatment. Figure 2.7 shows a typical treatment plant involving three steps with varying degrees of wastewater strengths. Primary treatment is mainly used to remove pollutants physically through either settlement or float in a clarifier tank. A typical primary treatment step removes over 60% of the SS from raw sewage and 35% of the BOD_5. However, primary treatment is inefficient against soluble pollutants, but these dissolved pollutants are removed in the secondary treatment step. In secondary treatment, dissolved organic pollutants are majorly removed in the form of BOD_5, which escapes the primary process. Also, the secondary treatment step typically involves disinfection and filtration using biological processes. Biological processes are naturally used to

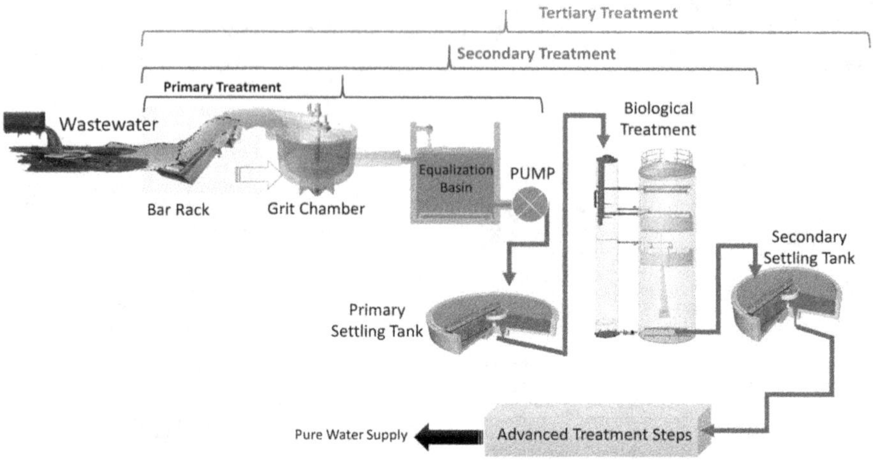

Figure 2.7 Different steps involving (i) primary, (ii) secondary and (iii) advanced wastewater treatment techniques and their specifications.

treat wastewater. However, secondary wastewater treatment steps are designed to speed up these natural processes. Nonetheless, secondary treatment may show restricted removal efficiencies up to 85% for BOD_5 and SS and this step does not remove significant amounts of inorganic contamination such as phosphorus, nitrogen, or heavy metals, and pathogenic bacteria and viruses [112–114].

On the other hand, tertiary wastewater treatment or advanced treatment processes are applied when the secondary treatment step proves inadequate to remove residual contaminants. However, advanced treatment techniques can be efficiently used for the removal of residual contaminants to 99% of BOD_5, bacteria, phosphorus, all SS, and up to 95% of nitrogen. Advanced treatment techniques involve several methods to produce sparkling colorless, clean, and odorless effluent indistinguishable in appearance from high-quality drinking water. Some of this broader technical categorization involves advanced oxidation, Fenton-type reactions, advanced filtration techniques like membrane-based separation processes or application of adsorbents such as activated carbon, the addition of advanced nanomaterial-based adsorbents, and so on in order to remove persistent organic pollutants and emerging contaminants. Among many advanced water treatment techniques, the addition of inorganic chemicals such as ferric chloride or alum improves the removal efficiency of excessive inorganic minerals like phosphorus. Further, nitrogen presence can also be controlled in the wastewater treatment plant with appropriate nanomaterials-based adsorbents or by introducing biological treatment steps. The biological treatment step generally is known as nitrification/denitrification process. However, this process requires specific bacteria to cause these two reactions. On the other hand, a chemical process known as ammonia stripping is used for fixing and removal of nitrogen in the form of ammonia by altering the pH (adding lime to convert ammonium ions to ammonia) (Davis, 2010) [63]. Interestingly, Hoekstra has defined the term water footprint (WFP) as an indicator of direct and indirect water status of availability and use. Even though WFP has not been widely discussed and acknowledged, however, in this water scarcity scenario, it is important to recognize the importance of the information on which how much water is required for daily consumption of common needs of the population [63, 115].

2.9 Biological Contamination in Wastewater

Biological pollutants are found in almost every type of water and wastewater with various quantities and types. The list of biological pollutants is increasing ever since first reports came on the contamination involving commonly found microorganisms such as viruses, bacteria, and parasites or pathogens that are responsible for waterborne diseases, such as cholera, dysentery, typhoid fever, hepatitis, polio, and schistosomiasis. For instance, the presence of bacteria like coliform found in domestic and municipal wastewater originated as fecal pollution is one of the causes of concern. Interestingly, wastewater plants containing biological contaminants can be termed a microbiological zoo that provides a conducive atmosphere for microorganisms like protozoa, metazoa, bacteria, and other microlife as depicted in Figure 2.8 [116].

Wastewater over time attains extreme deterioration due to the presence of microorganisms, which break down and remove organic material and nutrients resulting in high BOD. Microbiology of wastewater needs to be evaluated carefully to design effective and efficient treatment techniques. Microbiological understanding of microorganisms in wastewater plants is complex composition and constitutes change with time, temperature, and the extreme external stimulus in which wastewater treatment plants are operated. Interestingly, a large number of diverse microorganisms that are free-living exist as single cells or cell clusters in wastewater. Microorganisms being free-living, microbial cells can be distinguished from the cells of animals and plants as the plant-origin are not able to live alone in nature but only exist in characteristic groups. Generally, a single microbial cell provides a conducive atmosphere, carrying out its life processes of respiration, growth, and reproduction independently of other cells of a similar type or of different kinds. Following section of the chapter gives a detailed overview of the role of microorganisms or biological contaminants in different water and wastewater sources. Further, chapter gives a brief classification, their role as contaminants, lists the disease-causing microorganisms and the role of biological ingredients, additives, and micro-organisms in wastewater treatment. Interestingly, sections will include example and brief notes on groups of important microorganisms or biological contaminants like pathogens in water and wastewater treatment.

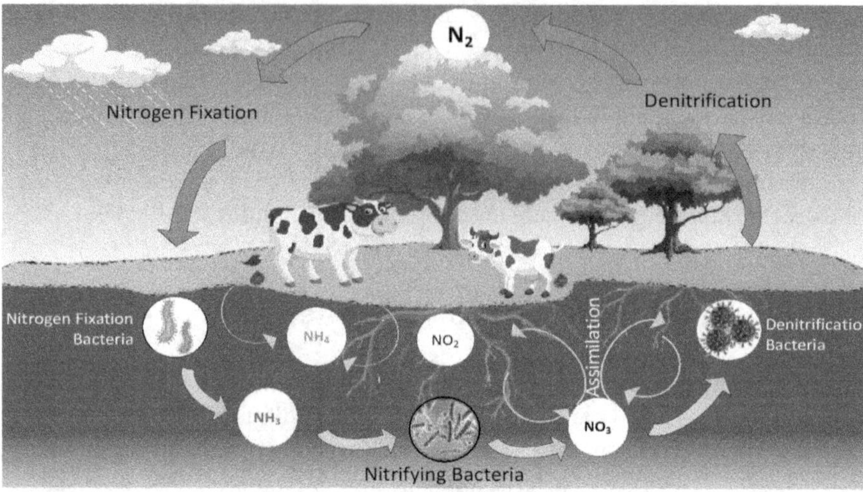

Figure 2.8 Ecological footprint and bacterial activities in the environment, which influences the nitrogen footprint, fixation, contamination, and recycling. *Source*: Dr_Microbe/Adobe Stock Photos

2.9.1 Taxonomy of Microbial Communities or Contamination

Taxonomy of microorganisms involves three main components: classification, nomenclature, and identification. Classification of micro-organisms allows the orderly categorization and further grouping. On the other hand, nomenclature refers to naming of these groups of organisms and this comes under the consensual agreement so that the same name is used unequivocally by everyone across the field. Altering the nomenclature may give rise to uncertainty and is subject to consensually agreed rules. These constituents together with the taxonomy of microbes make up the discipline of *systematics*, which is concerned with evolution, genetics, and speciation of organisms, and is generally referred to as phylogenetics. Microbial diversity in nature is staggering and microbes can be categorized into protozoa, fungi, and helminths. On the other hand, large subdivisions are finally classified into individual *species*, which is occasionally useful for recognizing a biological variant with particular properties [117, 118].

Despite their variety and complexities, all living cells can be categorized based on certain structural and functional characteristics into two main groups, namely prokaryotes and eukaryotes. In general, prokaryotes are structurally simpler and smaller than eukaryotes. However, presently accepted classification of life, there is one more extended domain making them three main categories, namely eukarya, bacteria, and archaea. Plants, animals, fungi, and protists are kingdoms in the eukarya domain. The domain bacteria comprises all the pathogenic prokaryotes as well as many of the nonpathogenic prokaryotes largely found in water and soil sources. Interestingly, the DNA of prokaryotes is generally in a single, circularly arranged chromosome and is not surrounded by a membrane. Therefore, these do not possess true nucleolus. Therefore, prokaryotes lack membrane-enclosed organelles, which are specialized structures, which carry different activities.

Microbes are environmentally, socially, and economically important species that have found a wide range of applications for centuries in a range of products and processes such as use of enzymes, probiotics, and biofuels like bioethanol and hydrogen gas. Presently, these products are used on an industrial scale, in various processes playing significantly important roles in medicine, health care, and food industry and cleaning up toxic waste. Microorganisms such as enzymes are mainly used in breaking down organic matter in wastewater. Through these important mechanisms, microorganisms perform a vital role and act as main natural engineers in regulating and governing all ecological processes. In many modern biotechnology processes, enzymes act as a universal catalyst and provide ecological transformations. On the other hand, through bioremediation process, enzymatic system uses microorganisms to degrade the various organic contaminants present in wastewater in soil and water environments.

Among many microbial constituents, bacteria and archaea that come from prokaryote entities constitute a wide group of very small unicellular organisms. On the other hand, microorganisms like fungi, protozoa, and algae belong to eukaryotes. Similarly, viruses which are noncellular entities do not fit into any organizational scheme of living cells (Figure 2.9). Even though viruses are genetic particles, which replicate but do not perform the usual chemical activities like other living cells [119–121].

2.9.2 Bacteria

Bacteria constitute a majority of prokaryotes. There are thousands of bacteria species present that are differentiated based on several factors such as their shape or morphology, biochemical activities, chemical composition, which is generally used for the detection through often staining reactions, nutritional requirements, and sources of energy such as sunlight or chemicals. Nonetheless, bacteria account for nearly about 95% of the microorganisms in wastewater. The single-celled microorganism's bacteria are mainly classified based on their response to oxygen. Aerobic bacteria utilize oxygen, which is added manually or mechanically to break down wastewater contaminants, converting it into energy. Interestingly, bacteria constitute the highest population of microorganisms in wastewater treatment plants (Gerardi, 2006; Mara and Horan, 2003). On the other hand, major human bacterial pathogens in wastewater include Escherichia spp., Vibrio cholerae, Salmonella spp., Shigella spp., Aeromonas hydrophila, Yersinia spp., Klebsiella spp., Mycobacterium spp., Leptospira spp., Legionella pneumophila, and Pseudomonas. Therefore, special attention is given to bacteria in surface water and wastewater as a contaminant as well as a source of energy for sustainable future, in particular, in the production and supply of safe and clean water sectors [39, 42].

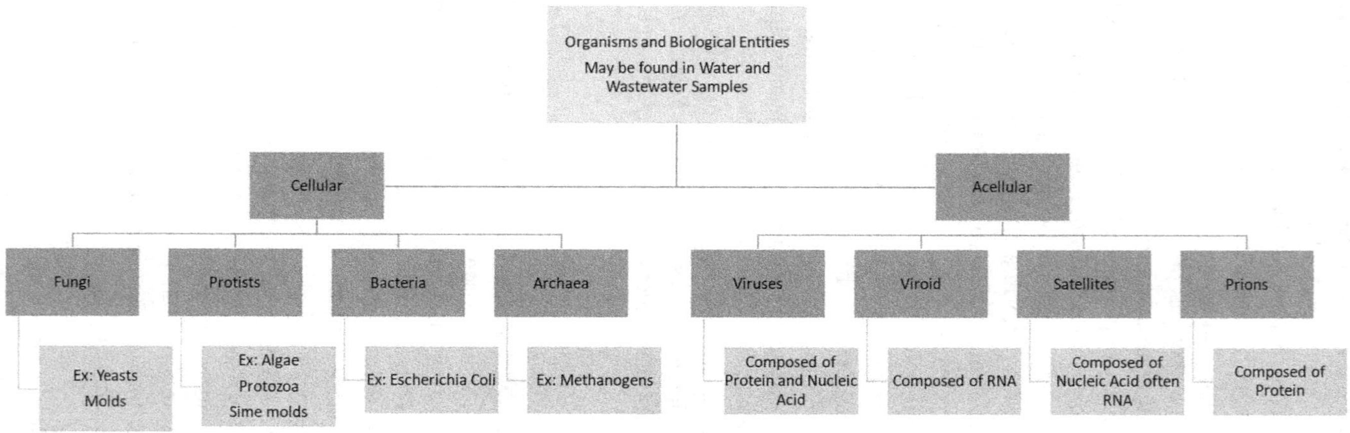

Figure 2.9 Details of biological entities enlisted by scientists in various ecosystems. Details of Bacteria Kingdom: in which major groups of bacteria archaea and eucary have been classified that are potentially found in water and wastewater systems.

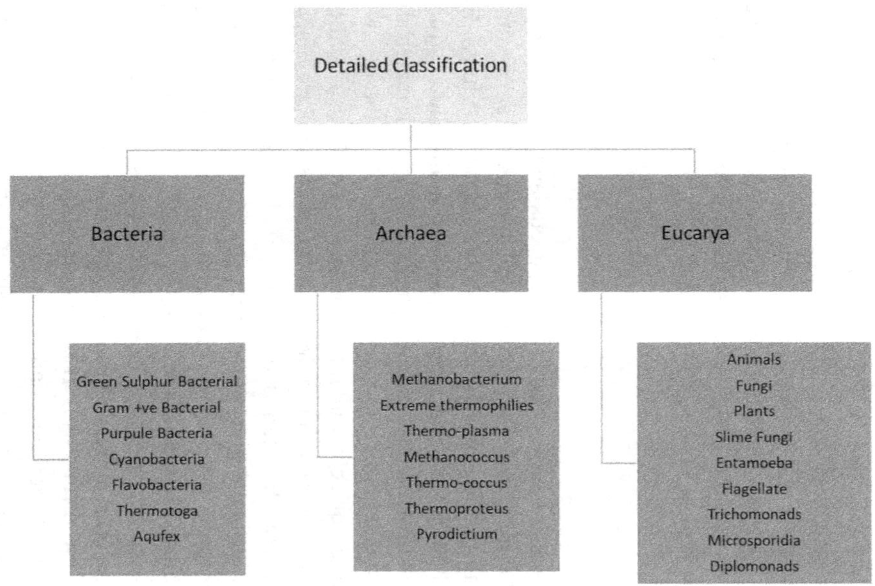

Figure 2.9 (Continued)

2.9.3 Archaea

Similar to bacteria, Archaea are single-celled microorganisms in their basic characteristics. However, these possess different chemical compositions but both belong to prokaryotes. Single-celled Archaea have no nucleus. However, these organisms carry a unique, tough outer cell wall and do not possess peptidoglycan in their cell walls. Archaea are evolutionarily different from bacteria and eukaryotes and these form the third domain of life. Interestingly, Archaea are obligate anaerobes that live in environments containing low-oxygen content wastewater, contaminated water, and soil. Archaea are mostly found living in extreme environments and these microorganisms carry out unusual metabolic processes. For instance, Archaea are known to be extreme species of all extremophiles that live in the frigid environments of Antarctica and others live in the other extreme boiling acidic springs of Yellowstone [122, 123].

2.9.4 Fungi

Fungi are eukaryotes whose cells have a nucleus and other membrane-bound organelles. In the microorganism world, a wide range of eukaryotic organisms exist including all plants, animals, protists, most algae, and fungi. Even though eukaryotes may be found in both single-celled and multicellular structures, these cells possess distinctive nuclei comprising specific cell genetic material surrounded by a unique envelope called the nuclear membrane. On the other hand, fungi primarily contain unique cell wall material, which is made up of chitin. Fungi are also nonphotosynthetic heterotrophic microorganisms and are obligate aerobes, which are organisms that can live or are found or survive and grow in an oxygenated environment. In general terms, fungi are multicellular. However, yeast is a form of fungi, which exists in unicellular form. More significantly, fungi can be usually categorized into four divisions: the (i) Zygomycota (bread molds), (ii) Chytridiomycota (chytrids), (iii) Ascomycota (yeasts and sac fungi), and (iv) the Basidiomycota (club fungi). These categories have been made based on their fungus reproduction mechanism. It is well established that fungi can reproduce following a variety of methods, namely budding, fission, and spore formation. Interestingly, the unicellular forms of yeasts and fungi are oval microorganisms that are larger than bacteria. In the real world, most commonly found fungi are molds, which form visible colonies called mycelia (cottony white or green growths found on bread and fruit). On the other hand, organisms known as slime molds possess both characteristics of fungi and amoebas. Fungi require half of the nitrogen amount that is required by bacteria to form normal cell material. This is the main reason why fungi reproduce and grow in nitrogen-deficient contaminated water and wastewater over bacteria [63] (Gerardi, 2006; Mara and Horan, 2003) [43]. In waste and wastewater systems, fungal communities play a critical role. These are the significant components of activated sludge. These microorganisms are used as the predominant decomposers in wastewater treatment plants or units due to their vastness of fungal biomass and diversity. This is particularly significant as fungi play a dominant role in biodegradation of organic matter and nutrient cycling processes (Evans and Seviour, 2012). A threadlike colony of bacteria such as Sphaerotilus natans and fungi like Fusarium aqueductum, protozoa, and other species can be found adhered on surfaces in sewage or wastewater treatment systems and plant and in natural waters exposed to biological contaminants or as a result of the discharging or discarding of partially treated wastewater from various sources. Nonetheless, microorganisms that are naturally found in the wastewater environment play a critical role in the wastewater treatment process. Presence of bacteria, protozoa, algae, metazoa, and fungi feed helps in breaking down organic material in wastewater [124, 125].

2.9.5 Algae

Algae are a group of eukaryotic microorganisms and they exist in either unicellular or multicellular forms. Generally, cell walls of algae are made up of a carbohydrate known as cellulose. Algae are composed of chlorophyll in their cell, which is used for performing photosynthesis. Algae are abundantly found in freshwater, salt water, soil, and in association with plants. The potential for growth and blooms for algae comes from nutrient pollution in water and soil sources and overabundance of the crucial plant nutrients like nitrogen and phosphorus. This overabundance potentially occurs due to the poor-performing STP. Interestingly, algae help wastewater treatment processes by producing and distributing oxygen in wastewater, which allows aerobic bacteria to degrade and break down organic matter in the water sources. These microorganisms also help in taking up excess phosphorus and nitrogen in the process. Nonetheless, algae in contaminated water sources are a useful ecosystem indicator of health of an ecological system due to their ability based on algal species type, family, and abundance. Harmful cyanobacteria and algae are sometimes called blue-green algae that produce toxins, which cause diseases in both humans and animals. Nevertheless, algae are found in all kinds of water and cause water pollution in a number of ways. Enrichments of algae in water selectively stimulate the growth of algal species through organic effluents producing massive "blooms" or surface growths. Interestingly, certain algae found abundantly in organic matter or waste-contaminated water plays a significant role in self-purification of water bodies. However, some algae in polluted water may cause toxicity to fish and humans.

Nevertheless, algae plays a significant role in the food chain of aquatic life, thereby influencing food chain of aquatic life including fish. Algae contamination or abundance is known to cause bad tastes and odor on surface as well as contaminated water sources. Interestingly, a large number of algae that cause altercation in the tastes and odor also vary in type and existence. Among many algae, certain blue-green algae, diatoms, and colored flagellates such as Chrysophyta and Euglenophyta are known to pose serious issues in water supplies [126].

2.9.6 Protozoa

Another major microorganism species are single-celled protozoa, which are bigger in size and more complex than bacteria. Among many microbes present in wastewater, the most commonly found protozoa include flagellates, ameba, and ciliates. Interestingly, protozoa enhance the clarity of wastewater by consuming small, unsettled floc and free bacteria as shown in Figure 2.10. Most of the protozoa are free living and solitary. However, some do form colonies. Also, many of the protozoa are aerobic chemotrophs in nature. Nevertheless, some protozoans such as amebae and flagellates can survive anerobic conditions. Protozoa are reproduced by binary fission and often consume bacteria. These protozoa are required highly in wastewater and effluent treatment plants as they act as polishers in consuming and breaking down bacteria. In modern wastewater treatment plants, protozoa are most commonly used components, which play an important role in contaminated water and wastewater purification processes. Therefore, in modern man-made wastewater treatment plants, protozoa significantly play a role in improving the quality of product water maintaining the density of dissolved or dispersed bacterial populations by predation. However,

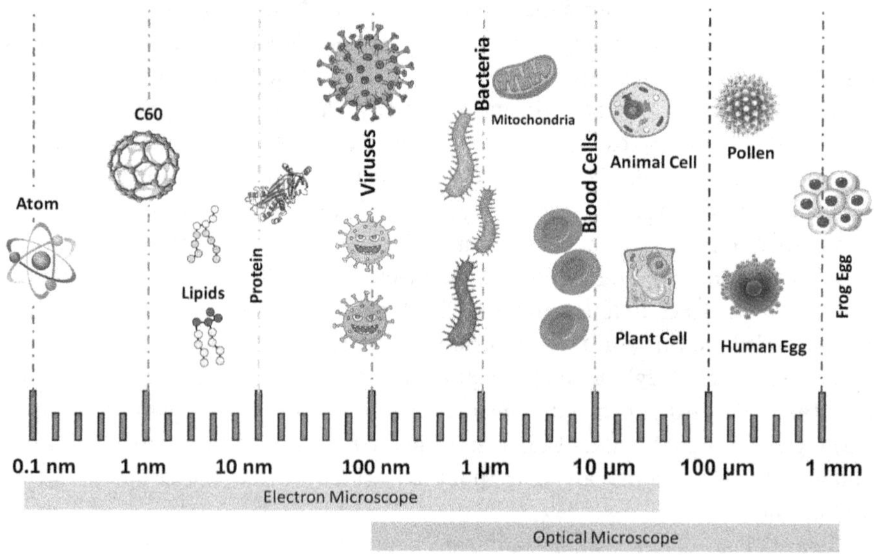

Figure 2.10 Size dependency of different waterborne contaminants, which potentially affect the removal efficiency of treatment techniques.

it is important to purify these microbes before human consumption as the protozoa present in drinking water may cause harmful effects on human health upon consumption or ingestion. Upon consumption, there are many diseases humans incur among many typical symptoms originating from the infection of protozoans, which include diarrhea or amebiasis (amebic dysentery), giardiasis (dysentery), acute gastroenteritis, abdominal pain, severe cramping, dehydration, and so on. Conventionally, boiling is commonly used to kill protozoa, pathogenic bacteria, and viruses [127–129].

2.10 Types of Microorganisms and their Environmental Factors

Microorganisms that mainly occur in the water and wastewater samples can be classified based on their energy and carbon sources. Carbon is the essential building block for cell synthesis in which energy is obtained externally from the cell to enable synthesis to proceed. On the other hand, if the microorganism utilizes organic material as a carbon source, it is categorized as heterotrophic. Nonetheless, autotrophs require only CO_2 to supply their carbon needs. Organisms that rely only on the sun as a source of energy are called phototrophs. Interestingly, chemotrophs obtain energy from either inorganic or organic oxidation/reduction reactions. Organotrophs utilize organic material, meanwhile, lithotrophs oxidize inorganic compounds.

Interestingly, microorganisms can be classified as *obligate aerobes* based on the ability to utilize oxygen as a terminal electron acceptor in a redox reaction in which microorganisms should have oxygen as the terminal electron acceptor. On the other hand, *Obligate anaerobes* are microorganisms that cannot survive in the presence of oxygen. Therefore, they cannot use oxygen as a terminal electron acceptor. Nonetheless, *Facultative aerobes* are an interesting class of microorganisms that will survive and grow in either the presence or absence of oxygen. Under anoxic conditions, which are also conditions of depleted DO, *facultative aerobes* use nitrate (NO_3^-) and nitrites (NO_2^-) as the terminal electron acceptor. Nitrate-nitrogen is converted to nitrogen gas in the absence of oxygen through the process known as *anoxic denitrification* [130].

Based on the survival atmosphere and temperature conditions, microorganisms can be further classified into four temperature ranges. (i) Psychrophiles are those that best survive and grow at a temperature below 20°C, (ii) Mesophiles grow best at temperature ranges between 25 and 40°C, (iii) Thermophiles best grow between 45 and 60°C, and (iv) Stenothermophiles are best grown above 60°C. On the other hand, the growth range of facultative thermophiles extends from the thermophilic range into the mesophilic range.

2.10.1 Size, Shape, and Structure of Bacterial Cells

Bacteria can be found in many sizes and shapes. However, large numbers of bacteria in diameter range from 0.2 to 2.0 μm and in length from 2 to 8 μm [43]. Most of the bacteria that are of medical importance range from 0.2 to 1.5 μm in diameter and from 2 to 5 μm in length (Parija, 2012). Nevertheless, bacteria are generally found in spherical coccus (meaning berries) shape, which means rod-shaped bacillus (meaning little staffs), and spiral as described in Figure 2.11. Figure clearly shows bacteria in various sources such as cocci, which are generally found in round shape but these bacteria can also be found in elongated, oval, or flattened on one side. When cocci reproduce upon division, the cells can remain attached to each other. Further, cocci which stay in pairs with each other after dividing are known as diplococci, similarly those that remain attached in chain-like patterns are known as streptococci. Cocci that divide in 2-planes and remain together for 4 are called tetrads, 3-three planes attached in cube-like groups of eight are known as sarcinae, and multiple planes in the form of grape-like clusters are called staphylococci. Interestingly, Bacilli divide only across their short axis and most bacilli look like single rods. Diplobacilli appear in pairs, streptobacilli appear like chains after division. Bacteria that appear like curved rods are known as vibrios. Remaining bacteria are classified as spirilla that have a helical shape similar to corkscrew, and fairly rigid bodies. Another class of spiral-shaped bacteria are mostly helical and flexible, known as spirochetes that move through

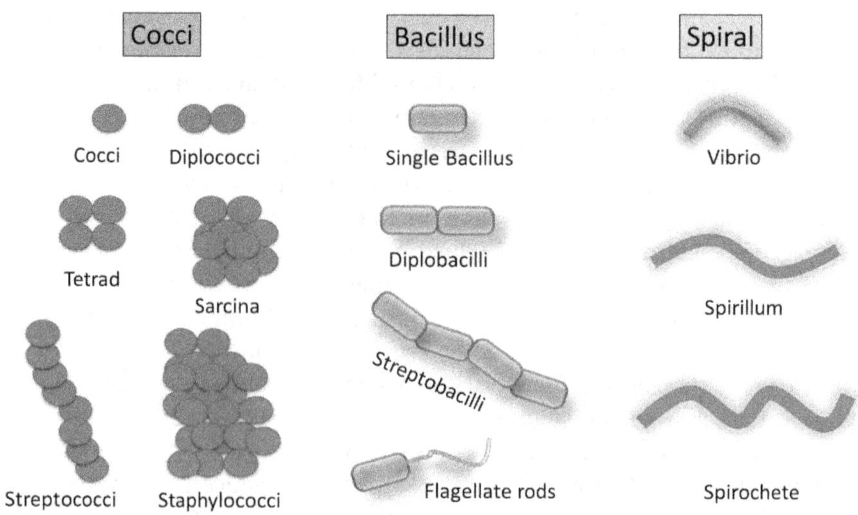

Figure 2.11 Schematic diagram representing different sizes, shapes, and structures of bacteria found in various sources.

their axial filaments resembling flagella within a flexible external sheath. In addition to these 3-basic shapes, bacteria can be found in star-shaped cells, flat cells, rectangular cells, and triangular cells.

Among many bacteria in the environment, majorly seven types of bacteria pose a serious threat to drinking water, namely (i) Escherichia coli, also known as *E. coli*, which cause vomiting, nausea, abdominal pain, and diarrhea upon consumption through contaminated drinking water. (ii) Campylobacter jejuni in contaminated drinking water can cause infections resulting in diarrhea, fever, cramping, and pain. (iii) Hepatitis A in drinking water causes jaundice, fever, stomach pain, dark urine, and fatigue. (iv) Giardia lamblia which is a parasite causes the infection such as giardiasis, cramps, gas, nausea, and diarrhea. (v) Salmonella are generally classified as pathogens and in contaminated water cause fever, diarrhea, headache, chills, and pain. (vi) Legionella pneumophila in contaminated drinking water can result in infection of shortness of breath, cough, and fever, muscle aches, which may require hospitalization or even cause death. (vii) Cryptosporidium, which is a protozoan, which functions like a parasite. Cryptosporidium spreads through contaminated drinking water and causes severe and painful diarrhea. Worryingly, Cryptosporidium can also be found in clean water supplies, which necessitates the regular water quality analysis [131, 132].

2.10.2 Bacterial Growth, Water Quality Indicators, and Analysis

Bacterial growth in various sources and environments refers to an increase in cell numbers. Bacterial growth in various ecosystems and environments is a complex process, which involves a number of anabolic and catabolic interactions, which further result in biosynthetic reactions causing cell division via binary fission mechanism. However, bacterial growth requires several conditions. However, microbial growth can be subdivided into two main categories, namely physical and chemical. Physical microbial growth mechanism requirements require conducive pH, temperature, and osmotic pressure conditions. On the other hand, chemical microbial growth mechanism requires macronutrients as well as micronutrients (Willey et al., 2016). Macronutrients such as carbon, nitrogen, hydrogen, oxygen, phosphorus, and sulfur largely help microorganisms in their compositional growth of bacterial cells, which are required in large quantities. Major micronutrients such as Ca, K, Mg, Fe, and Na, and minor micronutrients such as organic compounds, vitamins, and trace elements like Mn, Mo, Co, Zn. and Ni also help in growth and survival of bacteria in the specific environment. Bacteria consuming micronutrients are consumed in small quantities. For instance, methane-forming bacteria consume sulfur and trace elements like Fe, Co, and Ni to survive. On the other hand, Gram-positive bacteria consume calcium in large quantities for the synthesis of cell walls [43, 63]. In all cases, contaminated water sources or wastewater sources provide significant amounts of nutrients for bacterial and pathogenic contaminants to grow and multiply [133].

Most of the pathogenic microbiological contamination like bacteria, viruses, or protozoa in drinking water sources that are intended for human consumption are generally derived from human or animal waste. Quantity and the nature of pathogenic contamination in drinking water may vary in different samples, which causes most common symptoms upon consumption such as vomiting, diarrhea, and nausea. In extreme conditions, many pathogens can cause severe infections in lungs, kidneys, skin, nervous system, eyes, and liver, and these effects may lead to more severe and, in some cases, chronic or fatal. Nevertheless, worldwide, microorganism-contaminated water consumption kills millions of infants and adults every year. It is estimated that more than half of the global population contract diseases that are directly linked to contaminated water or polluted water sources including dysentery, cholera, typhoid fever, toxoplasmosis, and cryptosporidiosis. Waterborne diseases mainly originate from microbial contamination that kills thousands of children every day in developing countries. In the United States alone, it is estimated that every year more than 500,000 people suffer from several diseases that are termed as waterborne yearly, and more than 7 million population suffer from mild to moderate waterborne infections causing approximately 12,000 deaths a year that can be traced to bacterial infections and transmitted through contaminated water resources. Some of these diseases and their main causes are listed in Table 2.3 [134].

Table 2.3 The main bacterial diseases transmitted through drinking water.

Disease	Causal bacterial agent
Cholera	*Vibrio cholerae*, sero varieties O1 and O139
Gastroenteritis caused by vibrios	Mainly *Vibrio parahaemolyticus*
Typhoid fever and other serious salmonellosis	*Salmonella enterica* subspecies *enterica* serovar Paratyphi
	Salmonella enterica subspecies *enterica* serovar Typhi
	Salmonella enterica subspecies *enterica* serovar Typhimurium
Bacillary dysentery or shigellosis	*Shigella dysenteriae*
	Shigella flexneri
	Shigella boydii
	Shigella sonnei
Acute diarrheas and gastroenteritis	*Escherichia coli*, particularly serotypes such as O148, O157, and O124

The presence of microbial pathogen contamination is often complex and difficult to determine in various water samples due to several factors. Pathogenic contamination is generally in water samples occurring in low quantities, which necessitates the constant monitoring and regular analysis of water samples using "indicator" organisms such as thermotolerant, *E. coli*, and *fecal*, or *total coliforms*. This approach assumes the presence of bacterial indicator organisms, which is associated with the occurrence of microbial pathogens. Traditionally coliform bacteria have been used as an indicator group for most of the microbial pathogens. However, recent times have seen numerous shortcomings in the use of coliforms as indicators for waterborne disease, which led to the identification and adaptation of several alternative indicators. For instance, *Pseudomonas aeruginosa* has been adopted as alternative indicator for the traditional total coliforms, *Candida albicans* for *E. coli*, and *Bifidobacteria* as an alternative indicator for *Fecal streptococci*. Even though several indicators have been identified and thoroughly investigated, however, one in all-purpose indicator has not been adopted for general purpose. Interestingly, these indicators have been distinguished using different methods depending on enumeration procedures such as multiple tube (MT) dilution or most probable number (MPN) and membrane filtration (MF) [135].

The identification, detection, analysis, and enumeration of water and wastewater quality indicators presently rely on the use of important methods, namely (i) membrane filtration (MF), (ii) MT or MPN method, (iii) the standard plate count (SPC) or total bacterial count, and (iv) presence–absence (PA) test. It is important to note that in each method, specifically defined enzyme substrate technologies (DST) have been induced and are increasingly used in microbiological testing of both clinical samples of food and water monitoring. Nevertheless, each method and protocol possess their own advantages and disadvantages; however, methods can be optimized based on the type of water sample to be tested and analyzed [136].

In this membrane filtration method, a known volume of contaminated water sample is passed through a membrane filter of known filter pore size that retains bacteria on the filter surface. The bacteria colonies that grow after incubation on a conductive medium are presumed to have derived from individual bacteria resulting in a direct count of the number of bacteria in the original sample. The MF method is often used in the analysis of various types of contaminated water. On the other hand, MT or Most Probable Number Technique (MPNT) technique involves three main stages, namely (i) the preliminary stage that is presumptive due to the acid and gas, or growth/no growth in which reactions may result due to organisms other than those being selected. (ii) the second stage is the confirmed test, which is usually the extension of the testing step and (iii) the final stage is the complete test in which total coliform is analyzed. Although the final completed test gives the highest reliability, however, time for all tests to complete the analysis may be a time-consuming step [137].

2.10.3 Standard Plate Count

Generally, either pour plate or surface plate technique is used as viable facultative anerobic, aerobic, and heterotrophic bacterial count in water. This procedure gives detailed data on the bacterial colonies and population and their potential growth pattern. This is associated with the combination of incubation temperature and time used. However, it does not estimate the total bacterial count present in the water sources. Nonetheless, this method is frequently used for measuring bacterial populations. More significantly, this method has an important advantage in that it measures the number of viable cells. Further, a plate count is carried out through either the spread plate method or the pour plate method [138].

2.10.4 Presence–Absence Procedure

The presence or absence test (PA) method is often used to determine either presence or absence of pathogens or other traditional biological contamination indicator organisms in the water samples. This frequently used economical method is used for the pathogenic determination, which is also considered a simplified version of the multiple-tube procedure, which allows qualitative information rather than quantitative information. The PA test generally involves analysis of one large volume of test sample in a single culture bottle to produce qualitative information. This is also a simple process in which less utility logistics are involved compared to MPN or MF methods. The PA method gives accurate and precise count number, which is <1 mL coliform or *E. coli*/100 mL, which specifically satisfies the regulatory limits. Therefore, this technique is widely used for the quantitative determinations of respective organisms contaminated with minimal protocols [139].

2.11 The Role of Microorganisms in Wastewater Treatment Plants

The fundamental aim of the wastewater contaminated with biological contaminant treatment is to remove oxygen-demanding organic wastes and to remove nutrients, which limit the photosynthetic organisms in receiving waters that are limited in their growth. Pathogenic contaminated wastewater in treatment plants is a biological amplifier. The plants allow organisms in the form of sludge or biomass, which contains primarily bacteria to utilize various organic pollutants both as substrates and nutrients in the wastewater and convert those organic pollutants to new organisms resulting in nonpolluting wastes or less harmful contaminants.

On the other hand, there are three main biological water treatment methods that are in practice traditionally, namely (i) aerobic, (ii) anerobic, and (iii) anoxic. Aerobic biological water treatment method involves use of microbes and oxygen in a wastewater treatment plant or reactor to optimize the efficient formation of sludge or biomass. In this method, the microorganisms catalyze the oxidation of biodegradable organic and other contaminants in wastewater such as ammonia, and other organic contaminants, which generate less harmful by-products such as carbon dioxide, water, and sludge. On the other hand, anerobic treatment technique proceeds without oxygen and anoxic proceeds in an oxygen-deficient environment in the treatment plant, which is similar to aerobic treatment method. However, here during the use of microorganisms, additional oxygen is not required. These microorganisms utilize nonoxygen compounds to catalyze the oxidation of biodegradable organics resulting in production of harmless by-products. Nevertheless, these three methods can be effectively utilized as individuals or in combination to efficiently treat the contaminated water. In all cases, these methods result in formation of nonhazardous sludge that is easy to handle and dispose of at the end of the treatment process [140].

2.11.1 Secondary Treatment Processes

It is well known that a typical wastewater treatment strategy involves three phases, namely primary, secondary, and tertiary wastewater treatment techniques. Primary treatment methods generally involve physical removal of contaminants from wastewater sources. On the other hand, secondary treatment mainly involves biological treatment techniques briefly discussed in the previous section. However, the major objective of secondary treatment is to reduce or remove the biological oxygen demand that by-passes the primary treatment steps. This step provides further removal of fine SS. However, advanced secondary treatment often includes treatment of nitrogen and phosphorus as these nutrients have been known to cause adverse effects in humans upon excessive consumption through water sources. Further, the advanced secondary treatment technique involves the combination of aerobic oxidation, nitrification, denitrification, phosphorus removal, and activated sludge selector for effective treatment of all contaminants. The main objectives of different processes can be summarized as follows [141].

(i) Aerobic oxidation process is mainly used for the removal of BOD with the use of aerobic heterotrophic bacteria. (ii) Nitrification process involves the oxidation of ammonia in wastewater to nitrate in which aerobic autotrophic bacteria play a predominant role in carrying out nitrification. Here two genera are widely used in which *Nitrosomonas helps in oxidation of* ammonia to nitrite whereas *Nitrobacter helps in oxidation of* nitrite to nitrate. (iii) In denitrification reduction of nitrate to nitric oxide, nitrous oxide, and nitrogen gas are mainly carried out. Unlike the nitrification process, a large number of genera are utilized effectively for denitrification such as heterotrophic or autotrophic. However, *Pseudomonas* is the most commonly used species. (iv) In phosphorus removal process, phosphorus is first incorporated into cell biomass or sludge, which is subsequently removed through phosphorus removal step in which genera like *Acinetobacter*, *Nocardia*, *Aeromonas*, *Arthrobacter*, and *Pseudomonas* are generally used to proceed with the transformation process. (v) In the selector step, a bioreactor design is used that favors the growth of floc-forming bacteria in place of filamentous bacteria, which results in better biomass or sludge forming, settling, and thickening properties.

2.11.2 Activated Sludge and Membrane Bioreactor

Activated sludge formation process is one of the most commonly followed steps used in secondary wastewater treatment plants. The activated sludge is defined as the formation of particles in wastewater through the growth of organisms in aeration tanks. Aerobic bacteria are commonly used organisms for the formation of activated sludge. However, facultative bacteria along with fungi and protozoa can also be used for the purpose. Nonetheless, the exact composition and properties of bacteria can vary depending on the aerobic reactor design, type, environment, and type of wastewater and its characteristics. The microorganisms in sludge oxidize the organic carbon in the wastewater to produce carbon dioxide, water, and new cells. Further, as microorganisms grow, they result in the formation of aggregates of particles, which clump together. With time, these particles form and coagulate resulting in the settlement at the bottom of the tank leaving behind relatively clear supernatant liquid that is free from organic matter and SS. Further, this concentrated sludge is discharged into a safe dumping site and the clear liquid is either released to the river or sent to the tertiary treatment step in case further treatment is required. Interestingly, activated sludge step possesses several advantages such as this step yields high effluent treatment results, resistant to organic fouling, resistant to hydraulic shock loads, drastic reduction of BOD and pathogens, and so on. Nonetheless, activated sludge process has some disadvantages such as involvement of high process cost, high-energy consumption, high operating costs, and involves complicated chemical and microbiological reactions, which may create complex sludge formation and effluent treatment, which may necessitate the incorporation of further treatment or discharge steps [142–145].

On the other hand, membrane bioreactor (MBR) technique has emerged as one of the prominent wastewater treatment process alternatives to popularly used activated sludge technique. This method is proven to overcome several drawbacks of the conventional activated sludge treatment process. MBRs involve two basic treatment arrangements, namely (i) integrated systems that involve the immersion of active membranes in an activated sludge reactor and (ii) separate systems on the other hand involving a membrane system fitted outside the main reactor. Among these, integrated immersion techniques are the most commonly and widely used arrangement. Generally, in this method, membranes used in either flat sheet or tubular mode carry pore sizes ranging in the microfiltration or ultrafiltration regime from 0.035 to 0.4 μm. However, in both cases, membrane forms a barrier for impurities while selectively allowing pure water to pass across. This MBR Process can be efficiently integrated for the enhancement of steps like nitrogen removal, nitrification, and overall biological nutrient removal.

Nevertheless, MBR process provides several advantages over activated sludge process such as it requires less space for settling tanks or clarifiers compared to large-size sludge treatment facilities, thereby drastically reducing plant footprint by 30–50%. MBR produces high-quality effluent with higher volumetric loading rates. MBR requires less hydraulic retention time (HRT), and longer solid retention times (SRT). This method leads to less overall sludge production. An MBR can efficiently produce significantly higher sludge concentrations up to 10–20 g/L with a relatively small reactor compared to conventional activated sludge systems. Nevertheless, MBR technology also poses several disadvantages including higher capital costs and requires efficient membranes with low fouling propensity and high costs of membrane replacement [146].

Among these, membrane fouling in MBR system is a major drawback as it significantly reduces membrane performances and membrane lifespan resulting in an increase in operational and maintenance costs and overall low treatment efficiencies. Membrane fouling is attributable to deposition or accumulation of suspended particulates, colloids, organic solutes, or matter in activated sludge including biomass of microorganisms and cell debris, and sludge flocs known as "*foulants*" on the active surface of membrane. These foulants on the membrane surface result in the blocking or clogging of membrane pores leading to a drastic decline in the permeability of the membrane.

2.12 Future Prospective and Summary

Groundwater quality deterioration and concerns surrounding management is an ever-evolving topic in water sector. However, there are several important factors that greatly affect the ground, surface, as well as wastewater quality and subsequent treatment. Many of the ground and wastewater deterioration-causing contaminants have been discussed in this chapter. The major concern related to groundwater and wastewater treatment is the lack of guidelines. Even though several individual nations have adopted general water quality and treatment guidelines issued by the United Nations Water Project or WHO guidelines. However, water management practice that requires a specific framework and policy guidelines have not been listed comprehensively. In this situation, a determination of various water management entities, managers, and end users becomes vital. Therefore, this chapter has attempted to discuss the information available on both water treatment sector as well as the policy-related issues across the globe. This chapter has paid special attention to discussing the hydrogeological parameters that greatly influence the water quality such as geological formations, type of soil, porosity, specific yield including storage coefficient or effective porosity, type and sources of contaminants, hydraulic conductivity for actual water quality, and contamination levels. In summary, the chapter has highlighted the steps and protocols being framed and followed in maintaining water quality management that further involves consent scrutiny or approval from various stakeholders. These efforts mainly involve policy makers drafting guidelines that are mutually in agreement with water resource regulators, suppliers, monitoring agencies, water end users, and, at the end, stakeholders involving discharging hazardous, toxic, and dangerous substances to treatment plants or safe reserves. This chapter, therefore, provided a detailed account of the general water monitoring and regulatory framework currently followed throughout the world.

References

1 Sustainable Resource Management, Edited by: Chaudhery Mustansar Hussain and Juan F. Velasco-Muñoz Modern Approaches and Contexts, Elsevier, 2021, ISBN: 978-0-12-824342-8.

2 Boix, M., Montastruc, L., Pibouleau, L. et al. (2012). Industrial water management by multiobjective optimization: from individual to collective solution through eco-industrial parks. *J. Clean. Prod.* 22 (1): 85–97. ISSN 0959-6526.

3 Sotelo-Pichardo, C., Ponce-Ortega, J.M., El-Halwagi, M.M., and Frausto-Hernández, S. (2011). Optimal retrofit of water conservation networks. *J. Clean. Prod.* 19 (14): 1560–1581. ISSN 0959-6526.

4 Tan, Y.L. and Manan, Z.A. (2006). Retrofit of water network with optimization of existing regeneration units. *Ind. Eng. Chem. Res.* 45 (22): 7592–7602.

5 Alva-Argáez, A., Kokossis, A.C., and Smith, R. (1998). Wastewater minimisation of industrial systems using an integrated approach. *Comput. Chem. Eng.* 22 (Supplement 1): S741–S744. ISSN 0098-1354.

6 Hansen, É., Rodrigues, M.A.S., Aragão, M.E., and de Aquim, P.M. (2018). Water and wastewater minimization in a petrochemical industry through mathematical programming. *J. Clean. Prod.* 172: 1814–1822. ISSN 0959-6526.

7 Mekuria, W. and Tegegne, D. (2022). Water harvesting. In: *Reference Module in Earth Systems and Environmental Sciences*, 2e (ed. M.J. Goss and M. Oliver). Elsevier ISBN 9780124095489.

8 Topcu, S. and Kirda, C. (2005). IRRIGATION|environmental effects. In: *Encyclopedia of Soils in the Environment* (ed. D. Hillel), 267–273. Elsevier ISBN 9780123485304.

9 Arfanuzzaman, M. and Atiq Rahman, A. (2017). Sustainable water demand management in the face of rapid urbanization and ground water depletion for social–ecological resilience building. *Glob. Ecol. Conserv.* 10: 9–22. ISSN 2351-9894.

10 Kumar, D., Ram, S., and Srivastav, A.L. (2022). Chapter 11 – urban water quality and wastewater discharges in mega and metro cities of India – challenges, impacts, and management: an overview. In: *Current Directions in Water Scarcity Research*, vol. 6 (ed. A.L. Srivastav, S. Madhav, A.K. Bhardwaj, and E. Valsami-Jones), 223–244. Elsevier ISSN 2542-7946, ISBN 9780323918381.

11 Singh, N.S., Sharma, R., and Parween, T. (2020). Wastewater treatments plants and their technological advances. In: *Modern Age Waste Water Problems* (ed. M. Oves, M. Ansari, M. Zain Khan, et al.), 261–274. Cham: Springer.

12 Taniguchi, D.J. (1993). Recent advances in wastewater treatment applied to improve water quality. *Can. Water Resour. J.* 18 (3): 313–319.

13 Kesari, K.K., Soni, R., Jamal, Q.M.S. et al. (2021). Wastewater treatment and reuse: a review of its applications and health implications. *Water Air Soil Pollut.* 232: 208.

14 Kesari, K.K., Verma, H.N., and Behari, J. (2011b). Physical methods in wastewater treatment. *Int. J. Environ. Technol. Manag.* 14: 43–66.

15 Nourani, V., Elkiran, G., and Abba, S.I. (2018). Wastewater treatment plant performance analysis using artificial intelligence – an ensemble approach. *Water Sci. Technol.* 78 (10): 2064–2076.

16 Singh, A., Sawant, M., Kamble, S.J. et al. (2019). Performance evaluation of a decentralized wastewater treatment system in India. *Environ. Sci. Pol.* 26: 21172–21188.

17 Zamora, S., Marín-Muñíz, J.L., Nakase-Rodríguez, C. et al. (2019). Wastewater treatment by constructed wetland eco-technology: influence of mineral and plastic materials as filter media and tropical ornamental plants. *Water* 11: 2344.

18 Vaux, H. (2011). Groundwater under stress: the importance of management. *Environ. Earth Sci.* 62: 19–23.

19 Vergili, I. (2013). Application of nanofiltration for the removal of carbamazepine, diclofenac and ibuprofen from drinking water sources. *J. Environ. Manag.* 127: 177–187.

20 Chakraborty, B., Roy, S., Bera, B. et al. (2022). Evaluation of groundwater quality and its impact on human health: a case study from Chotanagpur plateau fringe region in India. *Appl. Water Sci.* 12: 25.

21 Li, J., Sun, C., Chen, W. et al. (2022). Groundwater quality and associated human health risk in a typical basin of the eastern Chinese Loess Plateau. *Water* 14: 1371.

22 Chen, F., Yao, L., Mei, G. et al. (2021). Groundwater quality and potential human health risk assessment for drinking and irrigation purposes: a case study in the Semiarid Region of North China. *Water* 13: 783.

23 EPA (Environmental Protection Agency) (2002). *National Recommended Water Quality Criteria: 2002*. Office of Water, Office of Science and Technology EPA-822-R-02-047.

24 Altenburger, R., Brack, W., Burgess, R.M. et al. (2019). Future water quality monitoring: improving the balance between exposure and toxicity assessments of real-world pollutant mixtures. *Environ. Sci. Eur.* 31: 12.

25 Busch, W., Schmidt, S., Kühne, R. et al. (2016). Micropollutants in European rivers: a mode of action survey to support the development of effect-based tools for water monitoring. *Environ. Toxicol. Chem.* 35: 1887–1898.

26 Encinas, C., Ruiz, E., Cortez, J., and Espinoza, A. (2017). Design and implementation of a distributed IoT system for the monitoring of water quality in aquaculture. Wireless Telecommunications Symposium (WTS), 26–28 April 2017, Chicago, IL, USA pp. 1–7.

27 Ahmed, U., Mumtaz, R., Anwar, H. et al. (2020). Water quality monitoring: from conventional to emerging technologies. *Water Supply* 20 (1): 28–45.

28 Yiping, W. and Chen, J. (2013). Investigating the effects of point source and nonpoint source pollution on the water quality of the East River (Dongjiang) in South China. *Ecol. Indic.* 32: 294–304. ISSN 1470-160X.

29 Weiner, R.F. and Matthews, R.A. (2003). *Environmental Engineering*, 4e. Elsevier ISBN: 978-0-7506-7294-8.

30 Worms, I.A.M., Szigeti, Z.A.-G., Dubascoux, S. et al. (2010). Colloidal organic matter from wastewater treatment plant effluents: characterization and role in metal distribution. *Water Res.* 44 (1): 340–350. ISSN 0043-1354.

31 Bunch, B.W. and Griffin, D.M. (1992). Rapid removal of colloidal organics from domestic wastewater. *Water Sci. Technol.* 26 (9–11): 2433–2436.

32 Garfí, M. (2022). Biological treatment of organic waste in wastewater – towards a circular and bio-based economy. *Water* 14: 360.

33 Wen, Y., Schoups, G., and van de Giesen, N. (2017). Organic pollution of rivers: combined threats of urbanization, livestock farming and global climate change. *Sci. Rep.* 7: 43289.

34 Schaider, L.A., Rodgers, K.M., and Rudel, R.A. (2017). Review of organic wastewater compound concentrations and removal in onsite wastewater treatment systems. *Environ. Sci. Technol.* 51 (13): 7304–7317.

35 Matilainen, A., Vepsäläinen, M., and Sillanpää, M. (2010). Natural organic matter removal by coagulation during drinking water treatment: a review. *Adv. Colloid Interf. Sci.* 159 (2): 189–197.

36 Soh, Y.C., Roddick, F., and van Leeuwen, J. (2008). The impact of alum coagulation on the character, biodegradability and disinfection by-product formation potential of reservoir natural organic matter (NOM) fractions. *Water Sci. Technol.* 58 (6): 1173–1179. https://doi.org/10.2166/wst.2008.475.PMID: 18845853.

37 Matilainen, A., Gjessing, E.T., Lahtinen, T. et al. (2011). An overview of the methods used in the characterisation of natural organic matter (NOM) in relation to drinking water treatment. *Chemosphere* 83 (11): 1431–1442. https://doi.org/10.1016/j.chemosphere.2011.01.018. Epub 2011 Feb 11.PMID: 21316073.

38 Ghernaout, D. (2020). Natural organic matter removal in the context of the performance of drinking water treatment processes – technical notes. *Open Access Libr.* 7: 1–40.

39 Pandey, P.K., Kass, P.H., Soupir, M.L. et al. (2014). Contamination of water resources by pathogenic bacteria. *AMB Expr.* 4: 51.

40 Barrantes, K., Chacón, L., Morales, E. et al. (2022). Occurrence of pathogenic microorganisms in small drinking-water systems in Costa Rica. *J. Water Health* 20 (2): 344–355.

41 Steven, J.A.C., Thorn, R.M.S., Robinson, G.M. et al. (2022). The control of waterborne pathogenic bacteria in fresh water using a biologically active filter. *NPJ Clean Water* 5: 30.

42 Some, S., Mondal, R., Mitra, D. et al. (2021). Microbial pollution of water with special reference to coliform bacteria and their nexus with environment. *Energy Nexus* 1: 100008. ISSN 2772-4271.

43 Tortora, G.J., Funke, B.R., and Case, C.L. (2010). *Microbiology: An Introduction*, 10e. San Francisco: Pearson Benjamin Cummings.

44 Georgiades, K. and Raoult, D. (2011). Defining pathogenic bacterial species in the genomic era. *Front. Microbiol.* 1: 151.

45 Najafpour, G.D. (2015). Chapter 19 – biological treatment This case study was partially written with contributions from:Roya Pishgar, School of Civil Engineering, Babol Noshirvani University of Technology, Babol, Mazandaran, Iran. In: *Biochemical Engineering and Biotechnology*, 2e (ed. G.D. Najafpour), 557–595. Elsevier ISBN 9780444633576.

46 McKim, J.M. and Goeden, H.M. (1982). A direct measure of the uptake efficiency of a xenobiotic chemical across the gills of brook trout (*Salvelinus fontinalis*) under normoxic and hypoxic conditions. *Comp. Biochem. Physiol. C: Comp. Pharmacol.* 72 (1): 65–74. ISSN 0306-4492.

47 Pedersen, C.L. (1987). Energy budgets for juvenile rainbow trout at various oxygen concentrations. *Aquaculture* 62 (3–4): 289–298. ISSN 0044-8486.

48 Abdel Magid, A.M. and Babiker, M.M. (1975). Oxygen consumption and respiratory behaviour of three Nile fishes. *Hydrobiologia* 46: 359–367.

49 Yu, L.J. and Brooks, M.S.-L. (2016). Chapter 7 – food industry protein by-products and their applications. In: *Protein Byproducts* (ed. G.S. Dhillon), 120–132. Academic Press ISBN 9780128023914.

50 Gogate, P.R., Thanekar, P.D., and Oke, A.P. (2020). Strategies to improve biological oxidation of real wastewater using cavitation based pre-treatment approaches. *Ultrason. Sonochem.* 64: 105016. ISSN 1350-4177.

51 Hatfield, R. (1957). Biological oxidation of some organic compounds. *Ind. Eng. Chem.* 49 (2): 192–196.

52 Madan, S., Madan, R., and Hussain, A. (2022). Advancement in biological wastewater treatment using hybrid moving bed biofilm reactor (MBBR): a review. *Appl. Water Sci.* 12: 141.

53 Cui, B., Shaozhu, F., Hao, X., and Zhou, D. (2023). Synergistic effects of simultaneous coupling ozonation and biodegradation for coking wastewater treatment: advances in COD removal, toxic elimination, and microbial regulation. *Chemosphere* 318: 137956. ISSN 0045-6535.

54 Woodard & Curran, Inc (2006). 5 – waste characterization. In: *Industrial Waste Treatment Handbook (Second Edition)* (ed. Woodard & Curran, Inc), 83–126. Butterworth-Heinemann ISBN 9780750679633.

55 Gupta, S.K. and Sharma, R. (1996). Biological oxidation of high strength nitrogenous wastewater. *Water Res.* 30 (3): 593–600.

56 Mines, R.O. and Sherrard, J.H. (1997). Biological treatment of a high strength nitrogenous wastewater. *J. Environ. Sci. Health* 32 (5): 1353–1375.

57 Sengupta, S., Nawaz, T., and Beaudry, J. (2015). Nitrogen and phosphorus recovery from wastewater. *Curr. Pollution Rep.* 1: 155–166.

58 Lv, J., Hongjuan, W., and Chen, M. (2011). Effects of nitrogen and phosphorus on phytoplankton composition and biomass in 15 subtropical, urban shallow lakes in Wuhan, China. *Limnologica* 41 (1): 48–56. ISSN 0075-9511.

59 Ayed Radi Khanfar (2010). Dissolved nitrogen in drinking water resources in Al-Mahareth village of Assir – Saudi Arabia. *Saudi J. Biol. Sci.* 17 (3): 265–268. ISSN 1319-562X.

60 Rouse, J.D., Bishop, C.A., and Struger, J. (1999). Nitrogen pollution: an assessment of its threat to amphibian survival. *Environ. Health Perspect.* 107: 799–803.

61 Li, Y.-h., Li, H.-b., Xin-yang, X. et al. (2017). Fate of nitrogen in subsurface infiltration system for treating secondary effluent. *Water Sci. Eng.* 10 (3): 217–224. ISSN 1674-2370.

62 Wagner, S.C. (2011). Biological nitrogen fixation. *Nat. Educ. Knowl.* 3 (10): 15.

63 Davis, M.L. and Cornwell, D.A. (2012). *Introduction to Environmental Engineering*. McGraw-Hill Education.

64 Prendl, L. and Nikolavcic, B. (2000). Aerobic treatment of industrial waste water- experiencies with the dosage of nitrogen and phsophorus. *Water Sci. Technol.* 41 (9): 241–249.

65 Peter, A. and Sarfert, F. (1991). Operation experiences with biological phosphorus removal at the sewage treatment plants of Berlin (West). *Water Sci. Technol.* 24 (7): 133–148.

66 Sharpley, A.N., Daniel, T., Sims, T. et al. (2003). *Agricultural Phosphorus and Eutrophication*, 2e. Agricultural Research Service: United States Department of Agriculture.

67 Karapınar, N. (2009). Application of natural zeolite for phosphorus and ammonium removal from aqueous solutions. *J. Hazard. Mater.* 170 (2): 1186–1191.

68 Lam, K.L., Solon, K., Jia, M. et al. (2022). Life cycle environmental impacts of wastewater-derived phosphorus products: an agricultural end-user perspective. *Environ. Sci. Technol. 56* (14): 10289–10298.

69 Huang, H., Xiao, D., Pang, R. et al. (2014). Simultaneous removal of nutrients from simulated swine wastewater by adsorption of modified zeolite combined with struvite crystallization. *Chem. Eng. J.* 256: 431–438.

70 Sedlak, R. (1991). *Phosphorus and Nitrogen Removal from Municipal Wastewater*. New York: Lewis Publishers.

71 Egle, L., Rechberger, H., Krampe, J., and Zessner, M. (2016). Phosphorus recovery from municipal wastewater: an integrated comparative technological, environmental and economic assessment of P recovery technologies. *Sci. Total Environ.* 571: 522–542. ISSN 0048-9697.

72 Ban, Z., Wei, T., Ke, X. et al. (2023). Fate and distribution of phosphorus in coking wastewater treatment: from sludge to its derived biochar. *Sci. Total Environ.* 881: 163384. ISSN 0048-9697.

73 Gleisberg, D., Kandler, J., Ulrich, H., and Hartz, P. (1976). Eutrophication and wastewater purification. *Angew. Chem. Int. Ed. Eng.* 15: 354–365.

74 Preisner, M., Neverova-Dziopak, E., and Kowalewski, Z. (2020). Analysis of eutrophication potential of municipal wastewater. *Water Sci. Technol.* 81 (9): 1994–2003.

75 Mohammed, I.N., Bolten, J.D., Souter, N.J. et al. Diagnosing challenges and setting priorities for sustainable water resource management under climate change. *Sci. Rep.* 12: 796.

76 Reed, P.M. and Kasprzyk, J. (2009). Water resources management: the myth, the wicked, and the future. *J. Water Resour. Plan. Manag.* 135: 411–413.

77 Lund, J.R. (2015). Integrating social and physical sciences in water management. *Water Resour. Res.* 51: 5905–5918.

78 Eštoková, A., Wolfová Fabiánová, M., and Ondová, M. (2022). Concrete structures and their impacts on climate change and water and raw material resource depletion. *Int. J. Civ. Eng.* 20: 735–747.

79 Mo, W. and Zhang, Q. (2013). Energy–nutrients–water nexus: integrated resource recovery in municipal wastewater treatment plants. *J. Environ. Manag.* 127: 255–267. ISSN 0301-4797.

80 Knotts, J. (1999). A brief history of drinking water regulations, drinking water news for America's small communities, On Tap, inter. 8 (4): 1–24.

81 Verouden, N.W. and Meijman, F.J. (2010). Water, health and the body: the tide, undercurrent and surge of meanings. *Water History* 2: 19–33.

82 Ertsen, M.W. and Hoag, H.J. (2011). Editorial issue 1 volume 3, 2011. *Water History* 3: 1–2.

83 Wilkinson, T.J. (2010). Introduction. *Water History* 2: 85–90.

84 Bhatnagar, A., Hogland, W., Marques, M., and Sillanpää, M. (2013). An overview of the modification methods of activated carbon for its water treatment applications. *Chem. Eng. J.* 219: 499–511.

85 Mulhern, R. and MacDonald Gibson, J. (2020). Under-sink activated carbon water filters effectively remove lead from private well water for over six months. *Water* 12: 3584.

86 Arvelo, W., Pavia, A., and Braden, C. (2008). Chapter 63 – foodborne and waterborne disease. In: *Principles and Practice of Pediatric Infectious Disease*, 3e (ed. S.S. Long), 397–406. W.B. Saunders ISBN 9780702034688.

87 Nangmenyi, G. and Economy, J. (2009). Chapter 1 – nanometallic particles for oligodynamic microbial disinfection. In: *Micro and Nano Technologies, Nanotechnology Applications for Clean Water* (ed. N. Savage, M. Diallo, J. Duncan, et al.), 3–15. William Andrew Publishing ISBN 9780815515784.

88 Nataraj, S.K. (2022). *Emerging Pollutant Treatment in Wastewater*. CRC Press. ISBN: 9781032103242.

89 Zhang, S., Wang, Y., Lu, J. et al. (2021). Chlorine disinfection facilitates natural transformation through ROS-mediated oxidative stress. *ISME J.* 15: 2969–2985.

90 Jin, M., Liu, L., Wang, D. et al. (2020). Chlorine disinfection promotes the exchange of antibiotic resistance genes across bacterial genera by natural transformation. *ISME J.* 14: 1847–1856.

91 Public Health Service Review: The health consequences of smoking: 1969 supplement to the 1967 public health service review united states. Public Health Service. Office of the Surgeon General, 1–25.

92 FEPA (1991). *Proposed National Water Quality Standards*. Federal Environmental Protection Agency.

93 US EPA (2012). Guidelines for water reuse 600/R-12/618. Environmental Protection Agency.

94 UNESCO (2007). UNESCO water portal newsletter no. 161. Waterrelated diseases. http://www.unesco.org/water/news/newsletter/161.shtml.

95 USEPA (2011). *Exposure Factors Handbook*, 2011e. U.S. Environmental Protection Agency https://doi.org/EPA/600/R-090/052F.

96 WHO (2011). *WHO Guidelines for Drinking-water Quality*, 4e. Geneva, Accesed on 15th February 2024, ISBN: 978-92-4-154995-0: World Health Organization.

97 United States (1974). Safe Drinking Water Act. Pub. L. 93–523; 88 Stat. 1660; 42 U.S.C. § 300f et seq.

98 EPA Alumni Association: Early implementation of the Safe Drinking Water Act of 1974, by EPA of the Safe Drinking Water Act of 1974.

99 Chen, G.-Q., Wu, Y.-H., Tan, Y.-J. et al. (2022). Pretreatment for alleviation of RO membrane fouling in dyeing wastewater reclamation. *Chemosphere* 292: 133471. ISSN 0045-6535.

100 United States. Clean Water Act (CWA) section 502 (7); 33 U.S.C. § 1362.

101 Totenberg, N. (2023). The supreme court has narrowed the scope of the clean water act. NPR. Retrieved 26 May 2023.

102 World conferences introduction. www.un.org. Archived from the original on 10 March 2018 (accessed 15 February 2024).

103 Directive 2008/1/EC.

104 Directive 96/61/EC.

105 https://www.gsa.gov/reference/executive-orders (accessed 15 February 2024).

106 https://www.unep.org/ (accessed 15 February 2024).

107 https://www.unep.org/explore-topics/sustainable-development-goals/why-do-sustainable-development-goals-matter/goal-6 (accessed 15 February 2024).

108 Pakharuddin, N.H. et al. (2021). Water treatment process using conventional and advanced methods: a comparative study of Malaysia and selected countries. *IOP Conf. Ser.: Earth Environ. Sci.* 880: 012017–012028.

109 Zinicovscaia, I. (2016). Conventional methods of wastewater treatment. In: *Cyanobacteria for Bioremediation of Wastewaters* (ed. I. Zinicovscaia and L. Cepoi). Cham: Springer.

110 Brandt, M.J., Johnson, K.M., Elphinston, A.J., and Ratnayaka, D.D. (2017). Chapter 7 – chemistry, microbiology and biology of water. In: *Twort's Water Supply*, 7e (ed. M.J. Brandt, K.M. Johnson, A.J. Elphinston, and D.D. Ratnayaka), 235–321. Butterworth-Heinemann ISBN 9780081000250.

111 Redondo, J.A. (2002). Brackish-, sea- and wastewater desalination. *Desalination* 138: 29–40.

112 Kurniawan, T.A., Chana, G.Y.S., Lo, W.H., and Babel, S. (2006). Physico–chemical treatment techniques for wastewater laden with heavy metals. *Chem. Eng. J.* 118: 83–98.

113 Abu Qdaisa, H. and Moussab, H. (2004). Removal of heavy metals from wastewater by membrane processes: a comparative study. *Desalination* 164: 105–110.

114 Owlad, M., Kheireddine Aroua, M., Daud, W.A.W., and Baroutian, S. (2009). Removal of hexavalent chromium-contaminated water and wastewater: a review. *Water Air Soil Pollut.* 200: 59–77.

115 Margolis, S.A. and Huang, P.H. (2005). Water determination. In: *Encyclopedia of Analytical Science (Second Edition)* (ed. P. Worsfold, A. Townshend, and C. Poole), 357–363. Elsevier ISBN 9780123693976.

116 Alygizakis, N.A., Urík, J., Beretsou, V.G. et al. (2020). Evaluation of chemical and biological contaminants of emerging concern in treated wastewater intended for agricultural reuse. *Environ. Int.* 138: 105597. ISSN 0160-4120.

117 Ze, R., Fang, W., Xiaodong, Q. et al. (2017). Taxonomic and functional differences between microbial communities in Qinghai lake and its input streams. *Front. Microbiol.* 8: 2319.

118 Kuang, J., Huang, L., He, Z. et al. Predicting taxonomic and functional structure of microbial communities in acid mine drainage. *ISME J.* 10: 1527–1539.

119 Dong, X., Greening, C., Rattray, J.E. et al. (2019). Metabolic potential of uncultured bacteria and archaea associated with petroleum seepage in deep-sea sediments. *Nat. Commun.* 10: 1816.

120 Malard, L.A. and Pearce, D.A. (2018). Microbial diversity and biogeography in Arctic soils. *Environ. Microbiol. Rep.* 10: 611–625.

121 Flemming, H.C. and Wuertz, S. (2019). Bacteria and archaea on Earth and their abundance in biofilms. *Nat. Rev. Microbiol.* 17 (4): 247–260.

122 Inkinen, J., Siponen, S., Jayaprakash, B. et al. (2021). Diverse and active archaea communities occur in non-disinfected drinking water systems–less activity revealed in disinfected and hot water systems. *Water Res. X* 12: 100101. ISSN 2589-9147.

123 Khangembam, C.D., Singh, S.P., Chakrabarti, R., and Sharma, J. (2023). Study of effect of various temperatures on the abundance of ammonia oxidizing archaea and bacteria. *Indian J. Anim. Sci.* 88 (5): 626–632.

124 Novak Babič, M., Gunde-Cimerman, N., Vargha, M. et al. (2017). Fungal contaminants in drinking water regulation? A tale of ecology, exposure, purification and clinical relevance. *Int. J. Environ. Res. Public Health* 14 (6): 636.

125 Hageskal, G., Lima, N., and Skaar, I. (2009). The study of fungi in drinking water. *Mycol. Res.* 113 (2): 165–172. ISSN 0953-7562.

126 Abdel-Raouf, N., Al-Homaidan, A.A., and Ibraheem, I.B.M. (2012). Microalgae and wastewater treatment. *Saudi J. Biol. Sci.* 19 (3): 257–275. ISSN 1319-562X.

127 Syed, T., Batool, U., Aslam, M. et al. (2019). Bioremediation and decontamination potential of flagellate Poteriospumella sp. *Bioremediat. J.* 23 (3): 142–153.

128 Pauli, W., Jax, K., and Berger, S. (2001). Protozoa in wastewater treatment: function and importance. In: *Biodegradation and Persistence. The Handbook of Environmental Chemistry*, vol. 2/2K (ed. B. Beek). Berlin, Heidelberg: Springer.

129 Teel, L., Olivieri, A., Danielson, R. et al. (2022). Protozoa reduction through secondary wastewater treatment in two water reclamation facilities. *Sci. Total Environ.* 807 (3): 151053. ISSN 0048-9697.

130 Rajta, A., Bhatia, R., Setia, H., and Pathania, P. (2020). Role of heterotrophic aerobic denitrifying bacteria in nitrate removal from wastewater. *J. Appl. Microbiol.* 128: 1261–1278.

131 Chien, A.-C., Hill, N.S., and Levin, P.A. (2012). Cell size control in bacteria. *Curr. Biol.* 22 (9): R340–R349. ISSN 0960-9822.

132 Teimouri, H., Mukherjee, R., and Kolomeisky, A.B. (2020). Stochastic mechanisms of cell-size regulation in bacteria. *J. Phys. Chem. Lett. 11* (20): 8777–8782.

133 Wen, X., Chen, F., Lin, Y. et al. (2020). Microbial indicators and their use for monitoring drinking water quality – a review. *Sustainability 12*: 2249.

134 Cabral, J. (2010). Water microbiology: bacterial pathogens and water. *Int. J. Environ. Res. Public Health* 7: 3657–3703.

135 Cohen, J. and Shuval, H.I. (1973). Coliforms, fecal coliforms, and fecal streptococci as indicators of water pollution. *Water Air Soil Pollut.* 2: 85–95.

136 Scow, K.M., Schwartz, E., Johnson, M.J., and Macalady, J.L. (2001). Microbial biodiversity, measurement. In: *Encyclopedia of Biodiversity*, 2e (ed. S.A. Levin), 259–270. Academic Press ISBN 9780123847201.

137 Shariq, M., Singh, S., Farooq, U. et al. (2016). Presumptive Coliform count in water sample collected from different sites of a University, Moradabad, Uttar Pradesh, India. *Int. J. Sci. Stud.* 3 (12): 91–96.

138 Spinedi, C. and Gisin, M. (1990). Standard plate counts of drinking water: a comparison between incubation temperatures of 20 and 30 degrees C. *Int. J. Food Microbiol.* 11 (1): 93–102. https://doi.org/10.1016/0168-1605(90)90042-4.

139 Rice, E.W., Geldreich, E.E., and Read, E.J. (1989). The presence-absence coliform test for monitoring drinking water quality. *Public Health Rep.* 104 (1): 54–58. PMID: 2493663; PMCID: PMC1580284.

140 Akansha, K., Majumder, S., Misra, M., and Sachan, S.G. (2021). Chapter 4 – role of microorganisms in performance optimization of wastewater treatment plants. In: *Wastewater Treatment* (ed. M.P. Shah, A. Sarkar, and S. Mandal), 67–91. Elsevier ISBN 9780128218815.

141 Islam, R., Yu, R.M.K., Andrew-Priestley, M. et al. (2021). Secondary treatment phase of tertiary wastewater treatment works significantly reduces estrogenic load. *Water Res.* 200: 117257. ISSN 0043-1354.

142 Al-Asheh, S., Bagheri, M., and Aidan, A. (2021). Membrane bioreactor for wastewater treatment: a review. *Case Stud. Chem. Environ. Eng.* 4: 100109. ISSN 2666-0164.

143 Judd, S.J. (2016). The status of industrial and municipal effluent treatment with membrane bioreactor technology. *Chem. Eng. J.* 305: 37–45. ISSN 1385-8947.

144 Erkan, H.S., Turan, N.B., and Engin, G.Ö. (2018). Chapter five – membrane bioreactors for wastewater treatment. In: *Comprehensive Analytical Chemistry*, vol. 81 (ed. D.S. Chormey, S. Bakırdere, N.B. Turan, and G.Ö. Engin), 151–200. Elsevier ISSN 0166-526X.

145 Aslam, M., Charfi, A., Lesage, G. et al. (2017). Membrane bioreactors for wastewater treatment: a review of mechanical cleaning by scouring agents to control membrane fouling. *Chem. Eng. J.* 307: 897–913. ISSN 1385-8947.

146 Ligero, P., de Vega, A., and Soto, M. (2001). Influence of HRT (hydraulic retention time) and SRT (solid retention time) on the hydrolytic pre-treatment of urban wastewater. *Water Sci. Technol.* 44 (4): 7–14. PMID: 11575103.

3

Groundwater Contamination and its Treatment Techniques

3.1 Introduction to Groundwater Contamination

Groundwater is a precious freshwater source that exists underground in the cavities and saturated zones beneath the land surface. The upper surface of these saturated zones is known as water table. These water tables are specific underground boundaries that exist between the soil surfaces and the area in which groundwater saturates spaces between sediments and cracks in rocks. Interestingly, atmospheric pressure and water pressure are equal at these boundaries. Contrary to popular belief, groundwater does not form or originate from underground rivers. Nevertheless, with all-natural protections and conservatives in place, groundwater can still become contaminated in many ways as the surface water by the diffusion of the same pollutants that contaminate surface water. The major contaminants from the land leach into the groundwater sources and aquifers through soil and land contamination. Over a period of time, saturated land and soil contamination further leads to the percolation into groundwater layers and results in quality degradation. On the other hand, soil pollution, soil contamination, or land pollution occurs as a result of land degradation caused by human-made activities and the presence of xenobiotic or synthetic chemicals or other alteration substances in the natural soil environment either for the purpose of agricultural activities or deliberate disposal. Also, soil contamination is typically caused by agricultural and irrigation chemicals like fertilizers, domestic waste dumping, or improper disposal of waste industrial activity. The other major source of land and soil contamination occurs from fossil fuel-based chemicals such as petroleum hydrocarbons like discarded plastics, polynuclear aromatic hydrocarbons, for instance, benzo(a)pyrene, pesticides, and naphthalene; solvents; and inorganic contaminants like lead, chromium, and other heavy metals. Today, soil, air, and water pollution are directly correlated

Materials and Methods for Industrial Wastewater and Groundwater Treatment,
First Edition. S. K. Nataraj.

with the degree of industrialization, urbanization, and, in particular, direct production and disposal of chemical substances. Specifically, the concern over soil and land contamination stems primarily from further contamination of water sources surrounded or available immediately next to the pollutants causing severe health risks. This may follow as a result of water sources that come in direct contact with the contaminated soil or land mass, leaching from the contaminants, or through the secondary pollutants discarded near water supplies within the vicinity of contaminated soil. Soil or land pollution is largely associated with the presence of a series of pollutants generally found in the soil samples as follows [1, 2].

Among several reasons, accidental discharge of domestic and sewage spills causes major soil contamination, landfill, and illegal dumping, oil spills, direct discharge of partially treated or untreated industrial wastes to landfills or open surfaces, electronic waste disposal, coal ash discharge, contamination by metal and heavy metal containing waste like Pb due to vehicle exhaust, Zn and Cd caused by tire wear, contaminated rocks structures containing large amounts of toxic elements, incineration of fossil raw materials, which contaminates through air or soil, formation of microplastics due to extensive use and discard of plastic items, mining and activities heavy industries, leakage of contaminants due to corrosion of underground waste storage tanks including piping, fitting, and so on that are used to transmit the contents, acid rain, intensive farming, petrochemicals, construction activities, agrochemicals and mineral such as fertilizers, pesticides, and herbicides, industrial accidents, road debris, exterior metal-induced paints like lead-added paints, contamination leaking from drainage to surface water into the soil, ammunition, chemical agents, and other agents of war, waste disposal, oil and fuel dumping, nuclear wastes, other emerging pollutants, and so on [3–6].

Therefore, groundwater pollution or contamination occurs when unwanted, undesired, toxic, and hazardous substances and chemicals are discharged or disposed of, deposited on, or leached through the porous soil layers of the land surface above the groundwater as shown in Figure 3.1. This groundwater contamination occurs even in absence of industrial, human activities, and domestic pollution sources in the area as contamination may transmit from one point to another within the groundwater sources. Therefore, it is important to understand that the water may not always be free from contaminants and should be thoroughly tested before human consumption. Recent concerns on emerging pollutants like arsenic, fluoride, pharmaceutical ingredients, and rock minerals are largely found in the earth's crust as these contaminants are found in high concentrations in different parts of the geological locations throughout the world. On the other hand, mineral contaminants that are found excessively such as manganese, iron, ammonium, dissolved organic materials, modern agriculture

Figure 3.1 Schematic diagram showing various sources of surface water contamination that seep into groundwater and in long-term contaminants groundwater sources.

and irrigation booster chemicals, additives, substances, soil preservatives, and so on are also found in high concentrations in groundwater sources that are posing serious health, ecological, and environmental threats [7, 8].

It is well known that pollution can occur from two types of sources: (i) point and (ii) nonpoint. Point sources are the ones which can be accountable, and identifiable and can be traced to localized sources. Point source type contaminants that pollute groundwater include landfills, septic systems, oil storage tanks, or buried gasoline or industrial sources and accidental spills. On the other hand, nonpoint sources refer to those that are added or induced externally in the form of nutrients, pesticides, and fertilizers that enter the soil or landmass as a result of intense agricultural activities or operations [9].

In addition to this, the soil can be contaminated with biological species like bacteria, viruses, and chemicals from water that percolate downward. However, in some cases, these naturally occurring bacteria and viruses can also act as a cleaning agent that clean water sources. However, not all nutrients act as cleaning agents for both soil as well as water sources as effectively as modern techniques against both domestic as well as industrial contaminants, thereby altering the soil's ability to remove contaminants and chemicals. This is an important phenomenon as some soil layers allow water to quickly percolate down to the groundwater or aquifers. This means that if the percolating water is contaminated, it is less likely that contaminants will be removed. On the other hand, if the pollutants originate from an underground source, such as a septic system or waste storage tank systems that are very close to the soil and groundwater, which does not have enough and appropriate time to remove all harmful substances. Therefore, generally, the quality of the groundwater depends on the pressure, temperature, and other geological conditions that depend on how deep the groundwater is located, the type of rock, soil, and the residence time of the water (Figure 3.2) [10].

Generally, groundwater in their catchment moves very slowly between a few millimeters and a few meters every day. This also means that contamination of various types and natural substances in it tends to be concentrated and localized, in the vicinity of waste or close to the pollution source. However, contamination in groundwater and aquifers can spread within or to nearby water sources like

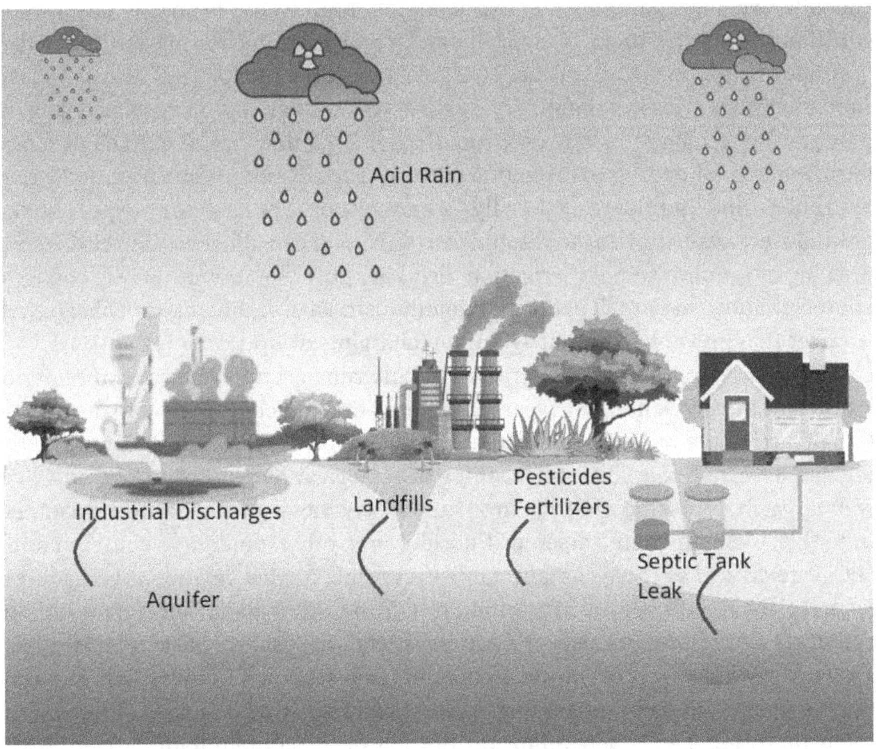

Figure 3.2 Several contaminants that seep into groundwater through soil or mismanagement of waste-dumping yards or waste storage reservoirs.

lakes and streams. However, groundwater pollution often is not noticed or identified until the water is characterized as already contaminated or the remediation process is already in place. In any case, reversing the groundwater contamination is an extremely expensive and difficult task and this process may take several days, months, and even decades before the water is considered usable again.

Globally, over 800 million population lack access to safe and clean drinking water. On the other hand, it is estimated that over a thousand children die due to the consumption of contaminated water. Majorly, contaminated groundwater and surface water result in life-threatening diseases like diarrhea, fever, vomiting, and so on. Poor-quality water also causes several issues associated with hygiene and sanitation. This situation is expected to worsen due to the increasing global water crisis due to threats from the impacts of population growth, climate change, and pollution. On the other hand, growing global population is expected to multiply the freshwater consumption capacities, which is another major cause of concern.

Therefore, groundwater depletion is expected to rise with every passing day. Globally, excessive pumping and extraction of groundwater primarily for human consumption is the main primary cause, which puts increasing stress on the groundwater sources. Other causes for groundwater depletion may also originate from inconsistencies in rainfall, drying up of wells, lakes, and streams and lack of groundwater recharge measures. Groundwater depletion can be defined as long-term water-level deterioration and decline as a result of sustained groundwater extraction and pumping. Globally, groundwater sources are experiencing groundwater depletion due to resource stressors that can affect aquifers, excessive drawing of groundwater for irrigation, drinking purposes, domestic use, changes in precipitation, snowmelt patterns, other industrial uses, and impermeable paved surfaces that prevent precipitation from recharging groundwater [11–13].

Another major groundwater stressor is increasing groundwater contamination issues and the ever-changing weather patterns that are contributing to deterioration of freshwater quality. Among many problems, groundwater salinity, decreasing and uncertain rainfalls, and leaching of geological minerals into groundwater sources such as arsenic, fluoride, iron, and heavy metals make natural resources unusable. Among them, arsenic, fluoride, and other emerging contaminants (ECs) are critical as these contaminants necessitate further treatment procedures to make them safe before consumption. On the other hand, increasing urban activities, agriculture, industrial discharges, and disposal of waste all can affect groundwater quality. Most of these contaminants originate from human-induced discharge of waste from modern agriculture and industrial activities [14–16].

On the other hand, depending on the nature of contaminant properties of groundwater can also be categorized into physical, chemical, and biological types that are being released into groundwater sources. Once the contaminants reach an underground water source either through soil erosion or dumping yards, aquifers get an infusion of contaminants as groundwater moves. However, many of the contaminants do not follow the flow of the groundwater current due to their varying physical or chemical properties. Nevertheless, it is possible to estimate and predict to some degree the transport within the underground water reservoir and aquifer of substances that move along with groundwater flow. For instance, both water and some or almost all contaminants can flow in the direction of the topography from recharge or catchment areas to discharge areas. Also, soils that are porous and permeable tend to transfer water and some or every type of contaminant mix with relative ease to an aquifer below. Nevertheless, generally, groundwater moves slowly and steadily, and so do infused contaminants due to which pollutants tend to remain concentrated in the form of a crest that flows in the same direction as groundwater. However, the size and speed of the crest depend on the extent and type of contaminants, their solubility, density, and the velocity of the surrounding groundwater as depicted in Figure 3.3 [17–20].

Newly added contaminants in the groundwater can move rapidly through cracks and fractures of rocks. These fractured rocks or cracks in the rock formations pose a unique problem connecting to groundwater sources and controlling leaching of contaminants in the underground reservoirs. This is due to the irregularities and randomness in the fractures that are spaced unevenly and do not follow the contours and profiles of the hydraulic gradient and the land

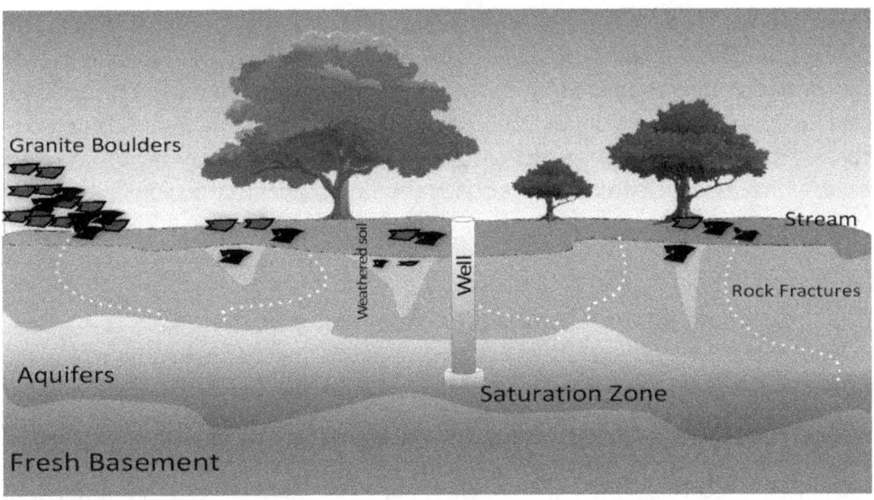

Figure 3.3 Various sources of contaminants that enter groundwater reservoirs such as rock minerals, soil contaminants, and so on.

surface. In addition to this, contaminants can also propagate, penetrate, and move into the groundwater reservoirs through macrospores, abandoned wells, animal burrows, root systems, and systematic fractures, holes, and cracks that provide pathways for contaminants. Additionally, contaminants enter the groundwater through the surrounding pumping wells and the potential for increasing the contamination originates from various water contamination zones, which have a land area bigger than the original recharge area, which draws into either the well or the surrounding aquifer [21].

On the other hand, many drinking water open wells, bore wells, and streams draw water from nearby lakes, streams, or rivers. These surface water sources can also contribute to groundwater contaminants. Nevertheless, recent advances in underground water recharge efforts have shown artificial infusion of various types of water into groundwater reservoirs. In particular, some wells are expected to receive various types and extent of contaminants during the artificial recharge process. Even though artificial recharge efforts increase the amount of water infiltrating into underground aquifers that often originate from irrigation, storm runoff, processed industrial processes, or treated sewage. Nevertheless, in many cases, artificial groundwater recharge efforts have resulted in dramatically increased concentrations of metals, synthetic chemicals, nitrates, and several other emerging pollutants. However, in some conditions, pumping can also result in groundwater contamination from another aquifer, which enters the one being pumped. This phenomenon is known as inter-aquifer leakage. Thus, accurately identifying and protecting the areas affected by well pumping is important to maintain groundwater quality. However, the greater the distance between a contaminated water source and a groundwater source, it is more likely that chances of contamination through natural processes decrease and thereby reduce the impacts of contamination. Also, processes like biological degradation, oxidation, and adsorption of ionic and metallic contamination through the binding of materials to soil particles may occur in the soil layers of the unsaturated zone, which drastically reduces the contamination concentration before it reaches groundwater [22–24].

3.2 Sources of Groundwater Contamination

Groundwater contamination can occur from various direct and indirect sources. In absence of aggravated industrial activities, groundwater contamination results as a result of numerous human activities and natural phenomena. Domestic activities, residential discards, municipal wastewater, commercial outputs, industrial discharges, and agricultural activities may severely affect groundwater quality. Contaminants can also reach groundwater from activities on the land surface activities like releases or spills from stored industrial wastes, structures beneath the water table like wells or contaminated recharge water, and sources below the land surface but above the water table like dumping yards, septic systems, commercial waste management sites, and leaking underground petroleum storage systems.

3.2.1 Natural Resources

Among these, natural sources comprise substances that originate naturally in rocks or soils, such as arsenic, fluorides, iron, sulfates, manganese, chlorides, or radionuclides that dissolve in groundwater. Other naturally occurring substances include decaying organic matter, which results in the form of suspended or dissolved nano-/microparticles in groundwater. However, the occurrence and the extent of contamination depend on the nature and types of rocks and their geological conditions. Even though many dissolved and dispersed contaminants may pose a health risk above permissible levels if consumed in excessive quantities, however, some may induce undesired physico-chemical characteristics to groundwater sources such as color, odor, and taste [25–27].

3.2.2 Septic Systems

One of the main reasons for groundwater contamination in many parts of the world is the effluent and leaking of waste from cesspools, septic tanks, and privies. It is estimated that one-fourth of all homes rely on septic systems or septic tanks to dispose of human waste. Even though each individual household septic tank may release a relatively small amount of waste into the ground sources, in situations where the large number or widespread use of septic systems can cause a serious contamination source. Most of these leakages or seeping of waste from septic systems may also originate from improperly designed and constructed septic system sites, or poorly maintained septic systems that may cause catastrophic damages may severely contaminate groundwater with viruses, bacteria, detergents, oils, nitrates, and chemicals. In addition to this, many commercially available synthetic organic chemicals such as 1,1,1-trichloroethane or methylene chloride used for cleaning septic systems may also cause serious groundwater contamination. These cleaners may also contaminate water supply systems such as wells and interfere with natural decomposition and decaying processes of septic systems [28–30].

3.2.3 Surface Water Sources

Groundwater contamination or pollution contamination mainly occurs due to the seeping of unwanted and discarded waste disposals from land surface or surface water sources into the groundwater over a period of time through natural processes or during the groundwater recharge process. Groundwater sources are found underneath the land surface including wells, springs, and various extended aquifers. The major contribution to the hydrologic cycle originates from rainfalls to the ground. After rainfalls, some water drifts toward lower ground altitudes along the land to tanks, lakes, and streams, whereas another portion of water evaporates into the atmosphere. A minor portion of it is taken up by plants, and a major proportion seeps into the ground based on the season and nature of soil or landscape. Also, groundwater can be described based on the nature of its origin in which the precipitation that has infiltrated or permeated the soil beyond the surface and accumulated or collected in empty and void spaces underground. The groundwater also originates from aquifers, which can be further classified into two types, namely confined and unconfined. Groundwater is stored in and moves gradually and steadily through geologic formations of sand, soil, and rocks called aquifers. Further, groundwater is found in two zones, namely (i) the unsaturated zone, which is located immediately below the land surface containing water as well as air in the open spaces and pores. (ii) The saturated zone is a cavity in which all the pores and rock fractures are filled with water that underlies the unsaturated zone. Further, nature of origin also plays an important role in determining quality and quantity of groundwater. Among these (i) Meteoric water, which is the main source of groundwater that is formed due to condensation and infiltration of atmospheric precipitation like sleet, hail, rain, etc., and infusion of lakes and river water. Interestingly, atmospheric precipitation and rain falling are distributed on the earth's surface in various types and ways, and in a pattern that enters the soil through infusion and infiltration. Even though a part of water returns to the atmosphere via evaporation, a major part of it flows into ground and surface water resources or runs off to lower levels. Therefore, water from rain and surface water apparently constitutes the bulk of groundwater, which is clearly from the fluctuation of levels of surface water and wells with the changing of seasons and weather patterns. (ii) Connate water that is also known as fossil water found in the entrapped sediments at the deposition on lake or sea beds. Connate water is classified into two types, epigenetic and syngeneic connate water. The epigenetic connate water is the one which enters the sediments from the basins into the rocks, which had been stored earlier or water condensed earlier, whereas the syngeneic connate water is the trapped water in the sediments. Connate water readily occurs in rock pores units with oil. Interestingly, water is contained in rock pores and sedimentary rock cavities under seas or ocean beds and lakes [31–33].

(iii) Magmatic water: Magmatic water is also known as juvenile water in aqueous phase form in equilibrium with other rocky minerals that are dissolved by magma deep within the earth's crust and released into the atmosphere during a natural volcanic eruption. This juvenile water is brought to the earth's surface from deep depths through the upward movement of intrusive igneous rocks. However, meteoric water is found on a large scale and in quantities compared to magmatic water and connate water [34, 35].

3.3 Waste Disposal to Groundwater Sources

This section discusses the major causes and origins of various waste sources that are disposed of or discarded into groundwater. Major reason for this uncontrolled and unmonitored waste disposal is ignorance and lack of social behaviors. This adverse effect of reluctant dumping or dumping of solid waste is due to ignorance of disposal protocols and lack of legal frameworks at disposal sites near the groundwater source. This practice of releasing or disposing of solid waste close to groundwater sources unsystematically or unscientifically is also one of the major reasons for increasing groundwater contamination. On the other hand, the water that is already present in the solid and semi-solid waste originated from the biodegradable material sources due to the infiltration of water either as one of the constituents or added during processing. This water in the waste that is generated or originates due to various sources mixes with the soil and causes contamination in groundwater upon leaking or contact. Overall, this process leads to contamination of groundwater and changes in the health parameters of the groundwater. The major concerns with the constant monitoring, evaluation, and risk assessment will depend on the universally adopted parameters largely agreed with the World Health Organization (WHO) recommendations. These physical and chemical parameters like pH, hardness, temperature, electrical conductivity, total suspended solids (TSS), total dissolved solids (TDS), and so on are further monitored using various advanced analytical techniques. Table 3.1 shows the list of contaminants and their origin generally found across the globe [36–38].

However, to control the pollution of groundwater through surface and activities, it is important to regulate the open area dumping of solid and semi-solid waste that needs to be prevented or prohibited by regulatory authorities. Nonetheless, the negative impact of dumping activity on groundwater sources clearly has shown high concentrations of total hardness, chlorides, TDS, electrical conductivity, chemical oxygen demand (COD), nitrates, and sulfates. Also, in many cases, leachate accumulated or collected from the designated site showed presence of high concentrations of heavy metals. Interestingly, it is noticed that concentrations of various contaminants in groundwater as well as surface water tend to decrease or dilute during the post-rainfall season and may increase in drought situations. Other important observations clearly reveal that landfills in highly populous communities or densely populated cities should

Table 3.1 List of various contaminants and their origin and severity which cause contamination to groundwater sources.

Contaminant category	Source of contamination	Severity
Agriculture	Fertilizer use or storage	Moderate
	Animal feedstock	Moderate
	Animal burial area	Moderate
	Irrigation source	High
	Manure pits and storage	Very high
	Pesticide and herbicide site or storage	Very high
Domestic and residential	Fuel oil	Moderate
	Household discharge	Very high
	Household lawns and fertilizers	Very high
	Furniture/refurbishment	Moderate
	Septic system	Very high
	Sewer lines	Very high
	Chemicals in swimming pool	Very high
Waste management sites and dump yards	Hazardous waste landfills	Very high
	Municipal incinerator	Very high
	Municipal landfills	Very high
	Open burning sites	Moderate
	Municipal sewer sites/lines	Very high
	Recycling facility	Very high
	Infrastructure facility	Moderate
Chemicals	Dry cleaners	Very high
	Cemeteries	Very high
	Gas stations	Very high
	Building infrastructure	Very high
	Laundromats	Very high
	Medical hospitals and institutions	Very high
	Scarp and junkyards	Very high
	Storage tanks	Very high
	Industrial processing	Very high
	Car washing	Very high

Table 3.1 (Continued)

Contaminant category	Source of contamination	Severity
Industrial sources	Asphalt plants	High
	Chemical manufacturing	Very high
	Electric/electronic Industry	Very high
	Foundries/metal processing	Very high
	Mining and mine drainage	Very high
	Petroleum and wood processing	Very high
	Toxic and hazardous waste	Very high
	Food/pharmaceutical industry	Very high

house a groundwater monitoring mechanism on a regular basis. Furthermore, groundwater in and around the surface land dumps and landfill sites should not be extracted or used for drinking water and household purposes unless it satisfies the specific quality standards. Nonetheless, indiscriminate and unregulated dumping of waste without appropriate solid waste management practices needs to be understood well [39–41].

3.3.1 Improper Disposal of Hazardous Waste

Recent awareness of waste disposal and environmental concerns have led to careful disposal of various types of waste. Nonetheless, in the past and to some extent till today, the hazardous waste is disposed of to the environment. Recent advances also show that hazardous waste is also properly regulated through licensed handlers or municipal collection protocols. Similarly, several protocols have also been followed to collect and manage industrial waste from liquid waste by reducing it to highly concentrated solid wastes as depicted in Figure 3.4. But it is important to note that many chemicals and synthetic substances should not be discharged directly into environment or treatment plants such as disposal of household septic systems, lawn and garden chemicals, including oils, medicines, disinfectants, photographic chemicals, paints and paint thinners, and swimming pool cleaning and disinfection chemicals. Similarly, many substances utilized as raw materials, during processing products or as by-products are also being disposed of either partially treated or without treatment. However, these industrial wastes should not be disposed of in surface water resources, drains, wells, or at the workplace as these activities may contaminate a drinking water source. These activities of improper disposal or handling of waste can be minimized by creating awareness, organizing education programs, and training water resource managers,

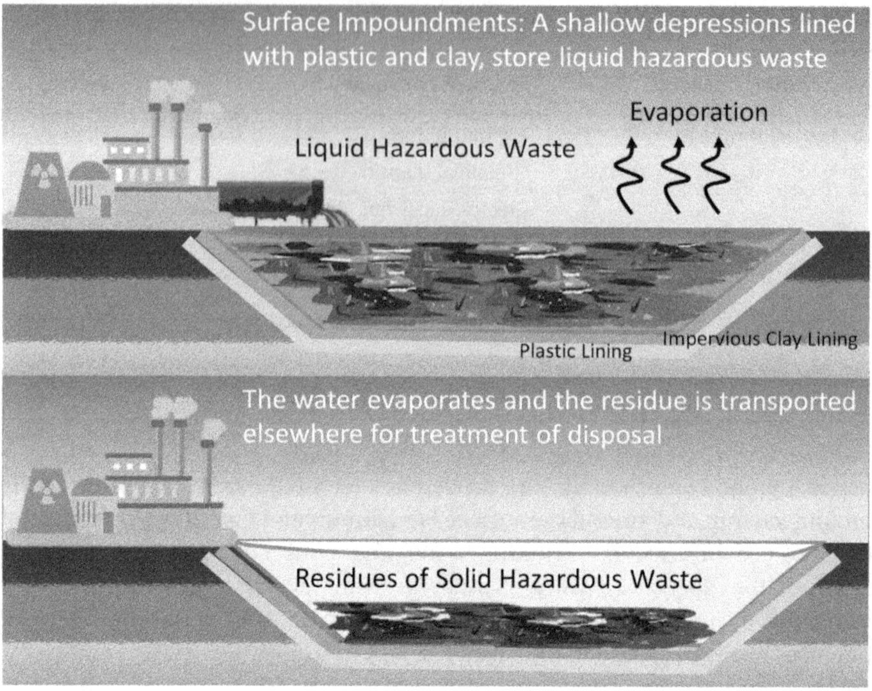

Figure 3.4 Liquid hazardous waste and solid residual hazardous waste generation and management protocols followed through advanced methodologies.

workplace employees, and end users to the extent of making proper use of water and disposal of all types of substances, chemicals, and by-products used on site [42–44].

3.3.2 Accidental Releases and Spillovers

The chemicals or commodity chemicals stored or on the way to transfer from one place to another for commercial purposes face risk of accidental release or spillover. Among many contaminant threats, (i) chemicals and substances such as petroleum raw materials, products that mostly depend on their transportation and storage mode of distribution through both underground and aboveground mechanisms or storage of these commodities until their distribution in various types of tanks commonly face vulnerability of leakage or spillover. For instance, many homes, offices, and commercial installations have underground fuel storage and heating oil tanks. (ii) Similarly, many commercial as well as municipal highway departments also install gasoline storage, fuel oil, diesel fuel, or chemical storage tanks, or facilities in on-site tanks. (iii) Industries contribute a major part of potentially hazardous chemicals and substances in storage tanks, which are either managed by licensed haulers or waste management authorities. (iv) It is expected that over 5 million underground storage tanks or various types and capacities exist across the globe that are always vulnerable to leakage or spill into the environment. (v) In a situation when an underground water or waste storage tank gets damaged or develops a leak that is generally expected or naturally occurs as the tank corrodes and ages. In this situation, it is expected to find all or most of the contents to migrate through the soil pores and cracks to reach the groundwater. (vi) Even though standard and high-quality storage tanks and catchment areas are designed and built following regulatory standards set by a state or federal government are less likely to undergo damage in a short time. However, in the long term, it is expected that almost every waste storage tank and slumping area is not foolproof. (vii) On the other hand, abandoned underground waste storage tanks and bins pose another serious problem due to their location, which is often unknown. Also, aboveground waste storage tanks can also pose a serious threat to groundwater sources in a situation of spilling or leak that may occur or degradation of decaying impermeable barriers. (viii) Another major threat to groundwater contamination is improper chemical storage, use of poor-quality containers, and sloppy materials handling, which can cause major threats to groundwater quality due to contamination. (ix) In the transportation chain, tanker trains, trucks, and tankers pose another chemical leakage hazard. Every year, it is estimated that approximately 16,000 chemical or oil spills occur from transportation trains, trucks, and storage tanks, which mostly occur when materials are either being transported or moved from one place to another. (x) Finally, another major water contamination occurs from marine oil spills, even though most of the time contaminate surface or oceanic waters, however, the site of an accidental spill, which contaminates often with high-density chemicals is often expected to reach river, lake, or sea bed. Also expected to reach directly soil through the surface flow increases the possibility of groundwater contamination [45–47].

3.3.3 Solid Waste Contamination from Landfills

Waste-disposed landfills house thousands of contaminants from various sources such as municipal, domestic, and industrial discharges throughout the globe. Hazardous and toxic chemicals are intended to be disposed of in solid and semi-solid waste landfills that are built and managed by expert managers. Major issue with domestic waste management is that it is not being regulated or monitored properly. On the other hand, domestic solids contribute significantly to soil contamination as most of the time, household waste is either dumped in landfills or thrown into open land and surface water resources. In many cases, many chemicals from these landfills tend to leach into the groundwater through precipitation and surface runoff. Therefore, to prevent the leaching of hazardous chemicals from landfills, it is important to have synthetic covering or liners or clay coating, leachate that contains contaminants, and secured waste collection systems in place to protect groundwater. Nonetheless, older landfills situated over aquifers or near surface water sources and waste dumps on permeable soils on shallow water tables increase the potential of leachate to contaminate groundwater sources. Also, closed landfills around or close to various water sources can continue to pose a groundwater contamination threat in a situation if such sites are not capped or properly maintained with an impermeable material layer to prevent the leaching of contaminants by leaking or precipitation [48–50].

3.3.4 Surface Impoundments

Surface impoundment is a method in which a unit is a designed man-made excavation on an already existing natural topographical depression or diked landscape formed mainly of earthen materials. However, modern techniques use highly impermeable man-made synthetic plastic and cement materials, which are designed to retain and accumulate liquids from waste disposal. Natural surface impoundments are relatively shallow lagoons or ponds strategically and systematically used by various end users like municipalities, and industries to dispose, treat, and store liquid wastes. Similar to man-made landfills, modern surface impoundment facilities are required to have impermeable liners. However, it is generally observed that many times these impermeable liners are also vulnerable to leaking [51].

3.3.5 Potentially Harmful Components of Common Household

Sewers, waste transportation, and disposal pipelines carrying wastes are vulnerable to liquid and fluid leaking into the surrounding soil and groundwater sources. Nevertheless, sewage and municipal waste consist of various types of biological contaminants like bacteria, viruses, pathogens, organic matter, heavy metals, inorganic salts, and various minerals like nitrogen and phosphorus. Also, pipelines carrying industrial waste chemicals, oil brine, and waste discharges have also been known to leak. This situation may aggravate when the pipelines carrying this waste get corrosive and prone to cracking, which leads to leakage of waste substances. For instance, pesticide and fertilizer industries handle millions of tons of pesticide and fertilizer constituents like insecticides, fungicides, rodenticides, herbicides, and avicides are used annually across the world in various loads and concentrations. These fertilizers and pesticides are not only used by farmers but also by businesses, sports grounds like golf courses, homeowners, utilities, and municipalities. In all cases, the type of fertilizer chemicals and their concentrations varied both in toxicity and hazardousness. Nevertheless, a number of such fertilizers and pesticides have been found entering various soil layers leading to contaminating groundwater. This process of contaminant percolation may be slow and some pesticides have been known to remain in soil for many days to many years and eventually enter groundwater [52].

Another major potential source of groundwater contamination and at large aquatic ecosystems is animal waste that releases and percolates into the soil and further to groundwater resources from farm feedlots. Several studies have confirmed the contamination of groundwater from feedlots, which has proven highly toxic over a period of time. Detailed surveys have examined the various types and depths of drinking water wells that contain minerals like nitrates, phosphates, pesticides, and so on in various concentrations and levels of toxicity. Also, bore well samples have also shown high-concentration contamination of agricultural chemicals generally used in farming including fertilizers, pesticides, herbicides, and other minerals. These groundwater samples have shown a significant percentage of several chemical concentrations that exceed the maximum permissible levels set by several health regulatory and monitoring agencies. Among many, the most commonly found pesticides in groundwater samples were metabolites of dimethyl tetrachloroterephthalate (DCPA), which is also commonly known as Dacthal or Atrazine, and is generally used for many utility easement weed-control purposes and in lawn care applications at domestic level [53].

3.3.6 Drainage Wells

Drainage wells are used in wet areas to help drain water and transport it to deeper soils. There are several types of drainage systems that allow comprehension disposal and transportation of domestic and municipal wastes. Among many municipal and domestic waste transportation and management systems, (i) surface drainage system sometimes referred to as a French drain removes excess wastewater from the land's surface through well-organized or systematically engineered channels or ditches. Sub-surface drainage systems are generally implemented beneath the top layer of soil or as underground pipelines or concrete channels. (ii) slope drainage systems on the other hand are built to allow water to flow from a systematic engineered structure in a downward direction with the aid of pipes that move through the slope. Since the installed pipeline is anchored to an incline and load under which high-pressure water flows, it makes them vulnerable to leakage and breakage. (iii) downspouts and gutter systems are another system of engineering structure largely used for municipal wastewater management systems. In general, the gutter system is a first defence against extreme stormwater and an emergency management system for stormwater oversaturation. However, systems are often used for domestic and municipal wastewater management. These systems are integrated with surface drainage structures along buried drain pipes, rain barrels, aluminum extensions, or other solutions. The purpose of these arrangements is to move wastewater away and route wastewater to other drainage systems on the street or sidewalk. Many modern gutter systems are connected to an underground sewer line using underground drains [54].

Nonetheless, understanding the types of drainage systems and the materials used for construction in detail often reveals multiple issues associated with overall wastewater leakage and potential groundwater contamination. Typically, lifespan of a construction material and its engineered structures to withstand high concentration of wastewater and their oversaturated flow patterns clearly show the potential of leaking several hazardous contaminants to groundwater. Further, improperly constructed drainage, gutter, and wells can result in groundwater contamination when contaminated surface or groundwater is introduced into the well. On the other hand, improperly abandoned wells can act as a conduit through which hazardous and toxic contaminants can reach an aquifer if the well casing is damaged or removed as part of a maintenance exercise due to corrosion. In many cases, lack of civic awareness people use abandoned drainage, wells, and surface water sources for the disposal of wastes such as plastics, kitchen waste, household garbage, motor oil, and so on. These contaminated wells then reach an aquifer that serves drinking supply underground water sources or wells. Similarly, abandoned exploratory wells such as gas exploration infrastructure, crude oil pipelines, coal mining infrastructure, or test hole wells that are generally uncovered and poorly marinated also act as a potential conduit for contaminants. Also, conventional drinking water supply wells that are regularly utilized are generally poorly maintained, which can result in groundwater contamination. This is also due to the use of poor-quality fittings, faulty construction materials like fittings and casings, inadequate covers, and inadequate maintenance or lack of concrete pads that allow external contaminated soil and water and other accompanying matter to flow into the well. This poorly maintained infrastructure also gets contaminants from surface runoff or septic systems or wastes from farm animals. Another segment through which groundwater gets contaminated is poorly constructed irrigation sources or wells. These wells can often allow contaminants such as fertilizers, pesticides, and residues of herbicides to be applied in the surrounding vicinity of wells on agricultural land. Conventional mining activities and abandoned mines can also contribute significantly to groundwater contamination. In all these cases, precipitation can lead to contamination through the leaching of soluble minerals from the mine wastes, which are also referred to as tailing and spoils into the groundwater. Both active and abandoned mining wastes are often characterized by minerals, acids, metals, sulfides, and so on. In some cases, abandoned mines, often small-to-medium scale in size, are used as waste wells and dumping pits. Finally, modern mining technologies sometimes use pumping techniques to keep them dry and these pumping methods can cause an upward migration and extraction of contaminated groundwater that is generally intercepted by a well.

3.4 Effects of Groundwater Contamination

Groundwater contamination can severely impact human health, aquatic ecosystem, socioeconomic development, and environment at large. For instance, many studies report clearly showing high concentrations of nitrate, arsenic, fluoride, heavy metals, and persistent organic pollutants are a health risk for the human population. Some of these are very critical for children and infants who are generally more susceptible to the effects of these contaminants than adults. For instance, higher concentrations of nitrate in groundwater sources that are consumed as drinking water cause "blue baby syndrome" in infant methemoglobinemia. Contaminated groundwater used, in particular, for drinking purposes can cause adverse effects on human health, which can also affect the food chain and thereby food production system. Irrigation and production from agriculture and horticulture with groundwater contaminated by heavy metals and wastewater containing energy contaminants can result in the accumulation of toxic elements in vegetables, cereals, and cause health risks to humans [55, 56].

Groundwater contamination results in poor drinking water quality, scarcity in water supply, degradation of surface water sources, high clean-up and treatment costs, increased costs for new and alternative water supply sources, and potential health issues associated with contaminated water sources. The consequences of degraded surface water or contaminated groundwater are often very serious. For instance, estuaries are impacted by the presence of high nitrogen in the groundwater sources, which has also resulted in loss of critical shellfish habitats. In some cases, contaminated groundwater is so severe that the water supply is generally abandoned as a main source of drinking water. In other cases, the groundwater can be treated to achieve standard quality to be used again if the contamination is mild or not too severe. Nonetheless, water quality monitoring is often a necessary and required protocol in case contaminated water sources are utilized for drinking purposes. As groundwater in natural conditions moves slowly, contamination often remains unaltered and undetected for long periods of time. This makes treatment of contaminated water supplies difficult if not impossible. In any case, groundwater treatment process is a cost-intensive process.

Groundwater contamination can also negatively affect the quality of soil, lands, and forests, which lead to soil quality degradation and decaying in land quality. For instance, many agricultural areas, particularly arid regions, face severe water crises and generally suffer from high groundwater salinity, which is one of the major factors that influence soil salinization. On the other hand, high concentrations of soluble salts and mineral contaminants in groundwater toxic metals, iron, manganese, and so on can accumulate in the root zone affecting vegetation growth. Further, groundwater contaminants can also be transported and transmitted through surface water-groundwater interactions resulting in deterioration of surface water source quality.

On the other hand, sustainable economic development essentially requires a balance between human demand and rate of renewal of natural resources. For uninterrupted life, human activities, and industrial growth, freshwater is probably the most necessary and valuable of the natural resources. However, chronic and ever-increased groundwater contamination may drastically reduce the availability of freshwater sources, thereby breaking the balance between water demand and supply. This is the main reason for reduced industrial growth, change in the agriculture produce, inflation, socio-economic crises, and even wars. Also, water scarcity induced by contamination may become a major factor that can be associated with conflicts among citizens in the near future, possibly slowing down the economic aspirations or delaying the socio-economic development of a nation. Therefore, groundwater contamination is at large not only an environmental issue but also a social-economical problem, which demands collaboration between researchers, innovators, nature scientists, economists, and social scientists.

3.5 Major Contaminants

Groundwater is the main source of freshwater for the global population for their daily needs, drinking purposes, food production, domestic utility, and industrial purposes. Among this, one-third of the global population depends on groundwater for freshwater needs for their drinking purposes. On the other hand, according to a report from the United Nations Environment Programme (UNEP), the largest 32 cities around the world have a population greater than 10 million known as "megacities," out of which 16 cities majorly depend on groundwater resources for life to sustain. Similarly, there are 657 cities in China of which nearly 400 cities are depending on groundwater as the main source for their water supplies. Likewise, India consumes more than 600 billion cubic meters or more than 150 trillion gallons of freshwater annually among which major portions are extracted from nearly 245 billion cubic meters drawn from groundwater aquifer sources. Therefore, around the globe, sub-surface water as well as groundwater has been the essential resource of freshwater to humanity.

Therefore, it is vital to closely monitor freshwater sources for life, ecology, and industries to survive and the environment on the earth. In this situation of extensive exploitation of natural freshwater sources, keeping water resources sustainable, effective, accessible, and efficient is a major concern for water managers, policy makers, scientists, and commercial entities working in a related field. Nevertheless, increasing urbanization, which demands modern farming, increasing industrial production consumes extensive water sources and looming climate change, which is drastically altering weather patterns posing significant threats to the quality and quantity of groundwater. Among many, hydrocarbons such as organic trace pollutants, toxic metals, pharmaceutical contaminants, pesticides, and other ECs are endangering human health, natural ecosystems, and long-term socioeconomic development. While groundwater contamination poses a substantial threat to human populations and environment, it also provides an opportunity for scientists and researchers to understand, evaluate, and explore suitable solutions to how our underground resources are formed, and aquifers evolved and exploited, which lead to contamination and pollution. This further helps policy makers and decision makers to understand and resolve the guidelines to mitigate the deterioration of groundwater sources through which to enhance the quality and quantity of these resources [57–59].

3.5.1 Inorganic Contaminants

There are several inorganic contaminants present in groundwater such as arsenic, cadmium, and iron that are dissolved in high concentrations and pose a big concern especially when groundwater sources are used for drinking or food production purposes. Even though contamination and pollution generally aggravate due to man-made activities, geological formations and natural activities also help in increasing concentration of unwanted and undesired chemicals and substances in various underground water sources. Human activities and industrial discharges, domestic waste disposal, and agriculture chemicals also leave several inorganic contaminants in groundwater sources. Furthermore, inorganic contamination could also happen due to agricultural activities such as the excessive use of fertilizers, plant nutrient chemicals, pesticides, and herbicides, which when used in excess migrate to groundwater and reach the water table. In any case, to make groundwater safe from these contaminants, which go into irrigation, drinking, or industrial purposes, requires different tests to ensure that it is suitable for these purposes. These inorganic contaminants cause health issues and environmental problems throughout the human and food chain if they are present in the groundwater with levels higher than the standard maximum contaminant level (MCL) recommended concentration. Table 3.2 gives a detailed description of different inorganic contaminants in groundwater sources, which directly affect human health and environmental risk above their MCL [60, 61].

Table 3.2 Inorganic pollutants presented in groundwater.

Contaminant	Source for the groundwater	Problems	MCL (mg/L)
Aluminum (Al)	• Groundwater passes through some kind of rocks. • Mines discharge.	If present in drinking water, it could cause turbidity increment besides water discoloration.	0.05–0.2
Antimony (Sb)	• Municipal waste disposal. • Industrial production and flame retardants such as glass manufacturing, ceramics and lead industry. • Fireworks or explosives.	Cause a change in cholesterol and glucose concentrations in blood in laboratory animals exposed to risky levels of antimony during their existence. Decreases longevity. Has a biochemical change in laboratory animals and toxic effect on neurobehavioral.	0.006
Arsenic (As)	• Industrial activities such as smelting of metals (zinc, lead, copper ore) • Using pesticides. • Naturally found in aquifers.	Liver, kidney and skin damage. Decrease blood hemoglobin. Chronic and acute toxicity. Can cause various forms of cancers. Hindrance of children's development.	0.010
Barium (Ba)	• Naturally it takes place in some kind of soils such as limestone and sandstones. • Landfill leachate. • Fertilizers and pesticides.	Cardiovascular and kidney diseases. Mental disorders. Metabolic syndrome.	2
Cadmium (Cd)	• Industrial and mining waste. • Phosphate fertilizers. • Landfill leachate.	High blood pressure. Replace zinc biochemically in the human body. Liver damage Destroy testicular tissues and blood cells (red).	0.005
Chloride (Cl)	• Industrial and domestic waste. • Saltwater intrusion.	Changes in drinking water taste. At high levels, it can deteriorate water heaters, municipal pipes, pumps, and works equipment.	250

Table 3.2 (Continued)

Contaminant	Source for the groundwater	Problems	MCL (mg/L)
Dissolved solids (DS)	• Naturally found. • Human activities such as landfill leachate, feedlots.	When presented, the water became unacceptable and objectionable to many. Affects the performance and life of water heaters.	500
Iron (Fe)	• Mining corroded metal, industrial waste. • Naturally found in sediments and rocks.	Changing water taste. Affect plumbing fixtures and cloth colors in laundries.	0.3
Lead (Pb)	• Industry, mining, gasoline and plumbing.	Affects babies' mental growth and can change red blood cell chemistry. Increase blood pressure. Probable carcinogen.	0.015
Zinc (Zn)	• Industrial waste, metal plating, is the major ion in sludge. • Naturally, it is found in mining areas.	Cause a change to the drinking water taste. Toxic to plants if exposed to high levels.	5

Inorganic contaminants such as toxic heavy metals, metal salts, and different types of nutrients generally constitute a major portion of inorganic pollutants that ultimately end up in the form of dissolved cations and anions. More significantly, some inorganic contaminants like heavy metals possess infinite lifetimes, for instance, cadmium always stays cadmium. It only can be attenuated through adsorption or other processes; however, no natural process essentially reduces the total mass of cadmium. Other inorganic nutrients such as ammonia (NH_4^+) or nitrate (NO_3^-) can be transformed through decay or degradation into other forms of inorganic nutrients N_2 or N_2O gas. Also, toxic metal contaminants such as lead (Pb) and chromium (Cr) pose serious problems and risky contaminants to soils, surface waters, sediments, and aerosols. This has been confirmed through various samplings and analyses in different urban environments.

Mining industry and abandoned mining activities are the main sources of toxic metals that occur as natural sources. However, active mining undertakings reinforce their disposal and release into the environment. Nevertheless, toxic heavy metals are less frequently found in groundwater sources due to their limited mobility. Heavy metal contamination mainly occurs in the form of cations such as Pb^{2+}, Cr(III), and Cr(VI) that are prone to adsorption, and many times these cationic pollutants precipitate as sulfides, lowly soluble oxides, or other minerals. However, it is well observed that the sediments in caves and conduits can act as reservoirs for toxic metals and other contaminants and in the events of high-flow events, these toxic inorganic contaminants transfer from sediment to active springs and freshwater sources. Among many toxic metal contaminants, cadmium (Cd^{2+}) is known to be more mobile and highly toxic than others. Similarly, mercury is extremely toxic and specific bacteria can convert it into highly mobile methyl mercury ($HgCH_3^+$) or volatile dimethyl mercury, $Hg(CH_3)_2$ [62, 63].

3.5.1.1 Arsenic

Arsenic is one of the most toxic metals that is the reason for large-scale geogenic groundwater contamination that orientate from various layers of soil formation as shown in Figure 3.5. Arsenic contamination in drinking water causes severe chronic health risks in humans. Long-term exposure to even low concentrations of inorganic arsenic in drinking water is fatal and known to cause human risk including thickening and discoloration of the skin, heart disease, cancer, nerve effects including numbness and/or pain, problems with blood vessels, high blood pressure, and interference with some important vital organs. High concentration of arsenic contamination is largely found in several regions of the world including Bangladesh, China, and the West Bengal region of India where millions of

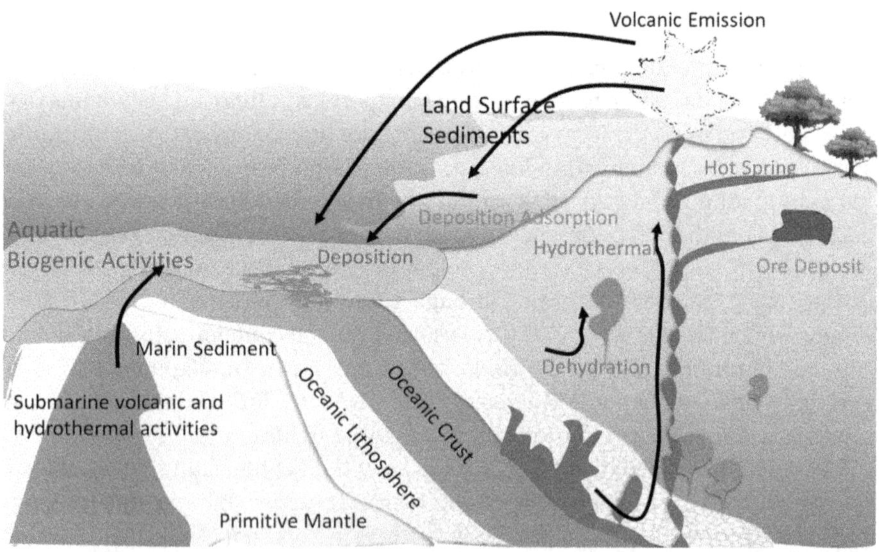

Figure 3.5 Infographic showing geologic arsenic redistribution cycle in which specific coding indicates the average concentrations of each geologic body. Red- and blue-colored letters indicate the releasing and fixing mechanism of arsenic in the cycle.

populations are severely affected and continue to get exposure to arsenic through contaminated drinking water. Interestingly, natural Arsenic contamination occurs in groundwater of inland basins in reducing alluvial aquifers, which is characterized by arid to semiarid, that is, flat, low-lying areas. In these areas, groundwater is slow and often these sources are extensively exploited through deep pumping wells. Interestingly, large springs are generally largely unaffected by arsenic contamination. However, high concentrations of arsenic contamination are also found in mining areas and earlier use of arsenic-containing insecticides has also been accounted for [64–68].

3.5.1.2 Fluoride

The high concentration of fluoride occurs in groundwater through natural as well as artificial sources, which concerns global population health potentially affecting hundreds of millions of people, primarily in the Global South. On the other hand, anthropogenic fluoride contamination in water sources occurs through several human activities like fluoridation of drinking water supplies, dental products, motorization, and industrialization, fluoride-containing pesticides, refrigerants, and fire extinguishers. Minor percentage of fluoride contamination is also accounted for by various airborne sources. Long-term exposure and ingestion of fluoride are known to cause severe health effects resulting in dental and skeletal fluorosis. Fluoride minerals are abundant in the earth's crust and the element fluorine forms fluoride (F^-) minerals that are naturally found in soil and aquifer sediments in extremely high concentrations that can lead to fluoride accumulation in freshwater as well as groundwater resources. The groundwater contamination due to fluoride mineral leaching is predominant among different water sources.

On the other hand, large urban populations consume or use fluoride-containing consumer products on a daily basis such as fluoride intake in the diet occurs primarily through food and drinking water. Even though fluoride has not been established that it is an essential element in the diet, even though moderate concentrations of fluoride in consumer products and personal care products (PCPs) like toothpaste and mouthwash can help prevent dental caries when used in limited concentrations , however, fluoride >1.5 mg/L is known to cause dental and skeletal fluorosis. Also, fluoride intake in any form >6 mg/d is known to cause severe crippling skeletal fluorosis. To address this serious situation and to protect against these potential ailments, the WHO has set guidelines for maximum permissible limit (MPLs) of fluoride in drinking, which should not exceed 1.5 mg/L. However, the WHO also recommends depending on the water scarcity and the available sources every national standard can take account of the total exposure to fluoride and establish a lower standard if the intake from all sources reaches 6 mg/d. For instance, the desired maximum fluoride concentration in drinking water recommended in India is 1.0 mg/L, whereas in the United States, it is 2.0 mg/L.

Naturally, high concentrations of fluoride are frequently found in aquifers of acidic igneous basement rocks, volcanic and geothermal rocks, sedimentary deposits and metamorphic rocks with low calcium concentrations, high pH or alkalinity, higher temperatures, and long groundwater residence times. At high pH conditions, the desorption of fluoride from natural clays having hydroxyl anions (OH^-) exchange with F^- in F-bearing minerals and bicarbonate (HCO_3^-) reacts with fluorite (CaF_2) to release fluoride. Even though, in this process, dissolved calcium can bind with fluoride and prevent it from dissolving in water sources again to form fluorite. Furthermore, higher temperatures such as in conditions of geothermal waters surrounding enhanced chemical weathering and prolonged groundwater habitation times offer more time for reactions to take place. As mentioned earlier, arid and semi-arid regions are usually likely to contain high fluoride groundwater due to higher pH, alkalinity, and longer residence times [69, 70].

Since fluoride is tasteless, odorless, and transparent in nature, presence of its contamination in a groundwater source can remain unnoticed and undetected until the contaminated water source is eventually analyzed thoroughly. Unfortunately, regular groundwater monitoring and analysis in general and fluoride contamination in particular are often not systematically tested in many parts of the world. Therefore, in order to analyze regions of areas of high fluoride concentrations, it becomes important to map and accelerate their discovery rates. Also, extensive analytical data also helps in obtaining high-resolution geospatial prediction mapping, which may help in creating data based on fluoride concentrations and the natural conditions related to fluoride accumulation in groundwater. Recent advancements in machine-learning approaches can also help in establishing statistical relationships between known fluoride contaminant concentration regions and environmental parameters. The resulting fluoride concentration model is then applied and used to match environmental variables to produce a future prediction map.

3.5.1.3 Nitrate

Nitrate (NO_3^-) is an inorganic substance largely found in areas where agricultural runoff, and fertilizers are extensively used, liquid waste discharged from septic tanks, in various types of manure, and dairy lagoons. Nitrate also occurs naturally in various vegetables and food produced at safe levels. On the other hand, rain or irrigation can also carry nitrate down to groundwater and other freshwater streams through soil. Nevertheless, nitrate is commonly found in groundwater, in the areas where agricultural activities having different N-compounds are extensively used as fertilizers or crop growth boosters. Nitrate in low concentration is not very toxic and harmful, but high concentration intake over a long period of time induces adverse human health risks. Therefore, US-EPA standard for nitrate in drinking water is set as 10 mg/L of nitrate measured as nitrogen in drinking water. Drinking water sources having MPL of nitrate at or below 10 mg/L is considered safe for consumption. Higher concentrations of nitrate (>10 mg/L) in drinking water lead to Blue Baby Syndrome in infants under six months of age and it is a disorder where a baby's skin turns blue. On the other hand, high nitrate concentration levels often go well with other types of agricultural pollutants such as fecal bacteria or pesticides, which deteriorate the general water quality. As nitrate is largely considered a nutrient in agricultural activities, sometimes increased concentrations in various ecosystems also influence positive natural growth. Effluent waters contain higher concentrations of ammonia (NH_4^+), which is frequently converted to nitrate under oxidizing conditions. Dissolved nitrogenous salts are natural water ingredients. However, elevated concentrations caused by human activities can render the water unusable. For instance, the use of road salt locally affects soils, landmass, and vegetation but rarely causes large-scale groundwater contamination. However, salt mining is another nitrate contamination source for groundwater. However, the hazards due to elevated concentrations are due to inappropriate agricultural and domestic water management practices [71, 72].

3.5.1.4 Heavy Metal Contamination

Among many modern pollutants that are categorized as one of the main emerging classes of contaminants, heavy metals are considered the most prominent toxicants in the environment. These are known for their toxicity, tenacity, and incremental nature that are detrimental to both living beings including humans and the ecosystem. Volcanoes and their eruptions break down rocks that contain deposits of heavy metals releasing huge quantities of toxicants and human activities such as mining, mineral extraction, and agricultural practices having lead-containing pesticides, and various industries are the main sources of heavy metals in the environment. Among these activities, rock weathering and volcanic eruptions are natural sources, whereas industries, agricultural practices, mining, mineral extraction, and various other infrastructure and developmental activities can be categorized as anthropogenic activities. In particular, anthropogenic activities mainly trigger the transportation of heavy metals in different ecosystems and largely into the environment further facilitating the distribution of heavy metals in biogeochemical cycles. There are several heavy metals found in nature and, in particular, in natural groundwater sources such as mercury, lead, iron, arsenic, cadmium, zinc, copper, and chromium and in many cases, the actual volume of these heavy metals is estimated to be more than $6\,g/m^3$. However, in low and trace amounts also, heavy metals possess unique properties such as bioaccumulation and environmental persistence. Among these, lead, mercury, chromium, arsenic, and cadmium have been the most commonly found heavy metals that are known to induce human poisoning. Exposure to these heavy metals for a prolonged period through water, air, and food can cause acute or chronic poisoning in living beings. Further, once the heavy metals reach aquatic ecosystems, these can lead to environmental serious biomedical issues in living beings, which adversely affect their health in both short and long terms [73, 74].

3.6 Organic Contaminants in Groundwater

Organic contamination in drinking water is mainly traced to both natural as well as human activities. Naturally, humic acid, plant residues, and animal decay are the main sources of organic contamination in various water sources. On the other hand, residues from various synthetic chemicals used in agricultural activities such as insecticides, herbicides, and fungicides have been widely used in conventional practices for pest control and agricultural yield production. On the other hand, humic substances, dye, petroleum, surfactants, pesticides, phenolic compounds, and pharmaceuticals are other main sources of organic contaminants that are significantly found in wastewater sources. Nevertheless, the major concern occurs from organic pollution in groundwater as these classes of contaminants are categorized as "emerging organic contaminants" (EOCs). The presence of organic contaminants in water over a period of time produces toxic chemicals, metabolites, by-products, and so on during disinfection.

The major concern from organic contamination worries the policy makers or decision makers is the uncivilized disposal or discharging of solid and semi-solid wastes having organic contaminants into the environment and water resources, which is also a pressing concern for researchers and innovators to find suitable solutions. The contamination and distribution of organic contaminants in groundwater are not only critical for health and environmentally related issues but also enable deeper studies to understand the nature and severity of problems in water resources and reservoirs. Additionally, ecological conditions also assist organic contaminants in contaminating groundwater from different sources like (i) solid waste disposal sites, landfills, sewer pipeline leakage, and storage tanks leakage, and (ii) discharges or overdoses from agricultural activities and farmyard drainage. Table 3.3 gives a detailed description of most commonly occurring organic pollutants usually found in groundwater sources, their origin, and potential health effects due to the exposure or consumption [75, 76].

Table 3.3 List of organic contaminants and their detailed description and potential risk involved due to prolonged exposure in groundwater.

Contaminant	Source for the groundwater	Problems
Volatile organic compounds (VOCs)	• Anthropogenic activities. • The pharmaceutical industry, dyes, polishes, inks, paints, disinfectants and spot removals industry • Crude oil industry.	Can cause damage and cancer in the liver, skin irritation, weight loss, nervous system damage and problems to the respiratory system.
Pesticides	• The use of herbicides, insecticides, fungicides, rodenticides and algicides.	It causes headaches, poisoning, cancer. Problems with the nervous system and gastrointestinal disturbance.
Plasticizers, chlorinated solvents and dioxin	• Solvents, pesticides, components of gasoline wood preservations.	Can cause cancer, problems in the nervous system, damage to the stomach and liver.
Pharmaceutical, antibiotics pollutants	• Confined animal feed operation facilities and feedlots. • The use of wastewater for groundwater recharge. • Municipal solid waste landfills. • Medical industrial activities.	The widespread use of antibiotics in the human and veterinary system caused a constant input of chemicals into the lifecycle, which caused the appearance of multi-drug-resistant bacteria.

3.7 Radioactive Contaminants in Groundwater

Many recent studies have shockingly discovered the presence of radioactive elemental (radionuclides) contamination in many drinking water sources, especially in groundwater. Specifically, radionuclides in water sources can be of great concern for human health as they pose severe toxicity and can be carcinogenic. To date, several radionuclides have been found in surface, sub-surface, and groundwater sources; among them 14C, 210Pb, 3H, 40K, 228Ra, 226Ra, 210Po, 232Th, 222Rn, and 234,235,238U are commonly found. On the other hand, thorium, uranium, and actinium are three NORM series of nuclear elements that contaminate water resources. Also, six radioactive contaminants were added to EWG's Tap Water Database, which also includes uranium, radium, and radon. Nonetheless, radium is most commonly found in two isotope forms known as radium-226 and radium-228, which has been largely found in contaminated tap water sources in several parts of the world. Radium is a naturally occurring radioactive element that is found in varying quantities in soil and rocks within the earth's crust. Also, small or trace amounts of radium have been determined in various groundwater sources and reservoirs. On the other hand, some radionuclides such as uranium have also been detected in rocks that are known to be present in the earth formed. Other than these, decayed radioactive by-products of those original radionuclides such as radon and radium have also been recently found in different water sources. As a result, these decayed or pristine radionuclides frequently occur in groundwater. Radiation is one of the intense and dangerous forms present in the form of electromagnetic rays or particles that are released from radioactive atoms. Upon exposure, radioactive contaminants release three types of radiation, namely alpha particles, beta particles, and gamma rays. Alpha radiation is relatively mild and unable to penetrate skin during normal exposure [77–79].

3.8 Groundwater Treatment Technologies

Groundwater contamination can be treated using both conventional as well as advanced water treatment techniques. Many conventional water treatment methods include physical treatment, and chemical as well as biological water treatment. Groundwater remediation is essential not only for cleaning contaminated sites but also to supply safe and clean drinking water. Groundwater reservoirs and aquifers play an integral role in maintaining the health of our ecosystem, soil, and crops. Groundwater sits just beneath the ground surface, which creates a complex situation, particularly treatment of contaminated sources becoming susceptible to spills and leaks. However, the treatment process of contaminated groundwater can be an expensive and lengthy affair. Today, in addition to conventional methods, different advanced water treatment techniques and remediation have been adopted to make groundwater safe as listed in Table 3.4.

Nevertheless, the first step to obtaining healthy and safe groundwater is to follow first with the environmental site assessment including site history and hydrogeology model evaluation in and around, which helps in identifying the risks to the quality of the groundwater. Once the site and risk assessment is done, detailed data analysis helps in identification, designing of remediation, and treatment techniques. Among many, conventional treatment methods are mainly categorized into three types, namely (i) physical, (ii) biological, and (iii) chemical methods.

Physical water treatment techniques are a set of widely practiced water treatment methods typically referred to as "pump and treat" techniques. These techniques involve pumping or extraction of groundwater by physically removing the water from the contaminated site from the sub-surface and treating the contaminated source through either chemical or biological means before returning it to the

Table 3.4 Types of reactive sources and their potential treatment techniques to remove radioactive contamination from groundwater.

Type of reactive media	Predominant remediation approach
Activated carbon products	Remediation by adsorption
Products made of amorphous ferric oxyhydroxides	Adsorption
Basic oxygen furnace slag (BOFS)	Sorption processes
Resins of ion exchangers	Adsorption
Limestone products	Precipitation
Zero-valent iron (ZVI)	Reduction then precipitation
Apatite products	Precipitation
Sodium dithionite	Reduction and precipitation
Sulphate-reducing bacteria	Microbiological degradation
Zeolites products	Adsorption
Sand beds or gravel beds with nutrients and oxygen	Microbiological degradation

original site. Similarly, air sparging or air stripping is used as another physical remediation tactic to treat contaminated groundwater. The sparging technique involves injecting pressurized air into the contaminated groundwater resulting in formation of bubbles that rise and cause pollutants to dissolve into a vapor state. Further, vapors containing contaminants are then extracted using vacuum systems which leave behind the treated and relatively clean groundwater [80, 81].

Chemical treatment techniques, on the other hand, are followed when all biological treatment techniques are ineffective in removing the contaminants from the groundwater. Among many chemical techniques, ion exchange, carbon absorption, Fenton reaction, chemical degradation, and oxidation play an important role in remediating groundwater sources. In the carbon absorption techniques, generally, ultra-high surface area activated carbon is used for the adsorption of semi-volatile and volatile organic compounds. In this process, organic contaminants are attached to the surface of the activated carbon, thereby removing contaminants and leaving behind clean groundwater. In the ion-exchange process, contaminated groundwater is cleaned by passing through granular ion-exchange resins, which facilitate the exchange of contaminants from ion-exchangeable mobile moieties in the media. On the other hand, the chemical oxidation process uses liquid or gas oxidants in the sub-surface to degrade or remediate organic molecules. Interestingly, these chemical methods are used individually or in combination to effectively treat groundwater [82, 83].

Biological treatments: Even though biological treatment techniques are widely used for treatment of domestic as well as industrial wastewaters, these techniques have also been applied for groundwater remediation due to their cost-efficiency and adaptability to treat a variety of contaminants. These biological remediation methods generally involve use of indigenous bacteria or introducing of biological material and matter to facilitate targeting the contaminants to either accelerate or speed up the breakdown of chemical waste in the groundwater. Similarly, other biological treatment techniques include bioventing, bioaugmentation, and biosparging. Bioaugmentation uses the introduction of strains that are capable of degrading contaminants in the groundwater, which also speeds up the degradation rate. On the other hand, bioventing uses microorganisms to biodegrade organic contaminants and biosparging involves pumping air and nutrients to increase the biological activity of helpful microorganisms in the groundwater. Biological treatments are generally effective against contaminants in groundwater having both inorganic and organic combinations and facilitate or help in accelerating remediation.

3.8.1 Inorganic Contaminant Treatment

Usually, underground water does not contain large quantities of suspended solids or particulate matter, which minimizes the use of physical water treatment techniques. Nevertheless, physical methods including filtration processes like membrane techniques can be very effective against most of the dissolved groundwater contaminants. Physical water treatment generally uses a variety of other filter media with different pore sizes. Similarly, physical adsorption methods are also being employed to capture physically absorbable contaminants from the groundwater. On the other hand, public water supply systems often use a series of water treatment steps that include disinfection, coagulation, flocculation, sedimentation, and filtration. Among these, some are chemical and others belong to physico-chemical techniques. However, coagulation, flocculation, and sedimentation processes that remove many of the particulate contaminants can be categorized as an important primary step in the water treatment process. On the other hand, dissolved organic carbon content and biological contaminants can be removed through disinfection techniques using either chemical or biological techniques. For instance, chemical reagent-assisted coagulation removes some of the dissolved substances using chlorine as a chemical disinfectant.

On the other hand, diverse and hazardous inorganic pollutants like heavy metals, nitrites, chlorides, phosphates, and other traces or impurities need to be addressed effectively to produce safe drinking water. Further, another set of inorganic pollutants like fertilizers, municipal wastewater containing surfactants, and numerous other toxic elements like agricultural discharges containing plant nutrients, radioactive traces and natural or geogenic seepages of elements such as selenium, fluorides, arsenic, and other heavy metals. These elements present even in trace amounts have proven threat to humans, aquatic organisms, and fauna and flora that are vulnerable to various infections, diseases, and even death [84, 85].

The efficient removal of these inorganic pollutants can be economically treated using a series of individual and hybrid treatment techniques based on the type of pollutant, concentration of impurities in groundwater, and environmental and geographical factors that may potentially hinder the efficiency of treatment methods. Among many technologies utilized currently, sedimentation, prechlorination, coagulation, precipitation, flocculation, Fenton process, adsorption, photocatalysis, incineration, ion exchange, ozonation, photoelectrochemical techniques, advanced methods like nanofiltration (NF), reverse osmosis (RO), membrane filtration, electrochemical methods, activated carbon technologies, ultraviolet light, and sono-degradation.

Nevertheless, inorganic contaminants can be efficiently treated using adsorption techniques that have been established as economical methods owing to their high efficiency and low-cost affordable prices. Among efficient adsorbents (as shown in Figure 3.6), numerous polymers from both natural and synthetic origins have been used as effective adsorbents derived from agricultural waste like vegetable/fruit

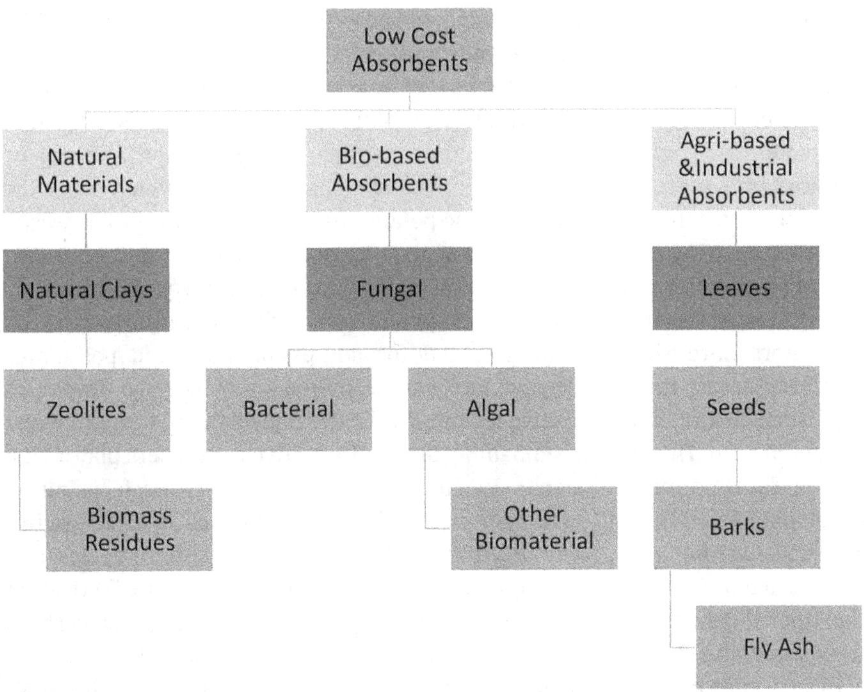

Figure 3.6 Different types of adsorbents available in the market or in developmental stage for the effective abetment of various inorganic contaminants in groundwater.

peels, bagasse, natural and modified clays, biomass-based activated carbon, sawdust, noble and precious metal-based nanomaterials, ferrite-based nanomaterials, nanocomposites, zeolites-based adsorbents, ionic bio-origin chitosan, industrial wastes like grits, dregs, and fly ash, and synthetic designer polymers like dendritic macromolecules have been utilized as inexpensive adsorbents for treatment and removal of different inorganic contaminants from groundwater sources. Further, nanomaterials and carbon-based microcleaners, and nanocleanse materials exhibit a large number of innovative and distinct physical, chemical, and structural properties that have great potential in treating inorganic contaminants. In particular, metal-organic frameworks, nano-zero-valent iron materials, carbonaceous-based and graphenaceous nanomaterials, biopolymers, microcleaners, hydrogel nanocomposite-based adsorbents, aerogels, composite aerogels, and xerogels have also been utilized as futuristic and emerging technologies for the treatment of contaminated groundwater and their purification. These nanomaterials are being utilized as effective adsorbents, and Fenton or photo-Fenton catalysts, which act as degradation media by virtue of their distinct structural, chemical, physico-chemical, electrical, and mechanical properties [85, 86].

Among many absorbents, low-cost biomass-based high-surface carbon materials can be an effective medium. These bio-based adsorbents can be obtained from biomasses like agricultural residues, industrial residues, and discarded bio-wastes, and abundant weed biomass can offer a sustainable solution for abetting and treating organically contaminated wastewater. With appropriate biomass extraction technique and controlled process parameters, these bio-gen adsorbents can be surface-tuned with required functionalities and characteristics, which allow the removal of harmful and toxic inorganic pollutants from potable as well as wastewater sources. These materials post-remediation also contribute toward minimization of secondary wastes at their disposal and maximizing recovery, and reuse.

Recent advances that are under development from various scientific groups and innovators have shown great potential for addressing inorganic pollutants. Scientists across material science and nanotechnology sectors have developed sophisticated and highly effective techniques for the remediation of groundwater contamination. However, sustainability and cost-effective parameters play a vital role in deciding material stocks. In this direction, researchers are focusing on developing or extracting precursor materials from abundant natural origins. These include biomass precursors, discarded or waste biomass, natural clay, and zeolitic materials that are not only considered abundant but also cost-effective. On the other hand, researchers can produce synthetic adsorbents (activated carbon) from agricultural waste, industrial waste, and household waste. Even though activated carbon, molecular sieve carbon, activated alumina, molecular sieve zeolites silica gel, and polymeric adsorbents are still considered a commercially viable alternative, however nanomaterial-based composite adsorbents either in the form of functional polymers, hydrogels, aerogels, xyrogels, modified nanoclays, metal-organic framework, and covalent organic framework and so on expected to replace the conventional materials stock in near future [87].

3.9 Organic Contaminant Treatment

In addition to various inorganic pollutants, recent studies have shown high concentrations of organic contaminants in groundwater. Diverse organic pollutants consist of traces or impurities of pharmaceuticals, endocrine-disrupting compounds, pesticides, perchlorates, halogenated and phenolic compounds, inorganic acids, herbicides, and numerous other compounds, which have been found in various groundwater sources throughout the world. Many of the recently found organic pollutants possess unique physico-chemical characteristics that were earlier either not found or analyzed. These pollutants are categorized as ECs that include PCPs including surfactants, detergents, pharmaceuticals, industrial compounds, pesticides, flame retardants, fragrances, secondary pollutants originating from primary water treatment materials and their by-products, as well as "lifestyle" compounds such as nicotine and caffeine. Recent studies reveal that ECs already have been known to induce adverse effects on aquatic ecosystems, especially on surface water sources and overall environmental health. However, recent studies have revealed high concentrations of pharmaceuticals, detergents, semi-volatile organic compounds, pesticides, herbicides, and numerous other toxic organic chemicals in groundwater supplies. Many of these organic pollutants have shown their origin in modern agricultural practices involving fertilizers, pesticides, herbicides, plant nutrients, growth boosters, and growth nutrients during irrigation activities to groundwater contamination. The other sources of groundwater organic contamination are also traced to industrial discharges, domestic and municipal waste dumping yards, and lancets. Other organic contamination origins include multifarious compounds, municipal wastewater discharges, natural or geogenic seepages, agricultural runoffs or discharges containing plant nutrients, biodegradable matter, pathogenic discharges, and biomedical waste disposal. These organic contaminants require advanced material stock and treatment techniques to effectively treat contaminated groundwater sources. Some of these techniques have been discussed in detail in the following section [88].

3.9.1 Pump and Treat Technique

Pump and treat method of water treatment refers to the process of pumping contaminated groundwater to a collection facility on the surface through wells, which are screened in the saturated zone and then subjected to treatment of

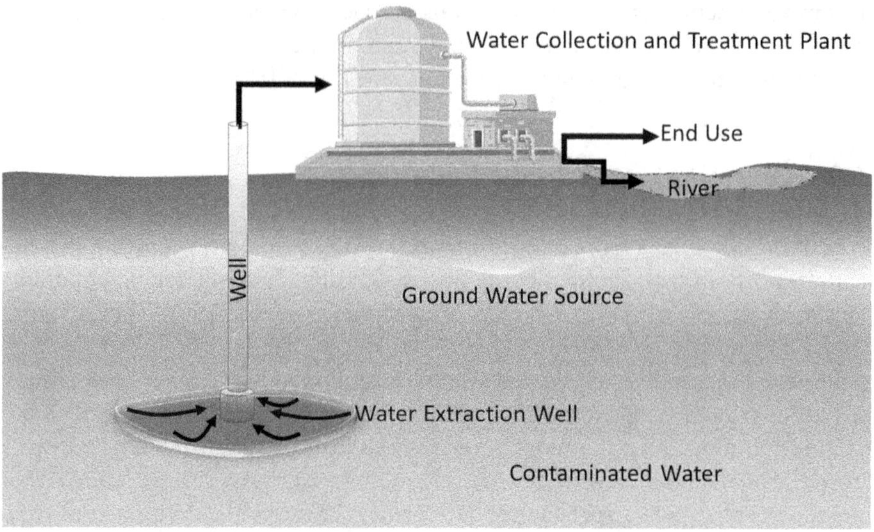

Figure 3.7 Schematic representation of pump and treat methods followed for the groundwater treatment.

contaminated groundwater with one or more technologies as shown schematically in Figure 3.7. Pump and treat generally involves the withdrawal of groundwater from a contaminated aquifer site using trenches or galleries, one or more extraction wells, and treating them as retrieved groundwater on above-ground treatment facility prior to discharge. This is one of the significant and popular procedures to remediate groundwater contaminated by dissolved organic/inorganic chemicals, organic solvents, semi-volatile organic compounds, PCPs, and fuel oil. This procedure also involves pumping water through a piped facility to ground lagoons or directly to treatment facility, which further treats the contaminated water using various methods like activated carbon or air stripping. At the final stage, the treated water on the ground facility is discharged either to further treatment facilities to make it fit for drinking purposes or be discarded to the nearest sewer system or re-pump to the sub-surface or groundwater aquifer sites as a measure of groundwater rejuvenation. Interestingly, this technique can be effectively used to treat large volumes of contaminated groundwater. However, the technique has several disadvantages such as high capital and operating cost, long operation time, and vulnerability of spreading removed contaminants into the freshwater ecosystem. This method has also proven ineffective against contaminants' precipitation or adsorption. This method is also proven limited to apply for low permeability material having silt clay or clay contamination. This method requires long-running operations, generates high quantities of secondary waste, and requires constant maintenance of pumping and treatment equipment to ensure clean and sustainable operations [81].

3.9.2 Air Sparging Procedure and Soil Vapor Extraction

Air sparging technique of groundwater treatment involves infusion or injection of air through engineered wells into the contaminated aquifer below sub-surface of groundwater table. The air sparging process involving high flow rate of air creates millions of bubbles to clarify contaminated groundwater sites and larger aquifer zones through two processes. First, infusion of air creates a turbulent condition in the polluted aquifer that increases the volatilization of the contaminants. The process of air sparging is considered one of the simple techniques widely used in remediating volatile organic contaminants (VOCs) from contaminated groundwater as shown in Figure 3.8. Air sparging is considered a fast, efficient, and relatively economical process in treating organic contamination from groundwater. The injection of high-pressure air at the lowest level or point of the contaminated groundwater site will clean up the polluted site by changing the state of volatile hydrocarbons to a vapor state. In this simple and unique physical treatment technique, pumping air under the saturated zone, pollutants are stripped out of the aquifer and oxygen is provided for the accelerated biodegradation of contaminants in the groundwater. Further, volatile organic mixed extracted air is then applied by vacuum extraction facilities for safe recovery or removal of toxic

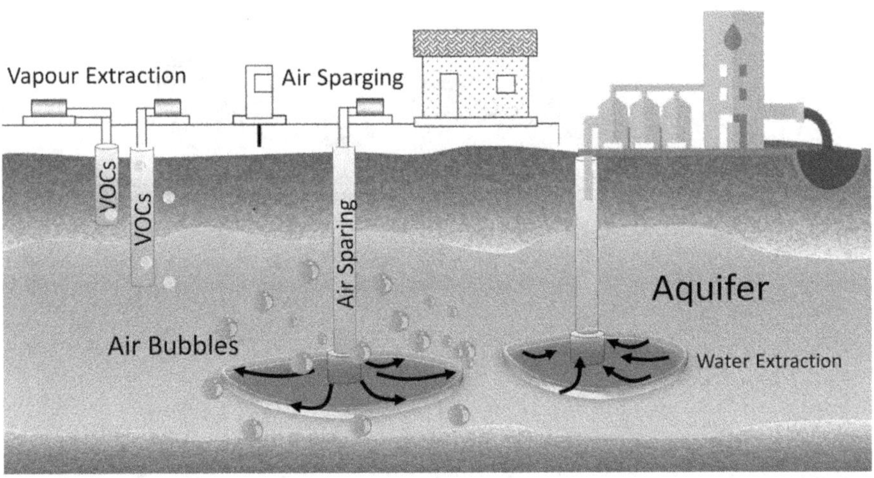

Figure 3.8 Schematic representation of air sparging treatment technique used for groundwater remediation.

contaminants. However, there are several limitations associated with this process such as high capital and operation costs associated with working with deep wells. Also, contaminants originate with soil heterogeneity that leads to an uneven treatment of the contaminated groundwater [89, 90].

3.9.3 Permeable Reactive Barriers

A permeable reactive barrier (PRB) is considered a low-cost and effective alternative for underground emplacement of pollutants or reactive moieties or substances through which a dissolved contaminant plume must move as it flows, typically operated under a natural gradient. In the PRB, the primary contaminant removal process is either physical like sorption or precipitation, or chemical techniques such as ion exchange, degradation, or biological. In this innovative underground water remediation technique, polluted groundwater flows through a reactive barrier or media, which will degrade or convert the contaminant into a less toxic or harmless or easily degradable moiety as shown in Figure 3.9. All the flow and transformation of the polluted water to treated water typically occurs under natural gradient flow. In this process, treated water exits the other side of the crossing reactive barrier. PRB is technically an in situ technology, which remediates groundwater contaminated with volatile organics and immobilizing metals. PRB is particularly designed to exchange or degrade through the reactive media designed with porous functional as well as natural materials such as sand

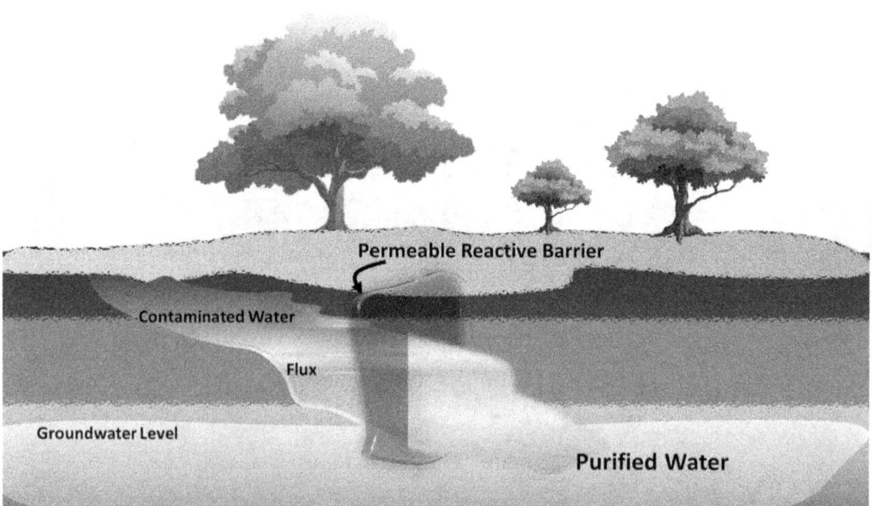

Figure 3.9 Schematic diagram illustrating the presence of permeable reactive barriers for remediating contaminated groundwater.

and polymeric resins. In the treatment barrier, contaminants are degraded or removed through the mechanism of adsorption, transforming into less toxic substances, degrading and precipitating out the targeted contaminants during water flow across barrier trenches. During the remediation, pollutants are passively removed, captured, or broken down while passing uncontaminated water to pass through the reactive barrier [91, 92].

PRBs are designed in many geometries such as (i) a continuous barrier, which is the most commonly used placement that consists of reactive media placed perpendicular to the contaminated plume of groundwater flow, (ii) PRB in funnel and gate configuration in which contaminant plume is directed to a treatment-filtering gate by two-sided impermeable barrier at contamination sites in which the soil is generally characterized as heterogeneous in nature in which placing the PRB plays an important role for the most permeable portion of the soil, (iii) PRB in radial filtration or caisson configuration in which the remediation filter is placed in a cylindrical shape of reactive barrier surrounded by coarse active material. In addition to this, a radial centripetal flow is applied to control the hydraulic gradient. The third-category reactive barrier has a long lifespan and has been known to perform better treatment efficiency by extending the contact time between pollutant and reactive material barrier.

On the other hand, there are several reactive materials that are being used as reactive materials to remediate contaminants, such as zero-valent iron (Fe^0), which is a soft reductant, which can easily treat heavy metal contamination. Zero valent iron can be halogenated and may halogenate hydrocarbon derivatives to control the type and extent of organic substance degradation. On the other hand, bio-sparging moieties or materials and slow oxygen-releasing compounds have the capability to treat groundwater contaminated with petroleum hydrocarbon moieties like nitrobenzene and aniline. This is carried out by utilizing the biodegradation step of these pollutants in/on reactive barriers. Similarly, vegetative substances are used as reactive species such as mulch to remediate perchlorates and chlorinated solvents. In this process, groundwater contaminants can also be precipitated on chemical reactive materials in the reactive barrier, for instance, granular activated carbon, iron/gravel, zeolites, iron/sponge, ferrous slats, fly ash, lime, organic carbon, iron/sand, phosphates, copper wool, and steel wool.

Interestingly, choosing a good reactive media plays a very important role in effective remediation of the water, which greatly depends on the following characteristics and parameters such as (i) Reactivity, which is the ability of reactive media to react or remediate contaminants and the equilibrium constant. The reactivity factors and functionalities of barrier materials are necessary to determine the desired time for the remediation, which is important to estimate the volume and size of the in situ reactive barriers; (ii) Stability is another important parameter of barrier materials required for any effective reactive material to be mechanically robust for a long period to survive during the remediate groundwater. Also, it is necessary to retain the physico-chemical integrity of the reactive barrier under the surface as a secondary precipitate. However, a major issue with PRB installation is that it is cost-intensive to replace or excavate a PRB with a new one; (iii) Cost and availability are other major parameters that are very important to make reactive media available widely at cost-effective prices; (iv) Hydraulic conductivity in which the reactive media must have or possess a permeability equal to or greater than the surrounding soil to ease the contaminated groundwater to flow within the receive barrier to achieve the groundwater remediation; (v) Environmental compatibility is another significant parameter in which reactive media to achieve compatibility to match the surrounding sub-surface soil environment by means of grain size. This is taken care of to match the grain size and functionality to achieve the goal and to avoid changes in the hydraulic conductivity of the soil.

3.10 Groundwater Contaminant Uptake Mechanism

Several researchers have proposed various contaminant uptake mechanisms from contaminated groundwater depending on the remediating media or technique. Among many phenomena, four important mechanisms can be broadly classified into physical, chemical, physico-chemical, and biological uptake mechanisms and these are specifically categorized as (i) adsorption and ion exchange, (ii) abiotic reduction, (iii) biotic reduction, and (iv) chemical precipitation. Most of the organic contamination remediation in groundwater occurs either following an individual mechanism or through two or more mechanisms [93].

3.10.1 Adsorption and Ion-Exchange Mechanisms

The adsorption and ion-exchange process of groundwater treatment are the process in which contaminant species in a different aqueous environment are selectively attached to a solid functional surface, which is referred to as adsorbates through the process of adsorption. Generally, the adsorption process that facilitates the interaction of groundwater contaminant moieties with the adsorbate's active site is considered robust or rapid and is generally required to be reversible. Adsorbents such as carbon-based active materials, graphene-based adsorbents, composites/nanocomposites, functional polymers, zeolite, zero-valent iron, and amorphous ferric oxyhydroxide (AFO) are the most commonly used adsorbents for various types of contaminant removal from groundwater. In all cases, the most significant physico-chemical property involved in adsorption efficiency is a high-to-ultra large surface area of adsorbent per gram, which provides optimal active sites to capture pollutants. On the other hand, the presence of active surface functionality available on adsorbents is another significant parameter that also determines the extent and recyclability of adsorbents. These functionalities significantly influence the adsorption of pollutants such as oxy-anions in groundwater like F^-, which is abundant in groundwater around the world and can be effectively adsorbed by metal oxide and metal-oxyhydroxide-functionalized carbon or polymers. Similarly, zero-valent iron (Fe^0) is the widely used and popular reactive media PRB, which has robust and high adsorption rate against contaminants from groundwater. Therefore, adhesion of toxic contaminant ions, or atoms, or molecules that occur in a liquid or dissolved solid state, is referred to as adsorption through the various chemical and/or physico-chemical interactions creating a thin film of the adsorbate on the adsorbent's surface as shown schematically in Figure 3.10 [94–96].

On the other hand, in ion-exchange process, remediation of inorganic substances and dissolved metals and heavy metals dissolved in groundwater is achieved through the ion-exchangeable site present in the resins or media. In the ion-exchange process, the ion which is a single atom or group of atoms possessing either positive charge upon loss of electrons or attain negatively charged after gaining an electron acts as exchangeable sites. When underground water containing ionic pollutants passes through the ion-exchange resins, contaminated substances are exchanged through the effect of metallic ions on the attraction of exchangeable ions in the resins. Interestingly, inorganic (zeolitic or modified

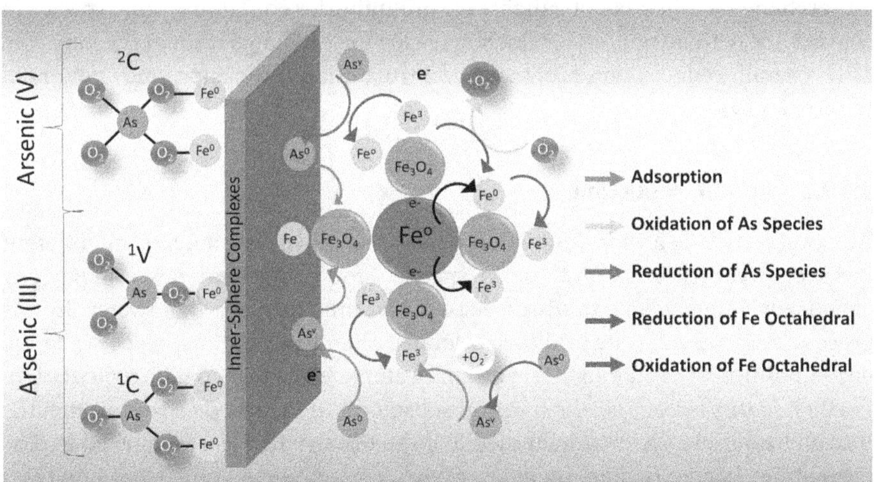

Figure 3.10 Schematic representation of various types of adsorption mechanisms found in various adsorption materials and media.

nanocomposite) and polymeric resins can be re-generated after being saturated or exhausted. In some cases, resins are irreversible in nature, which may be suitable for a single use. However, in advanced treatment materials and techniques, ion-exchange phenomena is a reversible reaction process in which repeated use of resins for abetment of ionic pollutants is used and resin's active site is regenerated by replacing with an identical ion on the immobilizing barrier. Conventionally popular ion-exchangers are made of natural source materials such as zeolitic compositions. However, there are several synthetic ion-exchange resins, which have been invented using polymers, organic/inorganic combination systems involving functional polymers, and nanomaterials-based ion-exchangers. Interestingly, ion-exchangers are effectively used to remove or the treatment of inorganic contaminants such as heavy metals, dissolved metals like Cr, and hardness-causing mono- and divalent ions like Mg^{2+}, and Ca^{2+} from polluted groundwater. Additionally, ion-exchange media is also used for the treatment of nonmetallic pollutants such as ammonia and nitrate. However, there are several limitations that exist in ion-exchange systems such as the oxidation of the soil, which when used in underground treatment plants could cause damage to the ion-exchange resin, which is expected to decrease remediation efficiency of the media. Another major concern associated with ion-exchange media is that the contaminants are physico-chemically held on the exchange media and not completely destroyed from the groundwater. That is, ion-exchange media only transfer ionic pollutants from one medium (in liquid) to another medium (on solid surface), which further needs appropriate disposal and management. Also, ion-exchange method is not suitable for groundwater containing ionic species in coexistence with oil or grease, which may clog the exchange resin in the long run. This overall reduces ion-exchange resin/materials efficiency and treatment economics [97–99].

3.10.2 Abiotic Reduction

Abiotic reaction is an iron reduction procedure, which is catalyzed by different types of organic compounds or in specific iron-sulfur minerals, often using chemically active agents to dissolve solid iron minerals in groundwater. In this process, suitable catalysts lead to accelerated chemical reactions that lead to the decomposition of contaminants in groundwater referred to as abiotic remediation. Further, in this technique, the harmful substances are reduced either through the immobilization in the treatment media of the reactive barriers or allowed to pass through the barrier making them less toxic. In this process, zero-valent iron (Fe^0) is the most commonly used reactive material in the abiotic remediation of groundwater. Upon the reaction, Fe^0 with the contaminants and low solubility minerals are precipitated out. For instance, the remediation of uranium (U) and chromium (Cr) contaminants from groundwater is removed as precipitation through the abiotic process. Equation (3.1) shows the ability of Fe^0 in abiotic process to reduce U(VI) to U(IV) in groundwater at moderate pH and high carbonate via the intermediate production of Uraninite (UO^2), which is a solid form of less crystalline product of uranium [100–102].

$$Fe^0 + UO_2\left(CO_3\right)_2^{-2} + 2H^+ = UO_{2[Solid]} + 2HCO^{-3} + Fe^{2+} \quad (3.1)$$

On the other hand, for the chromium Cr(VI) reduction to Cr(IV), Fe^0 is used as a catalyst as shown in Equation (3.2):

$$Fe^0 + 8H^+ + CrO_2^{-4} = Fe^{3+} + Cr^{3+} + 4H_2O \quad (3.2)$$

Further, Cr(VI) can be directly reduced to Cr(III) using various ferrous iron by introducing dissolved dithionite ions ($S_2O_2^{-4}$) to a contaminated aquifer, which can reduce the solid phase of ferric iron. On the other hand, dithionite oxidizes to sulfite (SO_2^{-3}) and Fe^{3+} is then reduced to Fe^{2+}. In this case, Cr(III) is to be separated out by precipitating in the solid form of Cr(III) and Fe(III) hydroxide in addition to the reduction of trace halogenated organic compounds with the help of Fe^{2+} as shown in Equations (3.3) and (3.4).

$$S_2O_2^{-4} + 2Fe\left(III\right)_{[Solid]} + 2H_2O = 2SO_2^{-3} + 2Fe\left(II\right)_{[Solid]} + 4H^+ \quad (3.3)$$

$$CrO_2^{-4} + 3Fe\left(II\right)\left[Solid\right] + 5H^+ = Cr\left(OH\right)_{3[Solid]} + 3Fe\left(III\right)_{[Solid]} + H_2O \quad (3.4)$$

3.10.3 Biotic Reduction/Oxidation

Abiotic reactions include chemical reductions or oxidations in anerobic or aerobic environments, respectively. Whereas biotic reactions are carried out by a diverse group of microorganisms, that is, biotic transformations are carried out in presence of and by bacteria and fungi in the environment, or reactions may take place within the bodies of plants and animals. Biotic reduction/oxidation can be effectively used either individually or in a combination of biologically and nonbiologically mediated processes, such as oxidation, reduction, and hydrolysis. In some cases, biotic reaction can be a completely biologically mediated process, especially in conjugation. Biotic reduction or oxidation is used when physical or chemical or any other physico-chemical remediation of groundwater techniques proves ineffective or shows little to no degradation efficiencies against groundwater contaminants. Then biological oxidation processes can be effectively used for the degradation of pollutants in groundwater. Using this process, several pollutants such as chlorinated solvents can be easily reduced through the biotic oxidation process. In this process, microorganisms perform a reduction process by utilizing organic contaminants as their main source of energy and the necessary materials to synthesize their cells. This bioremediation technique is a highly efficient water treatment process based on the degradation of types of contaminants using microorganisms. However, groundwater remediation efficiency in biotic processes depends on the working condition and environment of groundwater such as the pH, temperature, concentration of nutrients and electron acceptor ability of the media. More significantly, in biodegradation, it is essential that microorganisms use electron acceptors to accept any electrons liberated from contaminants and electron transfer facilitates the release of energy that is essential for microbes to survive or live for longer treatment cycles. Under aerobic conditions, which is a preferable route in the presence of oxygen, energy production from the overall process is higher compared to the process in which no oxygen is present. In addition to this, the oxidation rate of pollutants in groundwater is high. As the presence of oxygen in groundwater is generally less, in such cases, the anerobic microbe electron acceptors technique is utilized for the remediation. However, this method remediates groundwater contaminated with monoaromatic hydrocarbons, and is effective by using oxygen-releasing compounds in the reactive barrier [103, 104].

Nonetheless, the fundamental concept of biotic reduction/oxidation is to supply an electron donor in addition to the nutrient materials to be utilized using microorganisms to break down the contaminants. In this process, bio-origin leaf mulch, sawdust, and wheat straw can be used as electron donors. On the other hand, municipal waste can be utilized as a nutrient material. For instance, dissolved sulfate in groundwater or wastewater is a good electron acceptor, which can oxidate organic substances by consuming acidity coupling media by metal reduction as shown in Equations (3.5) and (3.6):

$$2CH_2O_{[\text{Solid or oganic}]} + SO_2{}^{-4} + H^+ = 2CO_2 + 2H_2O + HS^- \tag{3.5}$$

$$Me^{2+} + HS^- = MeS_{[\text{solid}]} + H^+ \text{ where Me} = \text{metal.} \tag{3.6}$$

3.10.4 Chemical Precipitation

Chemical precipitation process is a technique of converting solution into solid through the intermediate substance in the form of insoluble suspension and further separation out making it a supersaturated one. Therefore, this process consists of contaminant removal as hydroxides as shown in Equation (3.7) and carbonates in Equation (3.8) through the mineral precipitation resulting as a result of enhanced pH conditions. Here, in this process, contaminants are reduced

Table 3.5 Summary of different conventional and advanced remediation methods used for the treatment of groundwater contamination.

Type of reactive media	Predominant remediation approach
Activated carbon products	Remediation by adsorption
Products made of amorphous ferric oxyhydroxides	Adsorption
Basic oxygen furnace slag (BOFS)	Sorption processes
Resins of ion exchangers	Adsorption
Limestone products	Precipitation
Zero-valent iron (ZVI)	Reduction and precipitation
Apatite products	Precipitation
Sodium dithionite	Reduction and precipitation
Sulfate-reducing bacteria	Microbiological degradation
Zeolite products	Adsorption
Sandbeds or gravel beds with nutrients and oxygen	Microbiological degradation

to a less soluble species, and subsequently, insoluble species are retained as minerals in the reactive barrier. From groundwater sources, dissolved contaminants are chemically precipitated using limestone ($CaCo_3$) and apatite [$Ca_5(PO_4)^3(OH)$] as chemical agents.

$$Me^{2+} + 2(OH)^{-} = Me(OH)_2 \tag{3.7}$$

$$Me^{2+} + HCO^{-3} = MeCO_3 + H^{+} \tag{3.8}$$

Further, Table 3.5 gives the overall summary of conventional and advanced reactive media for groundwater treatment. However, selection and efficiency of these remediation techniques greatly depend on the nature of contaminants, their reactivity, geochemical process, and extent of their availability play a vital role in obtaining optimal and desired end results [105].

3.11 Conclusion and Future Perspectives

Conventional groundwater treatment typically involves aeration for the infusion of oxygen and the stripping of the volatile and semi-volatile organic compounds, sand filtration to remove particulate pollutants, and several biological processes have been utilized effectively. However, advanced techniques like

UV disinfection and membrane technologies have been adopted to improve the hygiene of extracted water. As sufficient availability of freshwater decreases, it is essential to protect sources, sites, and supply systems from potential ECs. Ever-increasing industrial activities and use of millions of synthetic products of which several of these are discharged into water bodies and into the environment without appropriate treatment and regulation. However, to keep the good health of humans and the environment, it is essential to regularly monitor, analyze, and adopt water quality protection measures like treatment techniques.

Conventionally, there are two main strategies that have been followed to treat groundwater, namely in situ and ex situ. In situ remediation technique is directly used in place or on-site, while ex situ means off-site and upon extraction of contamination from groundwater, which is further treated on the surface infrastructure or treatment facility. Whereas, in situ follows the cleaning of the groundwater where contamination occurs. The groundwater treatment facilities are now extensively used to remediate pollutants and various EC groundwater sources by applying various techniques that include both conventional as well as advanced processes to bring the water to standard drinking water qualities set by the various national and international regulatory agencies.

Among various advanced and modern techniques like membrane-based separation processes involving RO, NF, advanced adsorption materials like hydro/solvo/iono-thermal high surface carbon materials, nano cleaners, micro cleaners, hydrogels, polymer-based aerogels, nanocomposite-aerogels, and so on have been tried and tested for efficient treatment of contaminated groundwater. However, emergence of various persistent and toxic contaminants like fluoride, arsenic, active pharmaceutical ingredients, and constituents of PCPs like detergents, drug molecules, and so on have also been expected to pose serious challenges to existing conventional as well as advanced water treatment materials and processes.

On the other hand, future approaches also require protection of water zones from the exposure of emerging and industrial pollutants. These efforts also need to include regular chemical examination of the groundwater intended for mass consumption and need to be regulated to monitor maximum contamination levels to be under control. On the other hand, bacteriological toxicity levels in groundwater are also to be ensured by regular monitoring and disinfection. Even though groundwater contamination in individual sites or zones extracted regularly may not exceed extremely high concentrations, it can be simply treated and quality of water can be improved by regular cleaning and disinfection. However, adopting advanced materials and treatment techniques to remove sources of potential contamination would be the best choice.

References

1 Li, P., Karunanidhi, D., Subramani, T., and Srinivasamoorthy, K. (2021). Sources and consequences of groundwater contamination. *Arch. Environ. Contam. Toxicol.* 80 (1): 1–10.

2 Jones, J.B. and Mulholland, P.J. (2000). *Streams and Ground Waters, A volume in Aquatic Ecology*. Elsevier ISBN: 978-0-12-389845-6.

3 Mokarram, M., Saber, A., and Sheykhi, V. (2020). Effects of heavy metal contamination on river water quality due to release of industrial effluents. *J. Clean. Prod.* 277: 123380. ISSN 0959-6526.

4 Rehman, K., Fatima, F., Waheed, I., and Akash, M.S.H. (2018). Prevalence of exposure of heavy metals and their impact on health consequences. *J. Cell. Biochem.* 119 (1): 157–184.

5 Bexfield, L.M., Belitz, K., Lindsey, B.D. et al. (2021). Pesticides and pesticide degradates in groundwater used for public supply across the united states: occurrence and human-health context. *Environ. Sci. Technol.* 55 (1): 362–372.

6 Malla, M.A., Gupta, S., Dubey, A. et al. (2021). Chapter 7 – contamination of groundwater resources by pesticides. In: *Contamination of Water* (ed. A. Ahamad, S.I. Siddiqui, and P. Singh), 99–107. Academic Press ISBN 9780128240588.

7 Sharma, S. and Bhattacharya, A. Drinking water contamination and treatment techniques. *Appl. Water Sci.* 7: 1043–1067.

8 Mullenix, P.J. (2014). A new perspective on metals and other contaminants in fluoridation chemicals. *Int. J. Occup. Environ. Health* 20 (2): 157–166.

9 Qiong, W. et al. (2017). Current situation and control measures of groundwater pollution in gas station. *IOP Conf. Ser.: Earth Environ. Sci.* 94: 012005.

10 Cabral, J.P. (2010). Water microbiology. Bacterial pathogens and water. *Int. J. Environ. Res. Public Health* 7 (10): 3657–3703.

11 He, C., Liu, Z., Wu, J. et al. (2021). Future global urban water scarcity and potential solutions. *Nat. Commun.* 12: 4667.

12 Garrick, D. et al. (2019). Rural water for thirsty cities: a systematic review of water reallocation from rural to urban regions. *Environ. Res. Lett.* 14: 043003.

13 Krueger, E., Rao, P.S.C., and Borchardt, D. (2019a). Quantifying urban water supply security under global change. *Glob. Environ. Chang.* 56: 66–74.

14 Dillon, P. and Jiménez, B. (2008). Water reuse via aquifer recharge: intentional and unintentional practices. In: *Water Reuse: An International Survey of Current Practice, Issues and Needs* (ed. B. Jiménez and T. Asano), 260. London: IWA Publishing.

15 Chouhan, S. and Flora, S.J. (2010). Arsenic and fluoride: two major ground water pollutants. *Indian J. Exp. Biol.* 48 (7): 666–678.

16 Jha, P.K. and Tripathi, P. (2021). Arsenic and fluoride contamination in groundwater: a review of global scenarios with special reference to India. *Groundw. Sustain. Dev.* 13: 100576. ISSN 2352-801X.

17 Igboama, W.N., Hammed, O.S., Fatoba, J.O. et al. (2022). Review article on impact of groundwater contamination due to dumpsites using geophysical and physiochemical methods. *Appl Water Sci* 12: 130.

18 Abiriga, D., Vestgarden, L.S., and Klempe, H. (2020). Groundwater contamination from a municipal landfill: effect of age, landfill closure, and season on groundwater chemistry. *Sci. Total Environ.* 737: 140307. ISSN 0048-9697.

19 Kurakalva, R.M., Aradhi, K.K., Mallela, K.Y., and Venkatayogi, S. (2016). Assessment of groundwater quality in and around the Jawaharnagar municipal solid waste dumping site at greater Hyderabad, Southern India. *Procedia Environ. Sci.* 35: 328–336. ISSN 1878-0296.

20 Ya, X., Xue, X., Dong, L. et al. (2018). Long-term dynamics of leachate production, leakage from hazardous waste landfill sites and the impact on groundwater quality and human health. *Waste Manag.* 82: 156–166. ISSN 0956-053X.

21 Akhtar, N., Syakir Ishak, M.I., Bhawani, S.A., and Umar, K. (2021). Various natural and anthropogenic factors responsible for water quality degradation: a review. *Water 13*: 2660.

22 Meckenstock, R.U., Elsner, M., Griebler, C. et al. (2015). Biodegradation: updating the concepts of control for microbial cleanup in contaminated aquifers. *Environ. Sci. Technol. 49* (12): 7073–7081.

23 Burback, B.L. and Perry, J.J. (1993). Biodegradation and biotransformation of groundwater pollutant mixtures by Mycobacterium vaccae. *Appl. Environ. Microbiol.* 59 (4): 1025–1029.

24 Sims, G.K. (1990). Biological degradation of soil. In: *Advances in Soil Science. Advances in Soil Science*, vol. 11 (ed. R. Lal and B.A. Stewart), 289. New York, NY: Springer.

25 Miller, S.M., Spaulding, M.L., and Zimmerman, J.B. (2011). Optimization of capacity and kinetics for a novel bio-based arsenic sorbent, TiO2-impregnated chitosan bead. *Water Res.* 45 (17): 5745–5754.

26 Bissen, M. and Frimmel, F.H. (2003). Arsenic: a review – part I –: occurrence, toxicity, speciation, mobility. *Acta Hydrochim. Hydrobiol.* 31 (1): 9–18.

27 Nolan, J. and Weber, K.A. (2015). Natural uranium contamination in major U.S. aquifers linked to nitrate. *Environ. Sci. Technol. Lett.* 2 (8): 215–220.

28 Dilling, W.L., Tefertiller, N.B., and Kallos, G.J. (1975). Evaporation rates and reactivities of methylene chloride, chloroform, 1,1,1-trichloroethane, trichloroethylene, tetrachloroethylene, and other chlorinated compounds in dilute aqueous solutions. *Environ. Sci. Technol.* 9 (9): 833–838.

29 Li, B., Li, L., Lin, K. et al. (2013). Removal of 1,1,1-trichloroethane from aqueous solution by a sono-activated persulfate process. *Ultrason. Sonochem.* 20 (3): 855–863. ISSN 1350-4177.

30 Fawell, J. and Nieuwenhuijsen, M.J. (2003). Contaminants in drinking water: environmental pollution and health. *Br. Med. Bull.* 68 (1): 199–208.

31 Singh, R., Andaluri, G., and Pandey, V.C. (2022). Chapter 1 – cities' water pollution – Challenges and controls. In: *Algae and Aquatic Macrophytes in Cities* (ed. V.C. Pandey), 3–22. Elsevier ISBN 9780128242704.

32 M M Shahabudin and S Musa, Occurrence of surface water contaminations: an overview 2018 *IOP Conf. Ser.: Earth Environ. Sci.*140 012058.

33 Ritter, L., Solomon, K., Sibley, P. et al. (2002). Sources, pathways, and relative risks of contaminants in surface water and groundwater: a perspective prepared for the Walkerton inquiry. *J. Toxicol. Environ. Health A* 65 (1): 1–142.

34 Rasmussen, D.J. (2022). Magmatic water content controls the pre-eruptive depth of arc magmas. *Science* 375: 1169–1172.

35 Weis, F.A., Skogby, H., Troll, V.R. et al. (2015). Magmatic water contents determined through clinopyroxene: Examples from the Western Canary Islands, Spain. *Geochem. Geophys. Geosyst.* 16 (7): 2127–2146.

36 Ansah Amano, K.O., Danso-Boateng, E., Adom, E. et al. (2021). Effect of waste landfill site on surface and ground water drinking quality. *Water Environ. J.* 35 (2): 715–729.

37 Perry, R. (1982). Effects of waste disposal on groundwater and surface water. *Hydrol. Sci. J.* 27 (2): 273–283.

38 Gupta, N. and Nath, M. (2020). Groundwater contamination By E-waste and its remedial measure – a literature review. *J. Phys. Conf. Ser.* 1531: 012023.

39 Koul, B., Yadav, D., Singh, S. et al. (2022). Insights into the domestic wastewater treatment (DWWT) regimes: a review. *Water* 14: 3542.

40 Wild, S.R. and Jones, K.C. (1991). Organic contaminants in wastewaters and sewage sludges: transfer to the environment following disposal. In: *Organic Contaminants in the Environment*, Environmental Management Series, vol. 19 (ed. S.R. Wild and K.C. Jones), 133–158. Dordrecht: Springer.

41 Escher, B.I., Allinson, M., Altenburger, R. et al. (2014). Benchmarking organic micropollutants in wastewater, recycled water and drinking water with in vitro bioassays. *Environ. Sci. Technol.* 48 (3): 1940–1956.

42 Levi Anatolia, S.M. and Exposto, T.L. (2021). The impacts of hazardous and toxic waste management: a systematic review. *Interdiscip. Soc. Stud.* 1 (2): 103.

43 Fazzo, L., Minichilli, F., Santoro, M. et al. (2017). Hazardous waste and health impact: a systematic review of the scientific literature. *Environ. Health* 16: 107.

44 Marsili, D., Fazzo, L., and Comba, P. (2009). Health risks from hazardous waste disposal: the need for international scientific cooperation. *Eur. J. Oncol.* 14: 151–159.

45 Elisabeth, D., Ray, G., and Iska, B. (2021). Rebound and spillovers: prosumers in transition. *Front. Psychol.* 12: 636109.

46 Fang, D. (2019). Clean air for some: unintended spillover effects of regional air pollution policies. *Sci. Adv.* 5: eaav4707.

47 Coutts, A. (2022). Identifying communication spillovers in lab-in-the-field experiments. *J. Dev. Econ.* 157: 102845. ISSN 0304-3878.

48 Iravanian, A. and Ravari, S.O. (2020). Types of contamination in landfills and effects on the environment: a review study. *IOP Conf. Ser.: Earth Environ. Sci* 614: 012083-012091.

49 Xiang, R., Xu, Y., Liu, Y.Q. et al. (2019). Isolation distance between municipal solid waste landfills and drinking water wells for bacteria attenuation and safe drinking. *Sci. Rep.* 9: 17881.

50 Przydatek, G. and Kanownik, W. (2019). Impact of small municipal solid waste landfill on groundwater quality. *Environ. Monit. Assess.* 191: 169.

51 Johnson, B., Balserak, P., Beaulieu, S. et al. (2003). Industrial surface impoundments: environmental settings, release and exposure potential and risk characterization. *Sci. Total Environ.* 317: 1–22.

52 Seltenrich, N. (2014). Take care in the kitchen: avoiding cooking-related pollutants. *Environ. Health Perspect.* 122: 6.

53 Carpenter, R.A., Hollowell, R.H., and Hill, K.M. (1997). Determination of the metabolites of the herbicide dimethyl tetrachloroterephthalate in drinking water by high-performance liquid chromatography with gas chromatography/mass spectrometry confirmation. *Anal. Chem.* 69 (16): 3314–3320.

54 Ananth, M., Rajesh, R., Amjith, R. et al. (2018). Contamination of household open wells in an urban area of trivandrum, kerala state, india: a spatial analysis of health risk using geographic information system. *Environ. Health Insights* 12: 1178630218806892.

55 Fewtrell, L. (2004). Drinking-water nitrate, methemoglobinemia, and global burden of disease: a discussion. *Environ. Health Perspect.* 112 (14): 1371–1374.

56 Shaban, J., Al-Najar, H., Kocadal, K. et al. (2023). The effect of nitrate-contaminated drinking water and vegetables on the prevalence of acquired methemoglobinemia in Beit Lahia City in palestine. *Water* 15: 1989.

57 https://www.unicef.org/wash/water-scarcity

58 Boretti, A. and Rosa, L. (2019). Reassessing the projections of the world water development report. *NPJ Clean Water* 2: 15.

59 Figueroa, A.J. and Smilovic, M. (2021). Chapter 7 – groundwater irrigation and implication in the Nile river basin. In: *Global Groundwater* (ed. A. Mukherjee, B.R. Scanlon, A. Aureli, et al.), 81–95. Elsevier ISBN 9780128181720.

60 Apollaro, C., Fuoco, I., Criscuoli, A., and Figoli, A. (2022). Inorganic pollutants into groundwater: from geochemistry to treatment. *Geofluids* 9846802. 3 pages, 2022.

61 Kurwadkar, S. (2019). Occurrence and distribution of organic and inorganic pollutants in groundwater. *Water Environ. Res.* 91 (10): 1001–1008.

62 Adimalla, N., Li, P., and Qian, H. (2018). Evaluation of groundwater contamination for fluoride and nitrate in semi-arid region of Nirmal Province, South India: a special emphasis on human health risk assessment (HHRA). *Human and Ecol. Risk Assess.: An Int. J.* 25 (5): 1107–1124.

63 Jareda, G., Mahapatra, S.P., and Dhekne, P.Y. (2018). Water quality index, heavy metal pollution index and seasonal variation correlation of groundwater of Bailadila iron ore mine area and its peripherals: Dantewada district, Chhattisgarh, India. *Desalin. Water Treat.* 101: 7–16.

64 Masuda, H. (2018). Arsenic cycling in the Earth's crust and hydrosphere: interaction between naturally occurring arsenic and human activities. *Prog. Earth Planet Sci.* 5: 68.

65 Lima, I.Q., Ramos, O.E.R., Muñoz, M.O. et al. (2021). Geochemical mechanisms of natural arsenic mobility in the hydrogeologic system of Lower Katari Basin, Bolivian Altiplano. *J. Hydrol.* 594: 125778. ISSN 0022-1694.

66 Podgorski, J., Wu, R., Chakravorty, B., and Polya, D.A. (2020). Groundwater arsenic distribution in India by machine learning geospatial modeling. *Int. J. Environ. Res. Public Health* 17 (19): 7119–7136.

67 Keimowitz, A.R., Zheng, Y., Chillrud, S.N. et al. (2005). Arsenic redistribution between sediments and water near a highly contaminated source. *Environ. Sci. Technol.* 39 (22): 8606–8613.

68 Herath, I., Vithanage, M., Bundschuh, J. et al. (2016). Natural arsenic in global groundwaters: distribution and geochemical triggers for mobilization. *Curr. Pollut. Rep.* 2: 68–89.

69 Podgorski, J. and Berg, M. (2022). Global analysis and prediction of fluoride in groundwater. *Nat. Commun.* 13: 4232.

70 Jha, S.K., Singh, R.K., Damodaran, T. et al. (2013). Fluoride in groundwater: toxicological exposure and remedies. *J. Toxicol. Environ. Health B Crit. Rev.* 16 (1): 52–66.

71 Rahman, A., Mondal, N.C., and Tiwari, K.K. (2021). Anthropogenic nitrate in groundwater and its health risks in the view of background concentration in a semi arid area of Rajasthan, India. *Sci. Rep.* 11: 9279.

72 Wick, K., Heumesser, C., and Schmid, E. (2012). Groundwater nitrate contamination: factors and indicators. *J. Environ. Manag.* 111 (3): 178–186.

73 Nivetha, A., Sakthivel, C., and Prabha, I. (2021). Heavy metal contamination in groundwater and impact on plant and human. In: *Spatial Modeling and Assessment of Environmental Contaminants*, Environmental Challenges and Solutions (ed. P.K. Shit, P.P. Adhikary, and D. Sengupta), 233–246. Cham: Springer.

74 Mohankumar, K., Hariharan, V., and Rao, N.P. (2016). Heavy metal contamination in groundwater around industrial estate vs residential areas in Coimbatore, India. *J. Clin. Diagn. Res.* 10 (4): BC05.

75 Singh, G., Singh, A., Singh, P., and Mishra, V.K. *Organic Pollutants in Groundwater Resource*. Wiley 139-163.

76 Stuart, M. and Lapworth, D. (2013). Emerging organic contaminants in groundwater. In: *Smart Sensors for Real-Time Water Quality Monitoring*, Smart Sensors, Measurement and Instrumentation, vol. 4 (ed. S. Mukhopadhyay and A. Mason), 259–284. Berlin, Heidelberg: Springer.

77 Ma, R., Zheng, C., and Liu, C. (2018). Groundwater impacts of radioactive wastes and associated environmental modeling assessment. In: *Encyclopedia of Sustainability Science and Technology* (ed. R. Meyers), 1–12. New York, NY: Springer.

78 Marui, A. and Gallardo, A.H. (2015). Managing groundwater radioactive contamination at the Daiichi nuclear plant. *Int. J. Environ. Res. Public Health* 12 (7): 8498–8503.

79 Gamal Khedr, M. (2013). Radioactive contamination of groundwater, special aspects and advantages of removal by reverse osmosis and nanofiltration. *Desalination* 321: 47–54.

80 Al-Hashimi, O., Hashim, K., Loffill, E. et al. (2021). A comprehensive review for groundwater contamination and remediation: occurrence, migration and adsorption modelling. *Molecules* 26 (19): 5913–5941.

81 Speight, J.G. (2020). Remediation technologies. *Nat. Water Remediation*, 1st Edition 263–303.

82 Robles, I., Moreno-Rubio, G., García-Espinoza, J.D. et al. (2020). Study of polarized activated carbon filters as simultaneous adsorbent and 3D-type electrode materials for electro-Fenton reactors. *J. Environ. Chem. Eng.* 8 (5): 104414–104424.

83 Hafeez, M.A., Singh, B.K., Yang, S.H. et al. (2023). Recent advances in Fenton-like treatment of radioactive ion exchange resins. *Chem. Eng. J. Adv.* 14: 100461.

84 Mohan, D., Sarswat, A., Ok, Y.S., and Pittman, C.U. Jr. (2014). Organic and inorganic contaminants removal from water with biochar, a renewable, low cost and sustainable adsorbent--a critical review. *Bioresour. Technol.* 160: 191–202.

85 Soni, R., Bhardwaj, S., and Shukla, D.P. (2020). Various water-treatment technologies for inorganic contaminants: current status and future aspects. *Inorganic Pollutants in Water* 273–295.

86 Nathanson, J.A. (2023). Pollution control. *Encyclopedia Britannica*.

87 Chaturvedi, N.K. (2022). Comparison of available treatment techniques for hazardous aniline-based organic contaminants. *Appl Water Sci* 12: 173.

88 Xiang, Q., Nomura, Y., Fukahori, S. et al. (2019). Innovative treatment of organic contaminants in reverse osmosis concentrate from water reuse: a mini review. *Curr. Pollut. Rep.* 5: 294–307.

89 Adams, J.A., Reddy, K.R., and Tekola, L. (2011). Remediation of chlorinated solvent plumes using in-situ air sparging – a 2-D laboratory study. *Int. J. Environ. Res. Public Health* 8 (6): 2226–2239.

90 Qin, X., Huang, G., and He, L. (2009). Simulation and optimization technologies for petroleum waste management and remediation process control. *J. Environ. Manag.* 90 (1): 54–76.

91 Obiri-Nyarko, F., Grajales-Mesa, S.J., and Malina, G. (2014). An overview of permeable reactive barriers for in situ sustainable groundwater remediation. *Chemosphere* 111: 243–259.

92 Gomes, H.I., Dias-Ferreira, C., and Ribeiro, A.B. (2012). Electrokinetic remediation of organochlorines in soil: enhancement techniques and integration with other remediation technologies. *Chemosphere* 87 (10): 1077–1090.

93 Sethi, R. and Di Molfetta, A. (2019). Mechanisms of contaminant transport in aquifers. In: *Groundwater Engineering* (ed. R. Sethi and A. Di Molfetta), Springer Tracts in Civil Engineering. Cham: Springer.

94 Couper, J.R., Penney, W.R., Fair, J.R., and Walas, S.M. (2010). Adsorption and ion exchange. In: *Chemical Process Equipment (Revised Second Edition)* (ed. J.R. Couper, W.R. Penney, J.R. Fair, and S.M. Walas), 521–552.

95 Qasem, N.A., Mohammed, R.H., and Lawal, D.U. (2021). Removal of heavy metal ions from wastewater: a comprehensive and critical review. *NPJ Clean Water* 4 (1): 1–15.

96 Yunnen, C., Xiaoyan, L., Changshi, X., and Liming, L. (2015). The mechanism of ion exchange and adsorption coexist on medium-low concentration ammonium-nitrogen removal by ion-exchange resin. *Environ. Technol.* 36 (18): 2349–2356.

97 Lorenzen, L., van Deventer, J., and Landi, W. (1995). Factors affecting the mechanism of the adsorption of arsenic species on activated carbon. *Miner. Eng.* 8 (4–5): 557–569.

98 Samimi, M., Zakeri, M., Alobaid, F., and Aghel, B. (2023). A brief review of recent results in arsenic adsorption process from aquatic environments by metal-organic frameworks: classification based on kinetics. *Isotherms Thermodyn. Behav. Nanomater.* 13 (1): 60–72.

99 Hao, L., Liu, M., Wang, N., and Li, G. (2018). A critical review on arsenic removal from water using iron-based adsorbents. *RSC Adv.* 8: 39545–39560.

100 Ji, P., Park, J., Gu, Y. et al. (2021). Abiotic reduction of ketones with silanes catalysed by carbonic anhydrase through an enzymatic zinc hydride. *Nat. Chem.* 13 (4): 312–318.

101 Latta, D.E., Boyanov, M.I., Kemner, K.M. et al. (2012). Abiotic reduction of uranium by Fe(II) in soil. *Appl. Geochem.* 27 (8): 1512–1524.

102 Gao, Y., Zhong, S., Zhang, K., and Zhang, H. Abiotic reduction of organic and inorganic compounds by Fe(II)-associated reductants: comprehensive data sets and machine learning modeling. *Environ. Sci. Technol.* https://doi.org/10.1021/acs.est.2c09724.

103 Suzuki, I. (2011). Oxidation of inorganic sulfur compounds: chemical and enzymatic reactions. *Can. J. Microbiol.* 45 (2): 97–105.

104 Zhu-Barker, X., Cavazos, A.R., Ostrom, N.E. et al. (2015). The importance of abiotic reactions for nitrous oxide production. *Biogeochemistry* 126: 251–267.

105 Wang, L.K., Vaccari, D.A., Li, Y., and Shammas, N.K. (2005). Chemical precipitation. In: *Physicochemical Treatment Processes. Handbook of Environmental Engineering*, vol. 3 (ed. L.K. Wang, Y.T. Hung, and N.K. Shammas). Humana Press.

4

Advances in Desalination Materials and Technologies

4.1 Introduction to Seawater Salinity

Over 71% of the Earth's surface is covered with about 96.5% of water being held in oceans and seas. Even though water also exists in the form of water vapor in the air, sweet water in lakes, ponds, icecaps, and glaciers, ground as soil moisture, and aquifers altogether counted for nearly 2%. However, over 97% of the water found in the oceans and seas is salt water. This leaves less than 1% of the earth's water available to us for our daily water supply needs. Seawater typically consists of more than 35 g of dissolved salts per liter measurements, which is generally written as 35‰. However, the normal range of ocean salinity ranges from 33–37 g/L (33–37‰). Seawater mainly consists of two ions that are present often in seawater sodium and chloride. In seawater, sodium and chloride make up over 90% of all dissolved ions. Therefore, the concentration of salinity (dissolved salts) in seawater is about 35 parts per thousand (ppt); in other words, it is ~3.5% of the weight of seawater accounted for dissolved salts. Interestingly, an average of ~3.5% of salinity is accounted for in the five oceans of the world. However, regions like the Atlantic, the Mediterranean, and the Bay of Bengal have been characterized by much higher salinity levels [1–3].

On the one hand, the world distribution of freshwater sources is not uniform. While some places like high altitude snow-covered and sparsely populated Scandinavia, parts of southern South America, Alaska, northern Russia, and Canada possess abundant fresh surface water or groundwater. On the other hand, densely populated and highly industrialized areas are located in sites with low availability of freshwater resulting in a high degree of water stress according to the trade-off between demand for water and amount of water available. According to the United Nations, there are more than 2 billion people in the world who live in regions facing high water stress. It is expected to aggravate based on UNESCO's earlier projections considered during the period from 2017 to 2028. Further,

Materials and Methods for Industrial Wastewater and Groundwater Treatment, First Edition. S. K. Nataraj.

a greater demand for water is predicted not only in the field of agriculture where freshwater consumption is 70% of the demand worldwide but also for household, domestic, industrial production, and energy generation. Similarly, ever scary climate change phenomenon, increasing world population growth, contamination of freshwater sources, accelerated urbanization and population growth in cities, and expansion of public service networks are also expected to worsen global water stress. Therefore, the World Water Program (WWP) has estimated that by 2030, merely 60% of water demand will be available for consumption. Similarly, the Organization for Economic Cooperation and Development (OECD) estimated that by 2050, the availability of freshwater sources will be lowered by up to 55%, which is further predicted to decline by the end of the century due to which 40% of the global population will live in water stress regions [4–7].

Desalination is the process of removing salt from seawater or saline water to make it potable or drinkable using an appropriate material, method, or process. Traditionally, there have been two main techniques used in which boiling the water was the simplest method from which steam was trapped and condensed leaving behind the salt. Another technique passes the water through different types of filters that retain dissolved salt constituents and pass pure water across the filter. The desalination process is a cost-intensive technique. Due to these constraints, many countries are unable to afford modern technologies to produce freshwater. However, the increased research and innovation efforts have steadily improved the affordability of desalinated water from seawater resources. A seawater desalination process separates saline seawater into two components, namely a freshwater stream containing a low concentration of dissolved salts and a concentrated brine stream. This desalination process requires energy to separate salt from seawater and desalinate it into freshwater through various technologies for separation. Conventionally, two of the most popular and commercially successful technologies are in practice for seawater desalination: (i) multi-stage flash (MSF) distillation and (ii) reverse osmosis (RO) processes [8–10].

Although several traditionally followed desalination technologies have been successfully used to be a reliable source of freshwater from the sea, however, some of the processes have lately shown limitations in increasing the demand and supply of freshwater. Therefore, to fix these emerging issues, researchers and innovators have invested in a significant amount of research and development (R&D) activities in order to constantly improve the technologies, efficiencies, and sustainability to reduce the cost of desalination. In this direction, this chapter gives a detailed discussion on the current status, practices of desalination efforts, and advances in the desalination technologies and materials stock to produce sustainable, economical, and affordable seawater desalination to produce freshwater for larger populations. In addition to this, the chapter also gives detailed future prospects for the state-of-the-art seawater desalination materials and technologies under development.

Despite the presence of excessive amount of water on earth's surface ($1.4 \times 109\,km^3$) that includes 97.5% seawater having an average salinity of 35,000 milligrams per liter (mg/L) or parts per million (ppm) leaving behind merely 2.5% of earth's water suitable for direct use. However, 80% of the freshwater is stored in glaciers, and the remaining 20% is in lakes, rivers, and aquifers. With ever-growing urbanization and increasing population, global freshwater demand is expected to increase exponentially. As the demand for water is growing and water scarcity is worsening by expanding and intensifying around the globe, recent reports estimate nearly 40% of the global population is vulnerable to serious water shortages. This water scarcity is expected to exponentially rise to 60% by 2025. This situation may dynamically change due to various other reasons such as increasing industrial activities, construction, and operation of recreational aqua-parks, the building of large townships, special economic zones, and so on, which are largely considered a source of freshwater overexploitation and further contamination. These impending crises and statistics on freshwater shortages are expected to increase socio-economic conflicts within and among neighboring governments over the access, allocation, and sharing of water resources. Similarly, several states, regions, and locations across the world are facing depletion in their local water basin, which is forcing neighboring populations and communities to turn to alternative water resources, including water import, water recycling, and desalination [11, 12].

Desalination, which is also known as desalting, is a process of removal of dissolved salts from seawater and, in rare cases, removal of salt from the brackish or highly saline waters of inland seas, highly mineralized groundwater such as geothermal brine, and municipal wastewaters. This desalination process renders the separation of water from the solute moieties fit for human consumption, industrial applications, irrigation, and various other applications. Currently, existing seawater desalination technology requires high energy input, and in some cases, uses fossil fuels for the separation of salt from water, which makes the overall process expensive. Also, advanced desalination techniques like membrane-based separation processes use highly applied external pressure to break the osmotic barrier to facilitate water separation from salt contaminants. Due to this, desalination processes are currently used only in those regions where freshwater sources are scarce or not easily accessible. Therefore, a series of currently followed desalination techniques leave behind a large carbon footprint and greenhouse gas emissions. The currently followed desalination techniques also generate large amounts of brine wastewater due to which desalination plants pose significant environmental challenges. Nevertheless, the historic desalination process has been in practice for ages in various forms and extents [13, 14].

Historically, desalination was used by evaporating contaminated or highly saline water in different volumes. Aristotle described desalination processes as evaporation methods of seawater by Greek sailors of the 4th century BCE for survival. On the other hand, Arab literature that dates back to the 8th-century CE refers to a treatise on distillation. In recent modern (19th century) times, advances in science and technological development led to steam navigation, which created a demand for noncorroding water boiler technologies that revolutionized the desalination processes. These efforts were led by a ground-breaking innovation in England in 1869 that was granted the first patent in the desalination process and water recovery sector. Further, in the same year, the first water distillation plant was commissioned by the British government at Aden to provide a continued supply of freshwater to ships that dock or stop at the Red Sea port. On the other hand, in 1930, the first large stationary desalination plant was installed in Aruba, near Venezuela to provide fresh or reclaimed water for commercial purposes. By 2019, more than 20,000 desalination plants have been installed to produce a total of >100 m^3 of potable water per day that are operational across the world [15–17].

4.2 Desalination Process

In addition to salts, there are many chemicals dissolved in seawater that make it not only salty but bitter as well. Most of these dissolved salts and substances enter river streams and further rivers carrying chemicals accumulate the various chemicals and dissolved substances from rock and soil on their way. Among many salts, the main constituent of seawater is sodium chloride, or common salt, often just called salt. Most seawater contains nearly or more than 35 g of salt per every 1000 g (~1 L) of water. The general method of recording seawater and brackish water salinity is to measure the amount of salt dissolved in 1000 g or 1 L of water, which is generally termed as "parts per thousand" or ppt. Interestingly, most of the ocean possesses a salinity between 34 and 36 ppt. Further, due to the presence of high salinity, the properties of water change, which makes (i) seawater highly dense compared to freshwater; (ii) the freezing point of salty water drastically reduces, which necessitates keeping seawater in the freezer for more time than freshwater. Nevertheless, the salinity of the ocean and seas varies from place to place, particularly at the surface (between 34 and 36 ppt), but there are parts of the ocean where hardly any rainfall has a higher concentration of salinity as the warm dry winds cause high and accelerated evaporation. This accelerated evaporation removes water and rises into the atmosphere leaving behind salt and increasing the salinity of the seawater. For instance, the North and South Atlantic regions are characterized by high salinity as these regions experience strong winds and not much rain. Therefore, seawater is denser than freshwater [18–20].

On the other hand, some parts of the ocean receive heavy rains. This phenomenon adds a lot of freshwater to the surface of oceans, diluting the seawater, which reduces the salinity and so makes the seawater less dense. Seawater is also characterized as less saline near land where rivers meet oceans adding regular freshwater. For instance, the ocean near Antarctica has relatively low salinity even in some areas estimated to be below 34 ppt, and near the Arctic is even less than 30 ppt in some places. Thawing icebergs also contribute to the addition of freshwater that has broken off ice sheets formed over land that is not contaminated with salt. In this direction, the Baltic Sea, surrounded by Scandinavia, and northern Europe has very low salinity (~10 ppt). This is mainly due to the huge quantities of freshwater added from hundreds of rivers [21, 22].

Desalination is the process of selectively separating or removing excessive dissolved salts and other substances from the seawater. The desalination process drastically reduces the salt concentrations to below the permissible total dissolved solid (TDS) limits set by the World Health Organization's drinking water limit of 500 ppm. Desalination has been in use for centuries around the world. However, it has gained significance in the last few decades. Although, first references to desalination techniques and their use to extract freshwater are found as early as 300 BCE to 200 CE. In 320 BCE, desalination was extensively used by sailors among many. Alexander of Aphrodisiacs has been quoted that sailors boil seawater and use specially designed suspending sponges from sailors or vessels to absorb the vapor and then collect them as "sweet water" for drinking purposes. Whereas, in 1565, the French explorer Jean de Lery used the desalination technique during his journey to Brazil. However, in 1627, British philosopher Sir Francis Bacon proposed the use of sand filters to desalinate saline water. Boiling contaminated water and seawater to extract desalinated sweet water has been an ancient technique. However, large-scale application of boiling or evaporation techniques gained considerable significance in the last 300 years. One such effort was in the mid-1700s in which advances in steam-generation techniques allowed for the wide application of evaporation and condensation processes for desalination, which continued to dominate throughout the early 1900s. Further, major breakthroughs in advanced technology development led to the revolutionary transformation in the desalination process. In particular, the mid-1950s witnessed the development of membrane technology, which is considered the most efficient desalination technique. However, it was in the 1960s when Canadian inventors Loeb-Sourirajan patented an asymmetric RO membrane that drastically reduced the overall desalination process at a large scale economically, which also helped industries rapidly adopt and expand. Due to the innovation in membrane materials, processes, and overall technology, desalination of both seawater and brackish water has since grown rapidly around the globe. On the other hand, advances in nanotechnology and nanomaterial development in the mid-2000s further improved membrane performance and technology economics. It is estimated that in 2013, there were more than 17,000 active desalination plants that supplied nearly 8500 m^3/d to 300 million people in around 150 countries. By 2015, desalinated water production capacity increased to over 10,000 m^3/d and the supply of desalinated water from both seawater and brackish water is expected to increase to 20,000 m^3/d by 2050 [23–28].

Further, an increased water crisis in many parts of the world led researchers and innovators to explore the direct use of seawater to produce freshwater. In this direction, seawater became a major source of freshwater with the help of several desalination technologies. This has helped to some extent in reducing global water stress. Even though several desalination technologies exist in practice, however, many of these are categorized as energy-intensive processes, which restrict the wider use of technologies that may drastically reduce the water crisis in the world. Even though several countries and large populations of the world reside near large coastlines and suffer from a lack of freshwater resources, the high costs of desalination technologies restrict and restrain their massive use in many countries, societies, and regions affected by water scarcity. According to the International Desalination Association (IDA), there are over 16,000 operational desalination units/plants located in 177 countries accounting for production of over 95 million m^3/d of freshwater. It was recorded near 92.5 million m^3/d in 2017. Nevertheless, energy consumption is a major cause of concern in all these cases in which more than 50% of desalination plants depend on the electrical or thermal energy sources. Also, another major factor to make overall desalination plant operation sustainable and economical is the quantity of available input water, level of water contamination, the extent of treatment required, type of treatment technology used and the treatment plant capacity. However, recent advances in the production, distribution, and use of renewable energy also helped the desalination market adopt new forms of energy sources to operate treatment plants. Therefore, most of the recently commissioned desalination plants use renewable energy as a substitute or replacement for electrical energy, desalination units/plants/systems powered by renewable energies as a sustainable alternative to reduce overall operational costs in comparison to conventional desalination operations [29–31].

4.3 Types of Desalination Processes

Generally, advanced desalinations can be broadly categorized into two main types: (i) phase change technologies and (ii) nonphase change technologies. However, lately, several hybrid technologies for saline water desalination have been developed since the Beijing of the new century. Similarly, hybrid seawater and saline water desalination technologies cover a wide spectrum including the integration of advanced RO process in combination with other membrane processes, such as ultrafiltration (RO-UF), electrodialysis (ED) with membrane distillation (MD), and hybridization of RO or MSF distillation with other emerging desalination technologies. This section presents a review of the different commercial phase-change (thermal) desalination technologies and filtration-based desalination technologies that are currently practiced and also discusses a broad overview on the use of renewable energy sources in seawater and saline-water desalination technologies and at the end, future perspectives as a contributing factor to the sustainability of the water resource [32–35].

Nonetheless, modern desalination technologies are several decades old and improvements in materials and engineering technologies have led the desalination processes to become cost-competitive and robust in producing high-quality safe water for our growing demand. Post-World War II, many technologies found their ways to solving societal problems, which lead to innovation in various other technologies. One such technology was desalination technology where "desalting" saline water into drinking water. Subsequently, to regulate the technology to reach wider societal needs and industrial demands. One such act was adopted in the name of "The Saline Water Act" by the US Congress in 1952 to provide federal support for desalination. In this resolution, the US Department of the Interior provided funding in their initial efforts during the 1950s and 1960s for the development of desalination technology through the Office of Saline Water (OSW). Even though desalination technology gained momentum in the late 20th century due to the development of science and technology to a large extent, commercial-scale desalination efforts revived with great importance in the late 1990s and continued development in nanotechnology innovations contributed to technological improvements at the present time. Therefore, the first commercial-scale seawater desalination plant was established in 1961 in the United States at Freeport, Texas by Dow with the help of the US Department of the Interior, which had a 1 million gallons per day (mg/d) capacity [36].

Desalination is defined as the process that selectively removes or eliminates salts from water. Desalination processes can be applied to produce freshwater or sweet water from various contaminated or saline water sources of natural, municipal, groundwater, industrial, or commercial origin at large-scale applications. A desalination technique or a process essentially segregates or separates seawater or saline water into two parts: (i) one component, which has a low concentration of salt largely termed as treated water or recycled water or product water, and (ii) part in which a much higher concentration of salts is segregated than that of the original feed water, which is generally termed as simply as "concentrate" or brine concentrate. The three major types of technologies in practice across the

Table 4.1 The broad list of desalination technologies or processes that are categorized into mainly two types and subdivided into three types.

Phase change technologies	Nonphase change technologies	
Thermal technologies	**Pressure/concentration driven membrane technologies**	**Potential drive membrane technologies**
Multi-stage flash distillation (MSF)	Reverse osmosis (RO)	Electrodialysis reversal (EDR)
Multi-effect distillation (MED)	Nanofiltration (NF)	Electrodialysis (ED)
Vapor compression distillation (VCD)	Forward osmosis	Electro osmosis (EO)

world for both brackish water and seawater desalination can be broadly classified into thermal technologies, pressure driven/concentration gradient technologies, and potential-driven technologies as documented in Table 4.1. In all three cases, technologies require energy input in the form of heat, electricity, or potential to operate, extract water, and produce freshwater.

4.4 Phase Change Technologies

Phase change desalination technologies utilize the latent heat of vaporization of a solution, solvent, or working fluid to absorb heat/thermal energy in the evaporator cycle, which, however, releases the same energy during the condensation cycle. The phase change is defined as a transition of a matter between solid, liquid, and gaseous phases under the influence of external stimuli. Phase change typically involves a large number of energy inputs in comparison to the specific heat. It is, therefore, based on simple principles of boiling, evaporation, and condensation stages. In advanced units also, water is heated until it reaches the evaporation state, water is extracted in the vapor state and condensed in the next step, leaving behind salt to produce freshwater. Thermal desalination processes, often simply referred to as distillation, have been in practice for producing drinking water from brackish or seawater from ancient times. However, the medium to large-scale commercial plant rediscovery to serve the larger community and industrial needs is a modern practice. In the advanced plants, the energy required for thermal desalination plants to operate may come from steam generators, electricity supply, waste heat boilers, by the extraction of back-pressure steam from turbines in power stations, and other sustainable sources like solar plants, biomass origin, and so on [37–40].

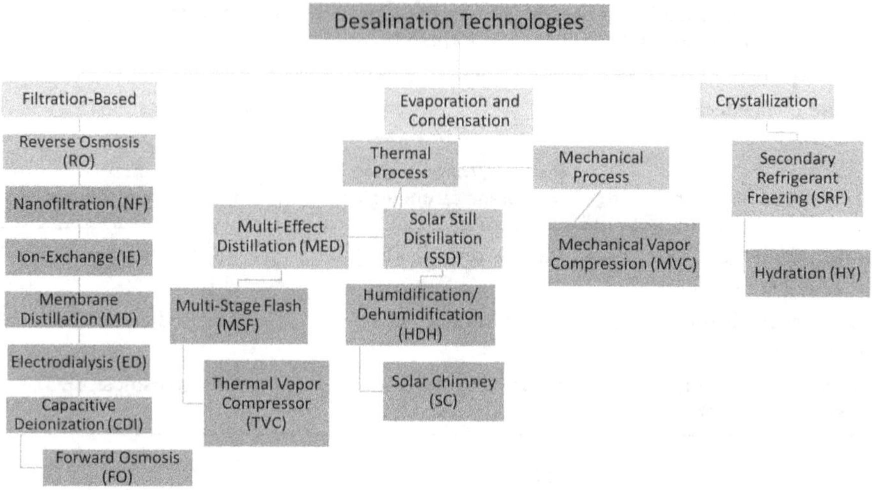

Figure 4.1 Classification of different desalination technologies available in the market and practice for seawater and brackish water desalination applications.

Specifically, water desalination processes can be classified into mainly three categories: (i) evaporation and condensation processes, (ii) filtration processes without phase change or membrane processes, and (iii) crystallization processes as shown in detail in Figure 4.1. Further, the evaporation and condensation process can be subdivided into mechanical processes and phase or thermal change processes. Mechanical vapor process or system that draws all the vapor from the evaporator and compresses it before returning it to the evaporator. To achieve this, the high pressure is used in the form of mechanical energy, which drives the compressor [10, 41–43].

On the other hand, the thermal processes proceed with the selective evaporation of water from seawater or brackish water source when it comes in contact with a heating surface or the surface, which influences the evaporation leaving the salts behind. Then the freshwater vapor condensed by cooling pipes resulting in high-pressure water of high quality and without salts. Nonetheless, phase change or thermal technologies are the very popular and commercially successful class of desalination technologies available in the market, which include (i) MED, (ii) MSFD, (iii) ED or compression (MVC), and (iv) Thermal Vapor Compression (TVC) as shown in Figure 4.2a–d. Similarly, the same segment can be further subdivided into three-solar energy-assisted technologies that directly use solar radiation as the main source of energy such as (i) Solar Still Distillation (SSD), (ii) Humidification/Dehumidification (HDH) and (iii) Solar Chimney (SC) as shown in Figure 4.3a–c. However, solar-assisted desalination technologies are currently not commercially available on a large scale. On the other hand, the thermal processes proceed with the selective evaporation of water from seawater or brackish water source when it comes in contact with a heating surface or the surface, which influences the evaporation leaving the salts behind. Then the freshwater vapor is condensed by cooling pipes resulting in high-pressure water of high quality and without salts. Nonetheless, phase change or thermal technologies are the very popular and commercially successful class of desalination technologies

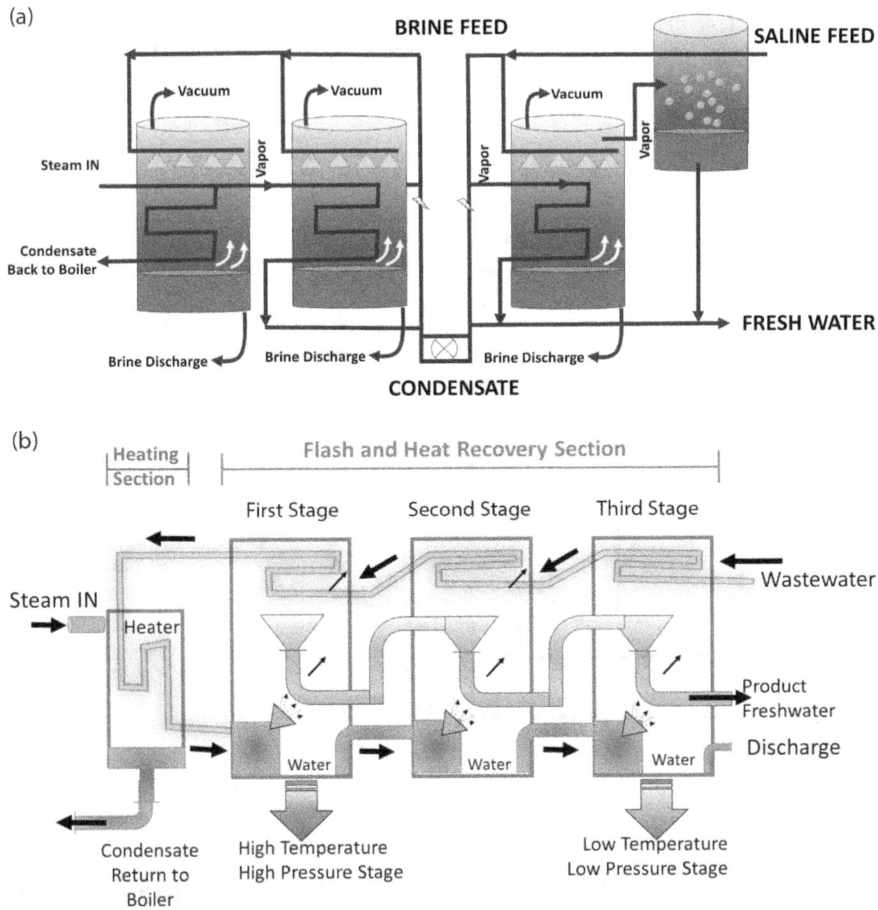

Figure 4.2 A schematic diagram showing different thermal desalination technologies including (a) multi-effect distillation (MED), (b) multi-stage flash distillation (MSF), (c) MVC, and (d) thermal vapor compression (TVC) available for desalination of seawater or brackish water.

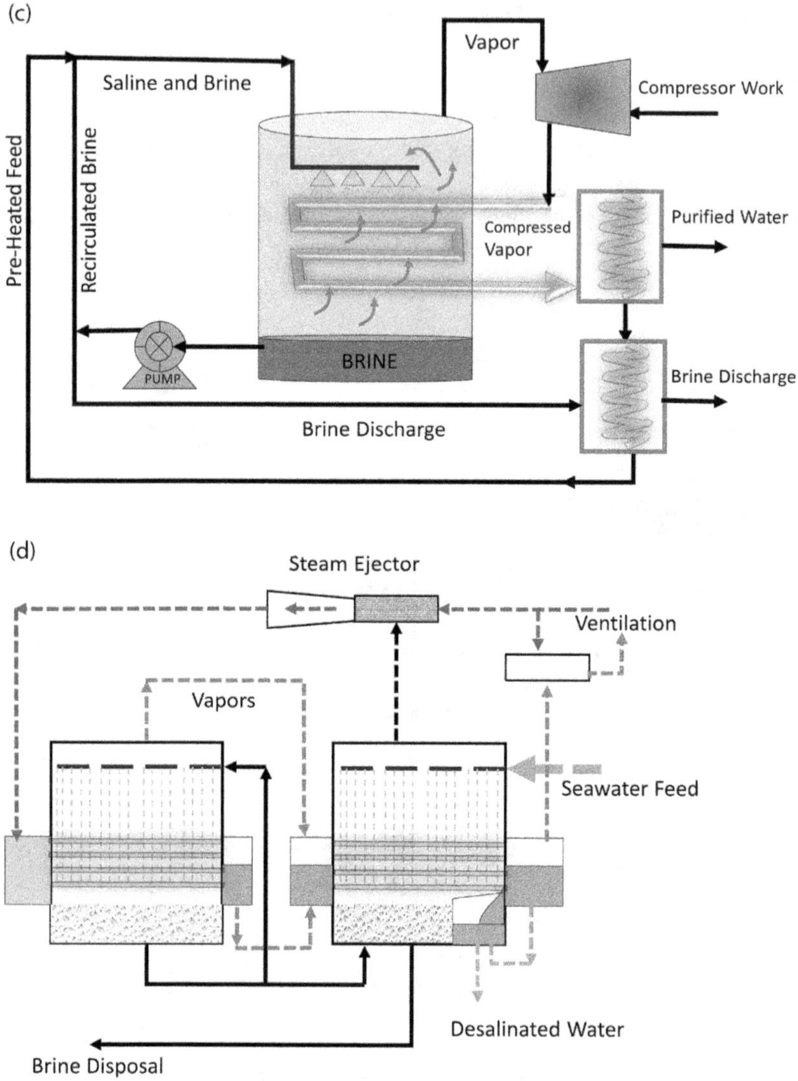

Figure 4.2 (Continued)

available in the market, which include (i) MED, (ii) Multi-Stage Flash Distillation (MSF), (iii) Mechanical Vapor Compression (MVC), and (iv) TVC as shown in Figure 4.2a–d [44–48].

Interestingly, thermal technologies, as the name suggests, involve the heat treatment or heating of saline water to produce steam or vapor and collecting the

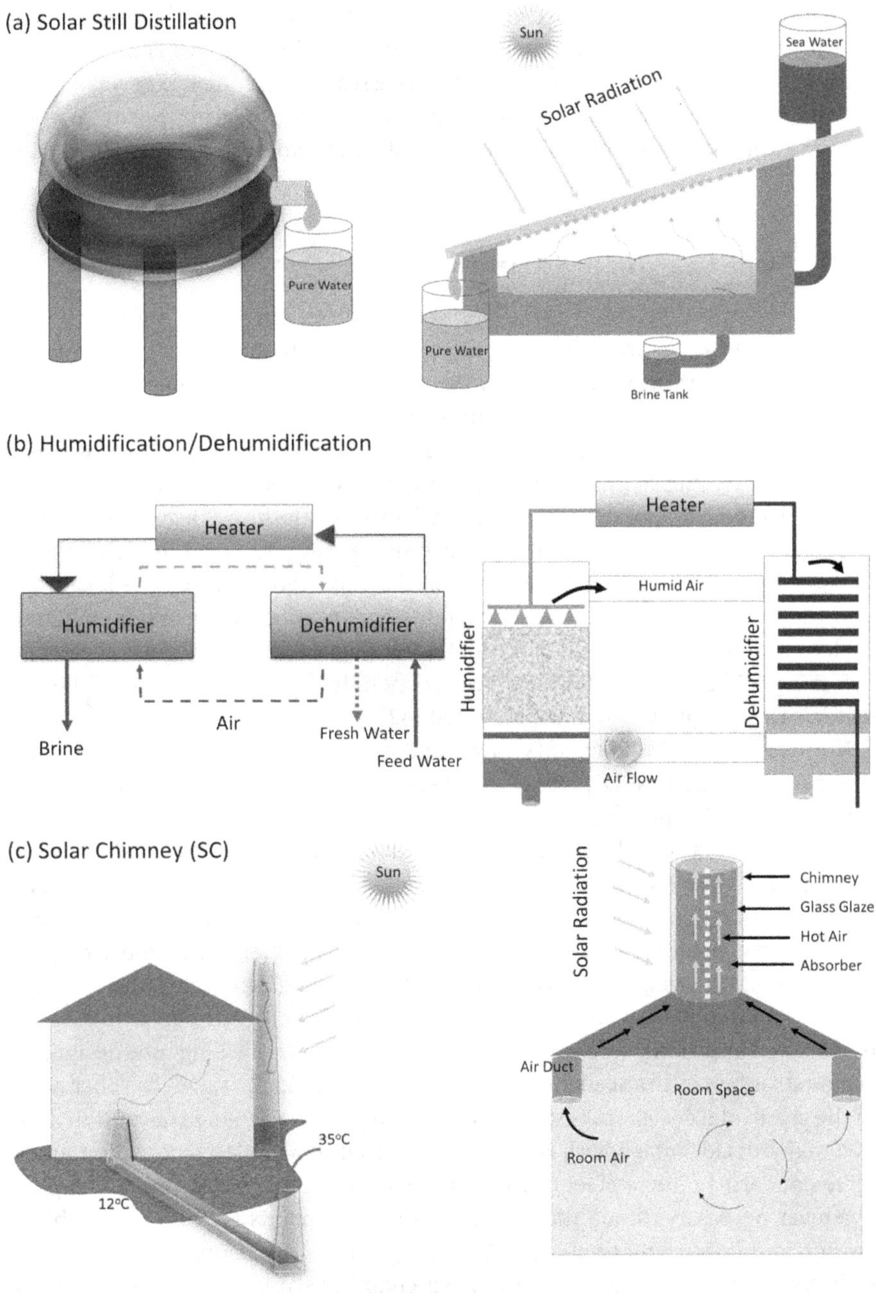

Figure 4.3 A schematic diagram showing different solar-assisted thermal desalination technologies including (a) solar still distillation (SSD), (b) humidification/dehumidification (HDH), and (c) solar chimney (SC) available for desalination of seawater or brackish water.

condensed vapor (distillate) to produce pure water. Nevertheless, thermal technologies in the past have rarely been utilized for brackish water desalination. This is because thermal desalination processes involve high-cost operations. However, most of the thermal desalination processes have been utilized for seawater desalination with economical freshwater production protocols such as MSF, MED, and VCD.

4.4.1 Multi-Stage Flash Distillation

The multi-stage distillation (MSF) process involves distillation through multi-stage or several chambers. In the MSF process, multi-stage involves a gradient pressure chamber in which each successive stage of the MSF plant operates at each progressive step in lower pressures. In this process, the feed saline water is first heated under high pressure, which further leads into the first "flash chamber," where the pressure is gradually released resulting in water boiling rapidly and thereby sudden evaporation or "flashing." In this "flashing," a part of the feed water continues to flow in each successive step as the pressure at each step/stage is lower than the previous stage. Further, the vapor produced by the flashing is then converted into freshwater through condensation using heat exchanger tubing, which is fixed and fitted through each stage. These tubes are cooled using incoming cold feed water. In this process, generally, a small percentage of the feed water is turned into vapor and condensed [42].

Nonetheless, the MSFD plants have been in use as early as the late 1950s. Even though advanced MSFD plants have been shortening their footprints, earlier versions of MSFD plants contain from 15 to 25 stages. However, even advanced plants are generally no larger than 15 mg/d in capacity. Today most MSFD plants can either have a "once-through" or "recycled" mode operation. In the "once-through" operation design, the feed water is passed through the flash chambers following the heater and further disposed of. On the other hand, in the recycled mode operation design, the feed water used for cooling is recycled and reused to increase productivity. Nevertheless, each of these operation mode processes can be designed as a "long tube" or "cross tube" structure. In the long tube design, the tubing structure is arranged in parallel to the concentrate flow. On the other hand, in the cross tube design, tubing is arranged perpendicular to the concentrate flow. The major disadvantage with MSFD plants is that they undergo corrosion unless high-standard stainless steel engineering material is used throughout the plant wherever necessary. In addition to corrosion, MSF plants are also vulnerable to erosion and impingement attacks in the long run. The erosion occurs due to the turbulence of the feed water in the flash chamber when the feed water travels from one stage to another. Nevertheless, cumulative distillation processes produce nearly 50% of freshwater worldwide with a desalination capacity of ~3.4 billion

GPD globally. Among all distillation processes, MSF plants contribute ~84% of the total freshwater production of all distillation plants' capacity. Interestingly, most distillation desalination plants are built where water scarcity is extreme and in those regions such as the Middle East region, where water dependency is on seawater and energy resources are available plentiful and inexpensive.

4.4.2 Multi-Effect Distillation

Similar to MSFD, multi-effect distillation (MED) technology has been in use since the late 1950s and early 1960s. This MED process uses a series of vessels (effects) that use the simple principles of water evaporation and condensation at reduced ambient pressure. In MED, a series of the evaporator (multi-stage) effects produce water vapor at progressively lower pressures. However, in this process, water boils at reduced/lower temperatures as pressure decreases; thereby, the water vapor of the first vessel or effect serves as the heating medium for the second, and so on. Further, the multi-effects (vessels) or the more the vessels are used, the higher the performance ratio. In this process, depending on the design, structure, and arrangement of the heat exchanger tubing, MED units can be subdivided into a vertical tube or horizontal tube, or vertically stacked tube bundles [10, 49].

4.4.3 Vapor Compression Distillation

Unlike MSFD and MED, the VCD process is either used in combination with other thermal processes such as the MED and, in rare cases, as a standalone process. In this process, the heat for evaporating the water is utilized from the compression of vapor, instead of the direct exchange of heat from steam produced in a boiler. VC units are built in a variety of configurations, designs, and structures depending on the size and footprint of the plant. Generally, in this process, a mechanical compressor is utilized to generate the heat for feed water evaporation. Nonetheless, the VC plants generally carry a small footprint in capacity and are often VC plants used at community or individual freshwater production capacities such as resorts, hotels, and, in rare cases, small-scale industrial applications [50].

4.5 Solar-Assisted Desalination Processes

Similarly, the same segment can be further subdivided into the three-solar energy-assisted technologies that directly use solar radiation as the main source of energy such as (i) SSD, (ii) HDH, and (iii) SC as shown in Figure 4.3a–c. However, solar-assisted desalination technologies are currently not commercially available on a large scale [51–58].

4.5.1 Solar Still Distillation

Solar distillation is one of the oldest and most traditional technologies that have been practiced for a long time. The first reported large-scale solar still distillation (SSD) desalination plant was established in 1872 at Las Salinas, Chile, which comprised solar stills of ~4700 m^2 total area, accounting for the production of over 23,000 L/d. Later, SSD processes gained increased attention during and after World War I when the industrial revolution led to the development of many technologies, processes, devices, and end products. During this period, the SSD process also evolved to be a standalone and economical technology suitable for arid, semi-arid areas, and remote locations where freshwater is scarce or drinking water shortage is a major problem. However, these regions and locations are blessed with high solar radiation.

Nevertheless, SSD is the process used to remove salts and other impurities using solar energy directly sourced from the sun to get potable or pure water for drinking purposes. Even though distillation is traditionally used for water purification using heat as a source, which is a relatively old technology for water treatment and purification, it is still considered one of the efficient methods that facilitate clean water to people. Unlike other methods, SSD uses a simple and energy-efficient route to produce drinking water, making it an economical alternative to clean drinking water. In its structure, the solar water distillation process involves the use of solar stills that segregate clean water by the process of vaporization.

There are several approaches through which solar-powered water distillation is achieved such as photovoltaic-powered RO. In the case of treating and purifying contaminated or polluted water, a sand filtration approach in addition to distillation can be adopted as a low-cost alternative. Therefore, the structure and design used in the solar water still distillation technique are solar still that distills water with the energy of the sun as an energy source that evaporates the water and then still accumulates and collects the condensate when the water is cool. However, there are several drawbacks to using solar energy for distillation as the process requires high initial capital cost and another dependency factor is the intermittent nature of the sun. Due to these limitations, the presently used designs and plant capacity of solar desalination water purification systems worldwide are still restricted to merely about 0.01% of the existing large-scale conventional distillation plants. However, SSD plants can effectively serve in their small- to medium-capacity installations in remote, isolated, or water-scarce small populations and poor communities [58–60].

4.5.2 Humidification/Dehumidification

Humidification/dehumidification (HDH) is another thermal desalination technology used to recover high-quality water imitating nature's rain cycle. The HDH plant involves the basic components, namely a humidifier, a dehumidifier, and a heater. In the humidifier component, the air is humidified using direct contact with salty water from the feed. In the dehumidifier section, hot moist air is infused in indirect contact with low-temperature steps having cold salty water. These simultaneous steps cause water vapor to condense, resulting in the production of freshwater of high-purity stream. Therefore, this simultaneous HDH system can be designed and configured on a large scale using multiple ways by varying and controlling, which stream gets heated, and monitoring whether streams are circulated in or through a closed loop or leave the system after one or multiple circulations or passes.

Among other configurations, water purification is achieved by varying the mass flow rate ratio through multi-staging. This configuration of water purification has the benefits of a balanced system extended by operating the plant under multiple mass flow rate ratios or multiple stages. There have been several approaches through which researchers have modelled single-stage and two-stage systems through a temperature pinch, used an enthalpy pinch to produce high-quality water production performance in single-stage, two-stage, and infinitely-staged systems, and further advanced studies have also extended the analysis to systems having up to six stages [61–64].

However, fixed-effectiveness or fixed-pinch modelling is highly beneficial in understanding the advantages of multi-staging in a treatment plant. These benefits can only be isolated and quantified when a fixed area is used. In another advanced approach, researchers modelled a system of fixed area and provided guidelines for the proper operation of single-stage and two-stage systems, including the proper direction of extraction. In addition, the effect of a single extraction on a system of fixed size has also been recently studied to produce an economical way of high-quality pure water.

In another approach, bubble column dehumidifiers use bubble columns as dehumidifiers in the HDH plants to make overall freshwater production costs economical. In a bubble column dehumidifier, condensation of humidity occurs at the surface of air bubbles instead of an expensive metal surface. Another set of innovators developed a design for a bubble column dehumidifier and evaluated it with experimental data. These innovators also further developed a correlation for the heat transfer coefficient of a cylinder that is submerged in a shallow bubble column as shown schematically in Figure 4.4. Further, in this design, the turbulence created by increasing the rising bubbles leads to high heat transfer coefficients, sometimes

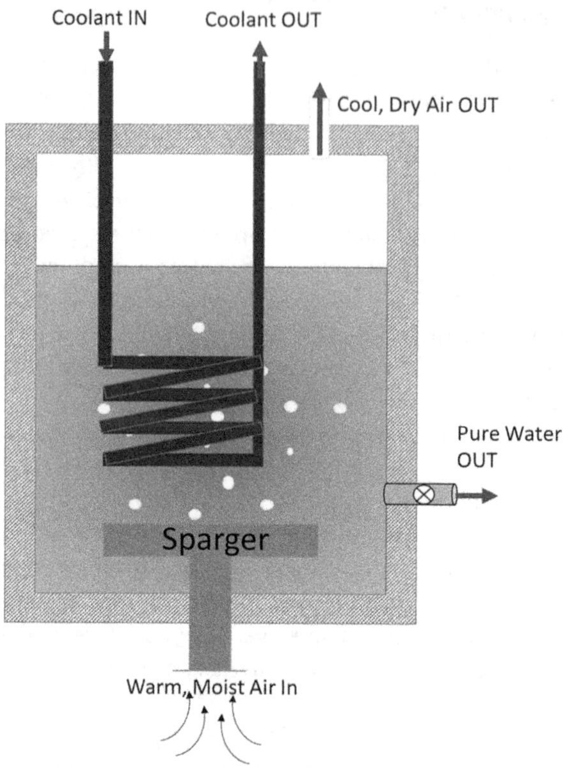

Figure 4.4 Schematic diagram of bubble column dehumidifiers used as an advanced distillation process.

even crossing $>5000\,W/m^2\,K$ at the surface of the cooling element, which reduces the material cost associated with condensation in the next stage.

4.5.3 Solar Chimney

A solar chimney (SC) is another type of solar-assisted water distillation technique that uses a passive solar heating and cooling mode of plant operation that can be used to regulate the heat and/or temperature of a building as well as provide ventilation. This mode of water purification unit also uses Trombe walls like solar walls, and SCs as means to achieve energy-efficient building design. Essentially, SC designs consist of hollow containers that connect the inside part of a building to the outside part of the building. A major advantage of using SCs is that these systems are simple, easy to install, and inexpensive means to acquire heat and ventilate a building. The establishment of a working SC first involves building a chimney and coated in some dark or black material. Black or dark-colored material or coating is used to minimize the heat loss absorbed from sunlight that usually reflects off of the chimney in the absence of these materials. However, dark materials have been adopted for absorbing more of the heat and ensuring retaining more of the heat is transferred to the air inside the building. Also, these chimneys are largely placed on a south-facing wall in case the home is in the Northern Hemisphere to make the overall process efficient [65–67].

The working mechanism in the SC process is very simple. In this process, heating a space using an SC is straight. In this wall-mounted design, when the solar radiation incident on the chimney at the side of the chimney, with this the column of air inside the chimney gets heated. In case the top exterior vents of the SC are closed, the heated air is forced back into the living space. This facilitates a type of convective air heating. In this mode, as the heated air cools due to convective flow in the room, it re-enters or pulls back into the SC, which gets heated up again. Therefore, when SCs are used for heating purposes, they operate like Trombe walls. However, cooling an inside chimney space using an SC is somewhat different from cooling using a Trombe wall. However, a roof overhang cannot be established in addition to an SC, two additional vents are provided to operate the plant efficiently. The first vent is present on the top of the chimney and the second is presented at the opposite end of the building making or providing an opening between the building and outside air, which allows ventilation. Interestingly, when solar radiation is incident on the side of the chimney, the column of air inside the chimney is again heated. The vent at the top of the chimney is therefore kept open so that heated air is not trapped and left for circulation. This leads to the heated air being pulled up and out of the chimney leading to pulling new air inside the chimney from the outside resulting in creating a sort of "draft," which provides fresh and cool air into the building.

Nevertheless, a relatively large, published literature base is available on renewable energy desalination processes, techniques, and methods that include details on the desalination plant operation. These studies account for detailed analysis and reliability of several solar-assisted and wind-powered desalination plant operations. However, there is considerably less literature available on the SC desalination process. Therefore, innovation and commercially viable plant operation in the sector of SCs for desalination are still in early stages. Nevertheless, there are several innovations that led to the advancement in technology. Among them, some initial SC desalination techniques have shown promising results and technology projections. However, at the commercial scale or operational scale, no single standalone or hybrid chimney-desalination configuration SC has yet shown itself to be the effective alternative choice. Nonetheless, researchers, innovators, and technologists are investing resources to make SC desalination technology a reality in the near future.

Traditionally, the thermal desalination processes have been used widely at great commercial scale with high success rate. Among many thermal desalination processes, MSF, MED, and MVC technologies share 87.3, 12.5, and 0.2%, respectively. On the other hand, other thermal desalination SS, HDH, SC, and freezing have not been commercially successful and have shown several limitations in the conceptual designs and experimental prototype scales.

4.6 Desalination by Filtration Processes

Desalination of seawater or brackish water can be achieved without phase change using physical filtration techniques such as the use of membranes or any other element or material to directly and selectively separate the dissolved salts from the water, which also requires high doses of applied pressure or energy. The physical method of desalination via the physical filtration process is achieved by removing salts and other minerals from seawater or brackish water. In these processes, the desalination process converts highly saline water potable that is suitable for consumption and irrigation. Among many physical filtration techniques, advanced materials and technologies involve majorly membrane techniques such as microfiltration (MF), UF, NF, RO, FO, membrane bioreactor (MB), MD, and ED. These processes either utilized high applied pressure or potential or concentration gradient as driving forces to selectively remove salt particles, bacteria, and other dissolved contaminants from the water following either size exclusion through membranes with different pore sizes or diffusion mechanisms [68–70].

In comparison to thermal desalination processes, physical filtration techniques like membrane-based separation processes are relatively advanced and do not require any phase separation to separate freshwater from saline water. Further, membrane-based separation processes and technologies can be subdivided into 2-broad categories: (i) ED and/or ED reversal (ED/EDR), and (ii) RO and, in rare cases, NF. ED and EDR processes are potential/voltage-driven membrane processes. In these processes, an electrical potential is applied to drive or influence the migration of ionic species through a semi-permeable barrier called ion exchange membranes leaving freshwater (dilute) behind as product water. The ED process was first commercially adopted in the 1960s, almost 10 years before the RO process came into existence. Interestingly, the ED process was initially conceived as a seawater desalination process. However, limitations associated with ion exchange membranes and the limit of the potential applied to the process restricted technology to be used efficiently in low saline feed water like brackish water desalination.

RO was the first modern filtration technique directly used for the extraction of high-quality drinking water from seawater, which was first commercialized in the 1970s. The development of cellulose and cellulose acetate (CA) membranes as a semipermeable barrier to separate dissolved salt from the rest of the water revolutionized water purification technologies. Currently, advanced innovations in the preparation and synthesis of novel membrane materials have transformed freshwater production from seawater. Nonetheless, RO is the most widely quick and physical separation method for desalination in the United States, Middle Eastern nations, and across the globe, in general, where the water scarcity is high and seawater is accessible in abundance.

4.6.1 Reverse Osmosis

Compared to all thermal phase change desalination techniques, reverse osmosis (RO) is an advanced and relatively modern process that was commercialized in the late 1960s. However, at present, the RO process is a widely used technique for seawater desalination across the world. Technically, RO desalination process is simple and straightforward. Pure water separation in RO is accomplished by applying pressure over osmotic pressure pushing seawater through a semipermeable membrane to remove higher salinity as shown in Figure 4.5a. RO process can be distinguished from other desalination processes as the technology offers a high purity of desalinated water as high as 99.99%. Every other desalination technology practiced conventionally lacks high-purity desalination efficiencies particularly when applied to seawater desalination applications. Also, due to extremely low pore size membranes and their unique distribution in their barrier structures, ranges below 1 nm in dimension retain most of the impurities including monovalent salt allowing only water molecules to pass through. However, for practical purposes, the RO process unit is fitted with several step pre-filter units to make the overall process economically viable and technologically feasible. Here, the RO process unit is designed such that seawater is passed through a series of pre-filters like sand wool filter to retain suspended particles, sediment filter to retain larger particulates, UF unit to retain bacterial and viral infection, and finally, RO unit to retain all the dissolved salts as detailed in Figure 4.5b [71–74].

In addition to this, the processing unit is also designed to comprise UV lamps to disinfect if any escaped biological contaminants are found in the permeate to ultimately achieve clean water in the end holding tank. Simultaneously, concentrated brine is rejected from the membrane surface containing almost every contaminant (~65 different contaminants that include chlorine, potassium, magnesium, lead, fluoride, chlorine, suspended solids, dissolved salts, and more) are collected as concentrate or rejectant or flushed down the drain. In this case, RO membrane plays a vital role in maintaining the quality and permeability of the product water. Also, physico-chemical properties of the RO membrane become extremely important as the process requires extremely high applied pressure (generally above 20 bar) to push water molecules through the ultra-nanoporous RO semipermeable membrane.

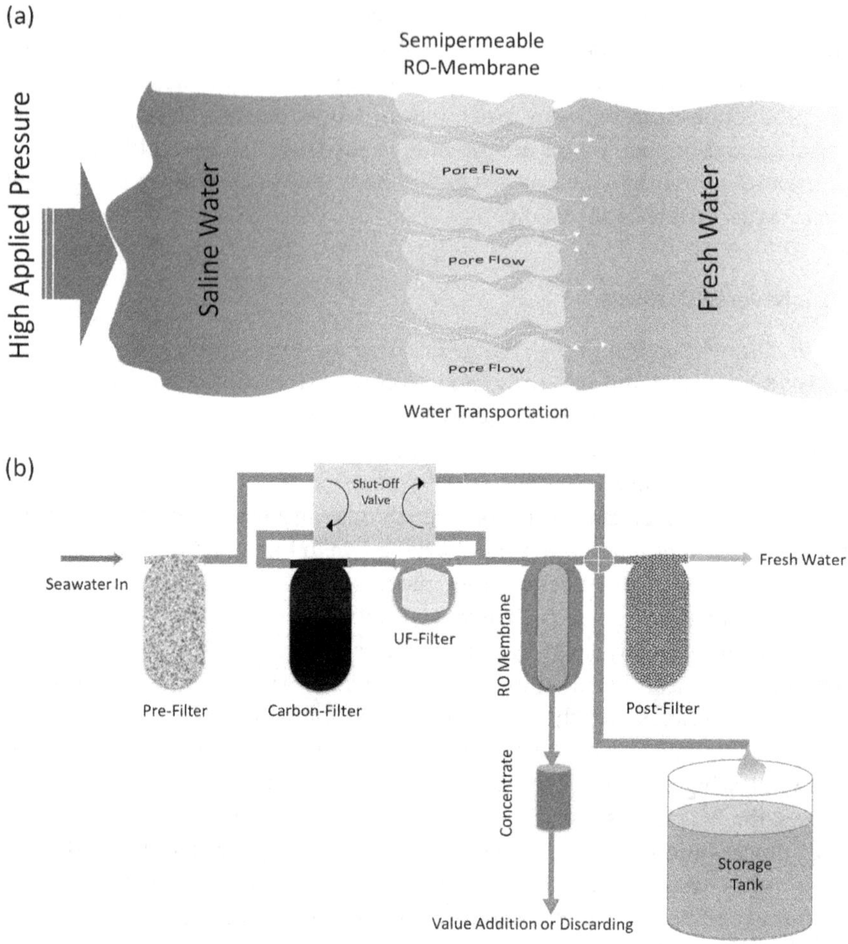

Figure 4.5 (a) Working principle of reverse osmosis processes widely used for seawater desalination and (b) actual working RO plant used for seawater desalination and its important components and steps involved in the process.

Nevertheless, there are typically four main steps involved in a commercially operated plant, namely (i) Sediment filter: This pre-filter step is designed to retain suspended particles, larger particulates, silt, dirt, and larger contaminants. This step is very essential as the sediment filter protects subsequent step filters from blocking, in particular, the dirt and larger particulates getting to the delicate UF and RO membranes that can easily be damaged by such contaminants. (ii) Carbon filter: The carbon filter step is incorporated in commercial plants to remove chlorine and other contaminants that affect the performance and life of both UF and

RO membranes in the subsequent steps in addition to improving the taste and odor of product water. (iii) UF step: This step is adopted in most of the large-scale commercial plants to retain biological and larger contaminants entering the RO step. Most of the fouling phenomena in the membrane processes, especially in RO units occur, from organic as well as biological contaminants. Therefore, the UF step helps in preventing organic/bio-contaminants from entering the next step. (iv) RO: This ultimate step involves a semipermeable RO membrane, which is designed only to allow water as a product to however retain or reject almost every contaminant. Among other post-treatment techniques, (v) Polishing filter or UV treatment is involved in a five-stage RO System. A final post-RO treatment will involve either a carbon filter to polish off the product water or improve the taste and odor in the water. In the case of UV filters/lamps, residual organic or biological contaminants are removed to induce taste and flavor to produce water.

Nonetheless, most of the domestic desalination units comprise a four- or five-stage process for producing optimal water quality. Even though RO processes became popular among utilization for municipalities in 1977, but enormous popularity in the last decade as there have been revolutionary changes occurring in the membrane preparation and RO membrane module fabrication improvements. Due to these reasons, RO has become increasingly popular among a larger population and an increased popular market because of safe, cost-effective, and easy-to-maintain characteristics involved in the technology. With suitable RO membranes and supporting components in place, a typical RO commercial either designed to serve a domestic desalination unit or seawater treatment can filter out impurities and produce outstanding quality drinking water. Most of the RO units collectively can remove contaminants to maximum extent, which cannot be achieved with any other conventional desalination technique or filtration processes. For instance, a typical RO desalination unit can reject chlorine (98%), sodium (85–94%), nitrates (60–75%), sulfates (96–98%), fluoride (85–92%), calcium (94–98%), lead (95–98%), barium (95–98%), phosphate (96–98%), pesticides (up to 99%), nickel (96–98%), arsenic (92–96%), mercury (95–98%), and so on [75–77].

4.6.2 Nanofiltration

The nanofiltration (NF) is another membrane-based separation process in which, other than monovalent ionic species, most of the other pollutants and contaminants can be separated from the feed solution. NF process uses less pressure to separate than RO process and is highly efficient in removing TDS contamination originating from the presence of ions such as magnesium, calcium, and sulfate. On the other hand, RO can also be effectively used for removal of the smallest ionic species like sodium and chloride. NF allows molecule size ranging from 1 to 5 nm to pass through and working pressure is substantially lower than RO process but comparatively higher than other high pore-sized membrane processes like UF and MF. Therefore, NF membranes effectively produce higher amounts of product water than RO membranes and show less hydrodynamic resistance for water molecules to pass through the filtration barrier. Nonetheless, the mass transportation mechanism and applications of the NF process a relatively similar to the RO process. NF process is a kinetic process in which the equilibrium condition between charge membranes plays an important role in determining the separation characteristic through the electric potential denoted as Donnan potential. These phenomena are similar to the basic principle of RO. Due to the existence of ionizable groups such as divalent dissociated ions interact with charged surfaces of NF membranes. Also, NF process produces higher permeate flux and selectivity and rate of pollutant (ionic contamination) retention, in particular, low molecular weight compounds can be enhanced by turning surface characteristics of the NF membrane. Therefore, the NF process is a highly productive and potentially better substitute for water purification having fed high TDS contamination. Nevertheless, the NF process can easily remove high molecular weight contaminants and charged molecules associated with monovalent salts simultaneously. The major disadvantage of NF membrane is its vulnerability to membrane fouling and NF membrane is not effective against monovalent cations. However, NF processes can be effectively used in industrial scale processes and technology applications such as electroless copper plating, treatment of heavy metals containing effluents, dairy produce processing such as cheese whey production, brackish water and groundwater purification having high calcium and magnesium TDS, and so on [69, 78–81].

4.6.3 Forward Osmosis

There are several alternative membrane-based desalination technologies, which use different driving forces to separate dissolved contaminants from contaminants such as osmotic gradient. Even though forward osmosis (FO) and membrane distillation (MD) are emerging membrane processes, which can also be listed as an alternative desalination process, the process utilizes semipermeable membrane as one of the major components to perform effective desalination. Therefore, FO and MD processes can be categorized as extended membrane filtration techniques [68, 82–84].

FO desalination process is also known as "direct osmosis," which uses a semipermeable membrane and a high osmotic-gradient or pressure-concentrated "draw" solution to drive the water molecules to separate from saline water to generate freshwater. Therefore, osmotic pressure originated from the concentration gradient between feed saline solution and highly concentrated draw solution as a "driving force" to separate freshwater across the semipermeable membrane

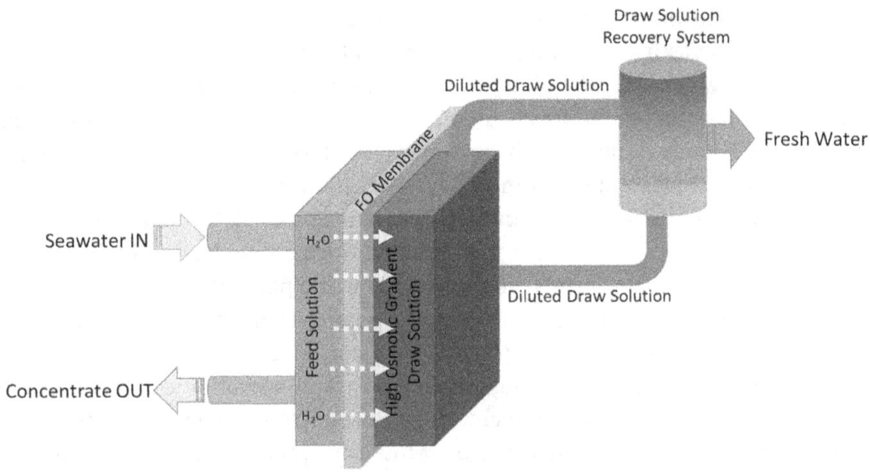

Figure 4.6 Schematic representation of forward osmosis (FO) seawater desalination process with detailed working parameters denoted.

instead of hydraulic pressure used in conventional membrane-desalination processes. A schematic of a typical FO desalination process is shown in Figure 4.6 in which freshwater passes through leaving behind dissolved salt through the membrane and diluting the concentrated draw solution. The draw solution receives only water molecules through osmotic diffusion from the feed saline water solution. Further, to make the overall process sustainable, diluted draw solution is further concentrated by a recovering system to produce freshwater. This allows the active draw solute/solution to be recovered and reused. On the other hand, development of suitable semipermeable membranes with high water flux, high mechanical stability, and low-fouling is very important to achieve optimal desalination performance. On the other hand, identification of appropriate draw solute/solution, and its mechanism for efficient recovery/recyclability also play key roles in the success operation of FO technology during the desalination process. Nonetheless, the draw solution should also possess high osmotic pressure and optimal viscosity, should be environmentally benign, and should also facilitate easily recoverable mechanisms. Therefore, all three components involving effective membrane, efficient draw solution, and its recovery and reuse parameters play a critical role in the development of technology with economic viability.

Even though the FO process uses nature-inspired osmosis phenomenon as the main driving force to achieve mass transportation, some breakthroughs in medium to large-scale commercial plants were achieved in the mid-2000s. A major breakthrough came from an innovation in developing easily recyclable and reusable concentrated draw solutions consisting of ammonium salts. In this study, researchers used ammonium salt along with carbon dioxide (CO_2) to generate high osmotic pressure, which was used for desalination of seawater. Upon dilution of the draw solution, the diluted draw solution was heated to remove ammonia and CO_2 leaving behind freshwater. This heat treatment step to separate from fresh product water extracted from seawater is achieved using the unique characteristics of the draw solute salts in the draw solution in which decomposition of active draw solute to ammonia and CO_2 gases is achieved. To make the overall process sustainable, heat treatment of product draw solute having high volume product water can be achieved using low-grade waste heat [85].

Ever since the RO desalination process was popularized for freshwater recovery from seawater sources, technology has gained considerable traction at commercial scale. However, due to the high capital cost and operation cost involved in producing freshwater from RO technology, several researchers and innovators have proposed alternative and less energy-intensive processes like FO technique. In one such operation, researchers designed FO process to perform seawater desalination with less overall cost comparisons to RO technology. These study results indicated that FOs could be economically viable and a better alternative. The FO process gained prominence with innovative product development from water companies, such as Modern Water, Oasys Water, and HTI Water Technology, which are involved in developing effective FO membrane and draw solutions for commercial FO technology. HTI developed the technology for draw solutions based on readily available concentrated draw constituents such as concentrated rehydration drinks, waste mud from drilling wells, oilfield-produced water, and potentially fruit juice concentrate. On the other hand, Oasys Water developed the FO technology utilizing concentrated solute constituents having ammonium carbonate salts, which act as easily removable constituents from diluted draw solution. Similarly, Modern Water developed their advanced version of FO technology, which was later known as manipulated osmosis desalination (MOD) for seawater desalination using a unique proprietary osmotic agent. The major advantage of FO technology originates from potentials of unlimited osmotic pressure that can be extracted from combination of various osmotic agents against either individual or in hybrid combination with various constituents. As there is no limitation for osmotic pressure to exist in FO technology, the process can be effectively used even for treating seawater to produce potable water or much-higher-salinity water to generate quality water [86–88].

4.6.4 Membrane Distillation

MD is a hybrid membrane-based desalination method that involves both thermal and membrane desalination concepts. In this process, vapor pressure is generated and further used as a "driving force" to pass evaporated vapor through the

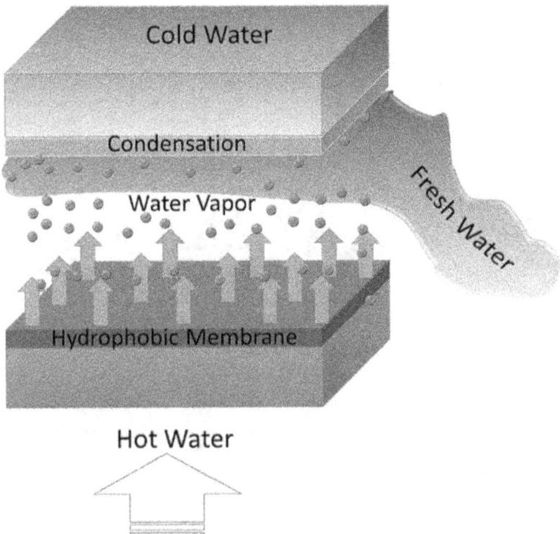

Figure 4.7 Schematic diagram and working of membrane distillation process used for desalination applications.

membrane to result in the separation of water from ionic constituents. Interestingly, hydrophobic membranes are used in this process to separate water vapor from feed solution to result in the production of liquid water leaving behind salts. In this case, the required vapor-pressure gradient across the membrane for the process is generated by heating the seawater or source feed water. A schematic in Figure 4.7 shows the working principle of the MD process, which has a temperature gradient design created across the membrane situated between the cold and hot sides of the unit. In this setup, the water vapor collected at the cooler side of the unit ultimately condenses to form freshwater. Therefore, the MD process thereby requires heat to produce freshwater from seawater feed through a membrane module followed by vaporization/condensation cycles [89–91].

4.6.5 Electrodialysis

Working of the ED process depends on the following fundamental principles, all salts dissolved through the dissociation mechanism in water are ions, either carry positively charged known as cations, or negatively charged referred to as anions. Since like charges repel each other and unlike charges attract, the ionic species migrate toward the electrodes with an opposite electric charge. To facilitate this migration without hindrance, suitable ion exchange membranes can be provided in a specific chamber design to permit selective passage of either anions (anion-exchange) or cations (cation-exchange). For instance, in seawater or brackish saline feed solutions, dissolved ions such as sodium (Na^+) and chloride (Cl^-) migrate to the

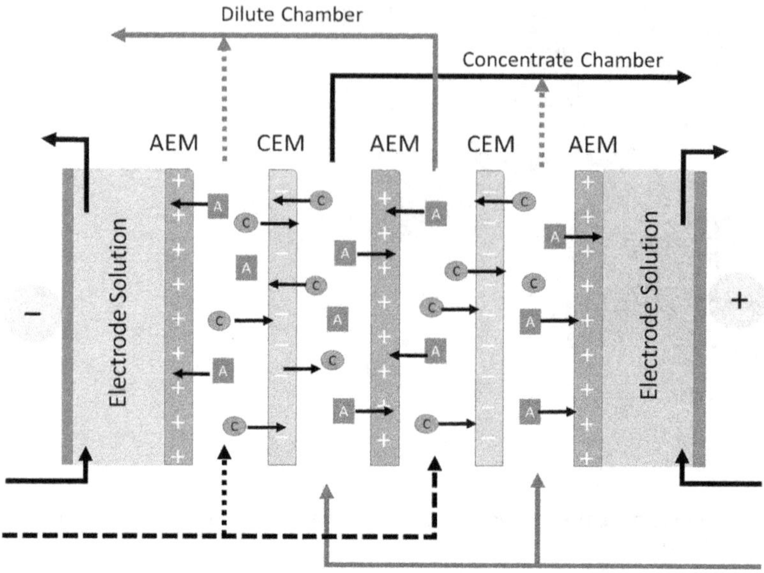

Figure 4.8 Schematic representation of electrodialysis working set-up and its constituents involved in the desalination process.

oppositely charged electrodes passing through their respective exchange membranes that either allow cations or anions to pass through (not both) as shown in Figure 4.8 [92–95].

In a typical ED set-up, membranes are generally arranged in an alternate pattern forming feed and concentrate chambers with an anion-selective membrane structured next to a cation-selective membrane. These patterns are formed by facilitating feed in between one anion-exchange membrane (AEM) and one cation-exchange membrane (CEM). Beside each of these membranes, a concentration chamber is provided to collect the exchanged ionic species. During this ion migration process, the dissociated ionic species in the feed chamber/channel get diluted, while concentrated brine is gradually built up near the electrode chambers or adjacent concentrate chambers. In this way, both concentrated and dilute solutions are formed in the channel/chamber provided between the alternating membranes. However, these chambers exist next to two membranes (AEM/CEM) that are referred to as a cell, and a typical small to medium size ED stack contains several cells. In large-scale ED units, from a few to several hundred cells are sandwiched together with electrodes that are known as a stack. In this stack, chambers are designed in such a way that feed water passes uniformly through all the cells simultaneously to facilitate a continuous flow of feed water, desalinated dilute water, and highly concentrated (brine) from the respective channels in the stack.

4.6.6 Electrodialysis Reversal

The electrodialysis (ED) reversal desalination is a potentially driven water desalination process in which electrical energy is applied to electrodes provided to pull dissolved salts through an ion exchange membrane to separate the water from the salts. First known EDR working plant was established in the early 1970s. Even though the EDR unit operates on a similar general working principle as an ED unit, except that both the product and concentrate channels are identical in their design and construction. The electrodialysis reversal (EDR) process creates two effluent streams: (i) a low salinity or dilute product water and (ii) a high concentrated brine or concentrate as shown in Figure 4.9. However, in the reversal process, the polarity of the electrodes is altered at constant intervals to control and reduce the formation of scaling, which leads to subsequent fouling. This constant switching of polarity at the electrode also allows the EDR process to achieve higher product water recoveries and reduce the volume of concentrate. In this arrangement, when at constant intervals of time, the polarity of the electrodes is reversed, which causes mobility of ionic pollutants attracted by the opposite direction across the ion-selective or ion exchange membranes. Immediately after the

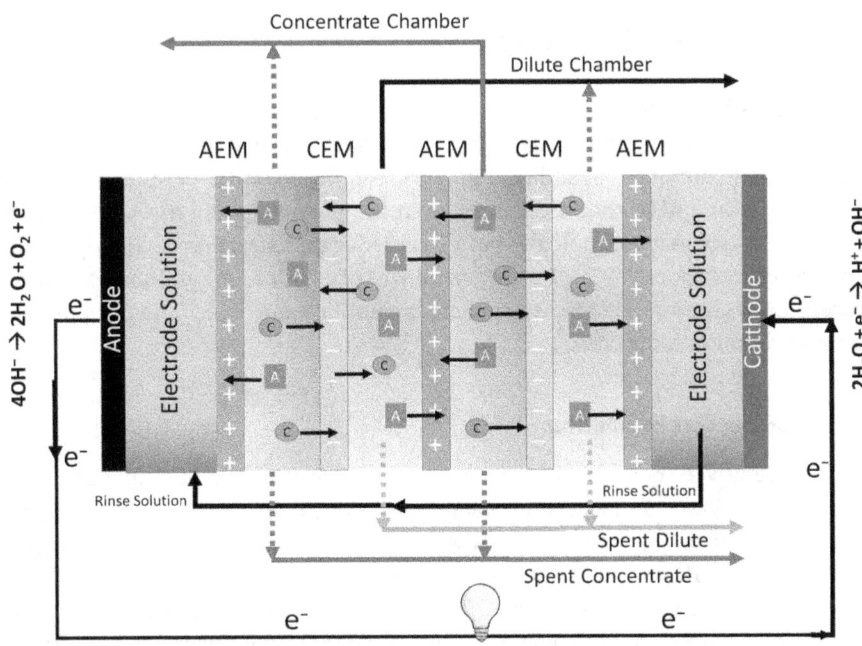

Figure 4.9 Working and structural depiction of electrodialysis reversal process used for desalination process.

polarity reversal, the product water is removed until the lines are flushed out and desired water quality is restored. In between, the flushing or bypass washing is provided just a few minutes before resuming the next cycle of water production. In this process, the flushing process is followed to prevent scaling or is useful in breaking up the potential sealants, slimes, and other deposits in the cells of a stack before they build up. On the other hand, flushing step also helps in drastically reducing the problem of membrane fouling and concentration polarization. Nonetheless, due to inherent characteristics and limitations associated with the potential driven process used in EDR units are generally used for the desalination of brackish water, rather than seawater or high-salinity-contaminated water. Nevertheless, initial studies of using EDR have also been successfully used for moderately saline water purification. However, a major disadvantage of EDR is the limitation in treating high-concentration saline water such as seawater. Therefore, EDR is mostly used for the treatment of brackish waters with low to moderate TDS and saline waters with a high scaling potential can also be treated effectively using EDR due to elevated levels of contaminants such as strontium (Sr) and barium (Ba). But, EDR is highly effective in high silica (SiO_2) feed waters [96–99].

4.7 Alternative Desalination Processes

Recent years have seen a series of innovations in the field of desalination and several alternatives came into existence other than thermal and filtration-based desalination techniques. In addition to two emerging seawater-desalination technologies (FO and MD) that are described in the previous section, the other two major innovative new technologies based on carrier gas extraction (CGE), dynamic vapor recompression (DVR), freeze crystallization, hydration, and ion exchange processes are also included to assess their capabilities and determine the suitability for seawater desalination.

4.7.1 Carrier Gas Extraction

CGE is a relatively new desalination process, which uses the principle of HDH phenomenon. The CGE technology uses atmospheric pressure and moderate temperatures as carrier gas to extract freshwater from high-salinity brines (<100 ppm salinity). The CGE was first developed by Gradiant Corporation, a spinoff company from Massachusetts Institute of Technology (MIT). In this process, the saline water is first heated, treated, and further sprayed onto porous and large surface area material, which then directly comes into contact with carrier gas that acts as dry air. Further, the air saturated with water vapor upon interface formation is

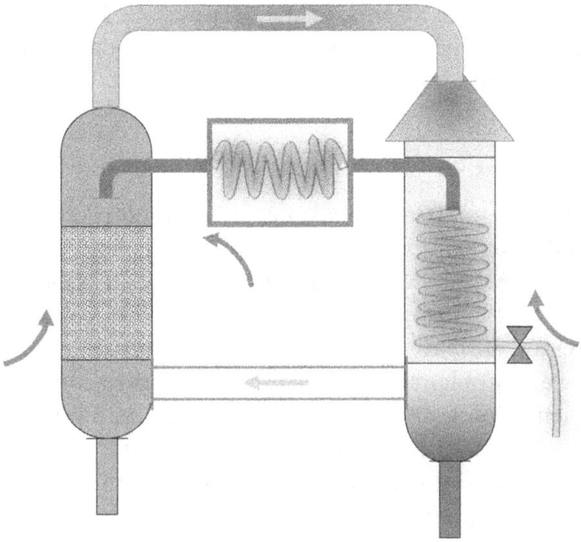

Figure 4.10 Schematic representation of carrier gas extraction process.

then processed in a multi-stage bubble column, which acts as both a heat and mass exchanger. In this step, the humid air is allowed to pass as bubbles through a series of pores/holes into water-filled trays. During this process of exchange, the water vapor in the bubbles turns colder and condenses to create more freshwater in the trays (Figure 4.10). The CGE technology has a potential to treat hyper-saline produce waters (>200,000 ppm TDS) from industrial sources like oil and gas production. The process in such a case requires rigorous pre-treatment to produce water while removing solid particles and oil/grease residual contents from source-contaminated water. Even though this novel CGE technology is currently being commercialized in various stages to treat produced water and extract freshwater required for hydraulic-fracturing operations in unconventional resources [99].

Nevertheless, the modular pilot plant that is in practice at a commercial scale can treat approximately ~10,000 B/D of product water. The technology has now been tested for its potential to be used for seawater desalination as well as to treat contaminated waters. The freshwater produced from CGE is of high quality, which can be compared to other well-established technologies, such as RO and thermal desalination processes. This technology can be effective in producing low- to medium-scale treatments in remote areas, and urban and semi-arid regions where reinjection/disposal facilities are rare or not available, making it one of the economical alternatives. On the other hand, suitable plant design and capacity building could also provide a better alternative solution for treating and desalinating industrial source water such as oilfield applications to meet the environmental regulations of zero liquid disposal (ZLD). However, CGE technology is in the evolution stage and at this stage is not a cost-effective solution for saline water or seawater desalination compared to RO.

4.7.2 Dynamic Vapor Recompression

The DVR process is another alternative and advanced desalination process developed by Salttech from the Netherlands, which involves the combinations of thermal distillation in a modular system (Figure 4.11). The DVR typically runs on a simple principle of mechanical vapor recompression in which water is evaporated at moderate temperatures by applying a vacuum followed by condensation to result in freshwater production. Here a heat exchanger is utilized to transfer the heat of condensation to the influent water stream to minimize the energy requirements. The key feature of DVR technology is that the system does not require any pre-treatment step, and the process operates without experiencing any type of scaling and fouling. This technology works with minimal energy requirements compared to those of conventional thermal desalination processes. However, the product water recovery is as high as 97%, making it one of the highly efficient techniques for the production of freshwater from saline water sources. During this process, remaining highly concentrated but low volumes of rejection of 3% are collected in the unit either in the form of high-concentration brine or crystallized salts. Another major advantage of this process is that it is capable of handling high-concentration salinity waters (more than 300,000 ppm TDS) including seawater, oilfield-produced water,

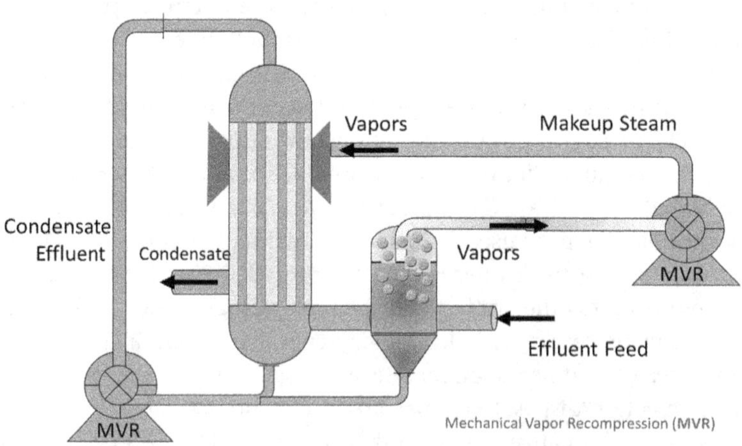

Figure 4.11 Schematics of working of dynamic vapor recompression desalination process.

fracture flow back, and so on. The process also provides high water recoveries from seawater compared with conventional membrane-based processes. On the other hand, the cost-competitiveness of DVR technology and plant footprint is relatively larger in comparison to commercially successful membrane-based desalination processes; however, advanced design approaches are expected to design concise and portable treatment plants in the near future. The DVR technology also has shown greater potential for the treatment of seawater as well as high-salinity contaminated water sources including oilfield waters, which requires minimal pre-treatment with high water recoveries. The process has also shown great potential in providing a ZLD solution with the help of existing membrane-desalination technologies to meet environmental regulations. Nevertheless, DVR has sown limited but very specific and industrially important applications at relatively small-scale applications, for instance, desalination of gypsum-contaminated brackish waters to produce drinking water. However, recent studies have demonstrated the potential scale-up of DVR technology to apply to a wide range of desalination applications. Also, the potential of DVR technology to remove or selectively treat specific salt ions from high saline brine solution is underway to understand the broader applicability of the process [100–102].

4.7.3 Freeze Crystallization

Ice crystallization is a complex coupled fluid and two-phase flow with heat transfer that initiates the ice nucleation leading to crystal growth. Freeze crystallization desalination (FD) is for the production of freshwater from saline source water in the form of ice crystals, which are then melted to get reusable water. Even though this technique was reported to be initially conceived and initiated way back in 1680 by Thomas Bartholinus. This was initially in practice to produce freshwater produced by the melting of ice, which was formed from seawater. Nevertheless, FD technology was first used at commercial scale in the 1950s by Karnofsky, Steinhoff, and Weigandt in which direct freezing was achieved with butane as a refrigerant for saline water desalination [103–105].

Potable water can be economically produced from brackish water as well as seawater using the FD process in which the nucleation is introduced by placing ice seeding. Further, the parameters are controlled to avoid the mixing of rejected salt from freshwater in the form of ice into the liquid phase. However, growth of ice crystals in this process becomes lower and in this condition, liquid subcooling is maintained at ~4 K. At the end, freshwater in the form of liquid is obtained through the post-treatment of ice if the FD process is not instigated ice-seeding as shown in Figure 4.12. Another major feature of the FD technology is that plant size can be altered and designed to suit the end application and capacity

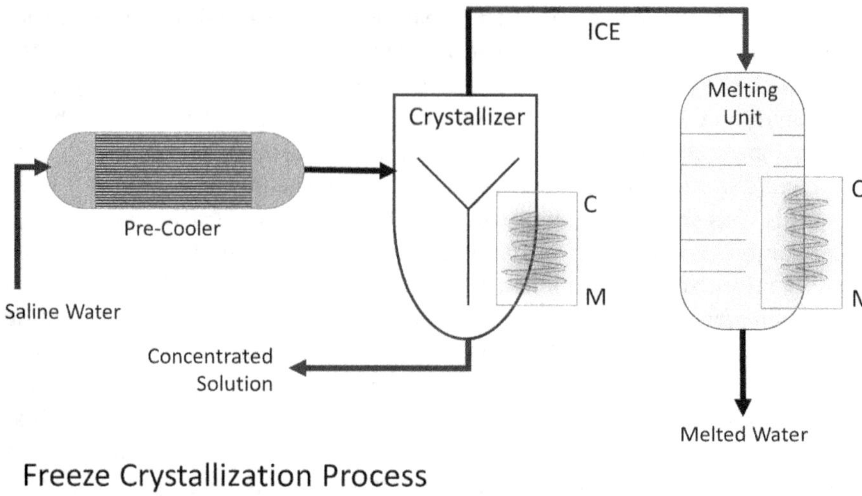

Figure 4.12 Schematic representation of freeze crystallization desalination unit and working.

requirements. Generally, the standalone advanced FD plant capacity is higher than the standalone suspension FD used for seawater desalination. However, recent advances in the FD technologies that are integrated with fractional thawing, falling-film, and block FD facilities have significantly improved the plant's desalination capacity than the standalone FD method. Freeze desalination processes possess several advantages compared to vaporization-based and membrane-based desalination processes. Energy consumption in the FD process is drastically less than other vaporization-based processes in which only ~1/7th of the latent heat is provided compared to other processes to achieve effective desalination from saline water required by the vaporization-based desalination processes. However, the energy requirement of standalone PFC- and SFC-based desalination is close to that of the RO method and the hybrid that is integrated with suspension FD with MD FD method uses less energy than the conventional RO method. Also, another major advantage is that the involvement of sub-zero temperature operations in FD decreases the risk of scaling and corrosion.

4.7.4 Hydration

For centuries, several desalinations techniques in practice, however, economical parameters and extent of technology feasibilities and their penetration factors are taking the front in process selection. Compared to conventional desalination, techniques such as membrane-based processes and thermal desalination techniques are cost-intensive and may be difficult to adopt for remote applications. Recently, attention has slowly shifted toward novel and advanced at the same time to simple processes like hydrate-based desalination (HBD) technologies. In this process, water is recovered in the form of vapor (gas) or Clathrate hydrate crystals. Clathrate in Latin means encaged, more precisely in desalination context, ice-like structures resulting from hydrogen bonding between water molecules at appropriate temperature and pressure conditions, which lead to the formation of so-called "cages" that encapsulate gaseous water molecules, namely the hydrate formers. Clathrate hydrates are crystalline special compounds having major species as water and minor portions as guest molecules. The temperatures at which Clathrate hydrate crystals are stable are typically above the freezing point of water, even though some guest molecules help in stabilizing hydrate crystals near ambient temperature [106–108].

There are three main crystalline structures proposed for Clathrate hydrates, namely structure I, structure II, and structure H. Every crystalline structure comprises a cage-like sub-architecture that is formed by the water molecules and Clathrate, which is a guest molecule within. The orientation of water molecules in the sub-architecture of these three-clathrate hydrate structures is shown in Figure 4.13. These crystalline structures perhaps differ significantly from the general hexagonal structure of solid water (ice). Structure I (sI) hydrates generally have unit cells comprising 46 water molecules with 6-cages, which can contain small gas molecules like ethane and methane. On the other hand, the unit cell of structure II (sII) hydrates contains 136 water molecules with 24-cages, which can hold relatively larger gas molecules like isobutane and propane. Further, the encapsulated gas molecules in the cages stabilize overall structure and, hence, allow the stable gas hydrate crystals to form. In this, any dissolved salt is left out

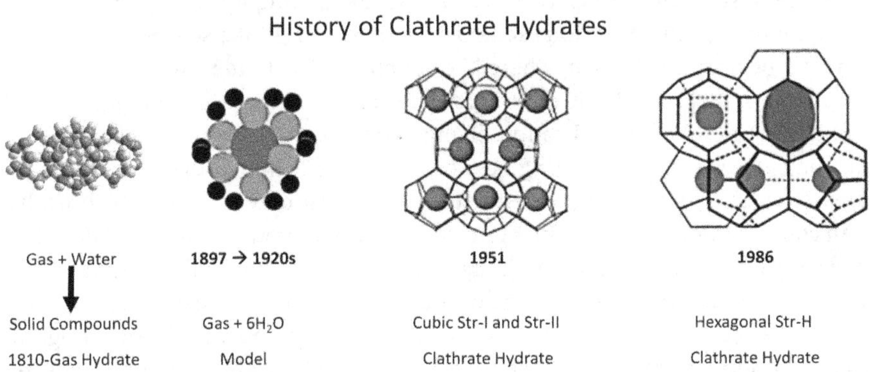

Figure 4.13 Schematic representation of cages of different forms of Clathrate hydrates structure.

from the hydrate crystals during hydrate formation, which is due to the size of the hydrate cages that ranges from 0.395 to 0.571 nm. When produced, hydrate crystals are separated from the saline or brine solution to produce freshwater. Here, the latter is produced upon the dissociation of the hydrate crystals caused by the heat stimulation or a pressure decline leaving behind the guest constituents, which can be reused in the subsequent process. With limited exposure, it is understood that the HBD process is a relatively energy-efficient method as its operating temperature ranges above the normal freezing point of water. Nevertheless, there are several major challenges that need to be addressed in this process such as the recycling of hydrate formers, slow-hydrate formation kinetics, and the hydrate-brine separation to make overall technology commercially viable at both domestic as well as industrial scale. Although hydration-based desalination technology in recent times has received significant attention across the scientific community, comprehensive cost, feasibility, and field efficiency trials need to be carried out over a long period of time.

4.7.5 Ion-Exchange

Ion-exchange desalination process is also generally known as "ion-exchange total desalination." Ion exchange water treatment method is one of the old technologies extensively used for producing ultrapure waters. This ion exchange total desalination is one of the oldest technologies mainly used for producing ultrapure waters. Ion exchange water softening or treatment concept is one of the modern concepts among all advanced water purification techniques, which is being used widely today as well in applications where quality is below 1 μs of conductivity [109–112].

Nevertheless, there are two distinctive techniques adopted in advanced ion exchange desalination units: (i) first, the method is uniquely equipped with the conventional cation and anion exchange charges, and (ii) the second is the use of a mixed bed ion exchanger where the anion- and cation-exchange resins are mixed in one step. The basic difference between two important configurations is that in anion-exchange and cation-change systems, charges need to be regenerated on surface site with existing network structure, while in the mixed-bed systems, charge is re-induced or re-generated with the use of resin media. Technically, both configurations produce the same quality output product water and quantity. However, regeneration of ion exchange site in both cases is a differentiating factor. Based on the end use and nature of source contaminated water or salinity, one can decide between two modes of the same technology. The mixed-bed configuration of an ion exchange system may be economically viable when the requirement is up to 1–2 m^3/h or 10–20 m^3/d water consumption. On the other hand, if the demand is for higher volume, then the separated cation-exchange and anion-exchange charge systems are highly recommended. However, in both cases, pre-treatment of raw water source or contaminated source of water is essential before feeding into the ion exchange system. The presence of good pre-treatment steps would lead to higher product quality and faster ion exchange in mixed-bed configuration.

Recently, the ion exchange process has been extended for the treatment of wastewater generated by steam boilers in power plants and treatment of boiler feed water heating plants, pharmaceutical industry, and component manufacturing. However, in absence of suitable high-performance exchange media and limited recyclability resulting in loss of traction for the ion exchange technology. Also, the chemicals such as acid used for the preparation of ion exchange charges on the resin structure, which impart main activity and functionality are becoming more expensive. Ion exchange systems also have material property limitations in which recycling reuse is a major issue. In such a situation, due to stringent environmental regulations, discharging used exchange resins or use of large amounts of the regenerating acidic and alkaline solutions restricts their affordability. Generally, large quantities of acid or alkali are required to regenerate ion exchange sites on a media, upon which waste is generated. Therefore, discharge of waste generated during the process into the waste collection reservoirs or directly into drain without neutralization is a serious challenge. These are the main uneconomical reasons why ion exchange desalination and wastewater treatment are fading. However, advances in nanotechnology, functional materials, and smart materials could revive the technology. Therefore, it is unlikely that ion exchange systems will replace RO technology for extended desalination applications [113–115].

4.8 Future Desalination Technologies

Advances in nanomaterials, nanoscience understanding, and ease with which nanotechnology helped innovators to design novel technologies, water treatment technology also greatly benefitted from the combination of events. One such class of nanomaterials are carbonaceous and graphene materials that are being tried in desalination applications. In this direction, several researchers have already experimented with several carbon-based systems like carbon nanotubes, graphene-based materials like graphene oxide and reduced graphene oxide nanosheets for selective separation, and purification of saline water sources. Graphene-based materials have been turned into stable filters either as standalone films or as composites in other carrier polymers. Arrangement of nanosheet-structure of the carbon has been further used for desalination applications.

4.8.1 Carbon Nanotube-Based Filters

The carbon nanotubes (CNTs) have been a fascinating class of materials that came into emergence in the last decade and have been widely studied for various applications. Several technology innovations developed using CNTs have shown potentials of materials. Therefore, researchers have used CNTs in water treatment applications as both standalone filters as well as in combination with other precursors. Many findings on desalination potentials of CNT-based filters have revealed that a desalination technology can be developed for long-term operations. In this approach, CNTs have been aligned and stabilized with the suitable binder or filler to result in a stable desalination membrane with well-organized and uniquely distributed pore structures. Typically, a membrane consists of CNTs and silicon to produce simple and low-cost seawater desalination filters. The hollow nature of CNTs in a pack can be ~50,000 times thinner than human hair, which therefore potentially act as pores in the membrane. Interestingly, upon alleging the nanotube bundles, the spaces between the CNTs are filled with either polymer or ceramic material to make overall structure stable and provide good mechanical strength. This also serves the purpose of binding in which adherence to CNTs is an important factor in achieving membrane characteristics. As prepared, aligned nanotube bundles having smooth and tiny holes (now can be termed uniform pore channels) in the membrane allow liquids (water) to selectively and rapidly pass through with exceptionally high flux while retaining larger ionic species and salt molecules. CNTs can be prepared with uniform and controlled hollowness having specific pore diameters, which can be leveraged to facilitate high permeate fluxes and highly selective rejection of salt ions through the nanotube hollow structure. Therefore, pores in CNTs can also be altered to reject ions selectively by use of size-controlled separation. Precise control over the pore structure and nature of binding can yield high product water permeability in CNT membranes, which is expected to drastically reduce energy consumption in comparison to conventional RO membrane-based desalination estimated by up to 75%. Therefore, several studies involving CNTs tested for their feasibility parameters in desalinating seawater have proven promising at various levels. Even though several attempts are underway to scale up the process of CNT-based membrane, nonetheless, the technology is still under rigorous scrutiny for R&D-related optimization and it may take some time for the road to develop technology to commercial scale.

4.8.2 Graphene Nanosheet-Based Desalination

Unlike CNTs-based membranes, graphene-based membranes or filter technology is designed to make use of graphene sheets with precise control over nano-sized pores that are useful for desalination applications. Graphene, which is an allotrope of carbon, consists of a single layer of atoms arranged in a hexagonal lattice nanostructure. The name graphene is derived from "graphite" and the suffix -ene, reflecting the fact that the graphite allotrope of carbon contains numerous double bonds. In this approach, sheets of graphene, which is a one-atom-thick form of elemental carbon, can be converted to ~1000 times thinner membrane compared to conventional desalination RO membranes. Due to these unique physical characteristics, the requirement for applied pressure to separate water from seawater drastically reduces. The pore size can also be controlled in graphene by creating holes/pores of specific size and dimension on the graphene sheets. This precise control over pore creation is a critical step in graphene-based technology development. However, due to the limitations involved in creating precise control over pore size and distribution is a quite challenging task. Nonetheless, researchers have demonstrated the feasibility and affordability of graphene-based membrane technology using lab-scale membrane extended their understanding through computer simulations. Interestingly, theoretical simulation and evaluation found that pore sizes in graphene sheets can be precisely controlled between 0.7 and 0.9 nm in diameter very effectively. This range of pores is very effective and plays a significant role in selective separation of water molecules, rejecting precisely even monovalent species like sodium ions. Again, graphene-based membrane technology is still in the early stages of R&D.

4.8.3 Capacitive Deionization

The capacitive deionization (CDI) is an electrochemical-based desalination technique in which salt upon dissociation electro-adsorption occurs into the high surface area electrode to result in desalination of feed saline water. Therefore, the main working principle is that charged electrodes are used to adsorb ionic species in an electrostatic adsorption phenomenon. So, CDI works on a selective electrostatic adsorption technique in which ions from seawater are electrochemically attracted onto charged electrodes. With controlled potential and other parameter optimizations, adsorption of ions present in feed water by the charged electrode plates significantly reduces the total salinity at the end of the process to produce treated pure or low-salinity water. In a typical CDI equipment, a pair of highly conductive and high surface area carbon electrodes are used, and through each set of electrodes, flow channels are created to pass feed or seawater. When direct-current voltage is applied to the electrodes, charged ions from flowing ionic sources of feed water are attracted to the appropriate electrodes to form an electric double layer. These adsorbed ionic species are then regularly removed by temporarily changing the polarity during the electrode regeneration step. The CDI has been very effective in desalinating the feed solution in low concentrations of TDS. Therefore, CDI technology has been successfully applied only for brackish-water desalination applications. Nevertheless, recently several researchers are investing resources to make it suitable for seawater desalination technologies through the modification of technology. In this approach, desalination of seawater on a larger scale is achieved by designing small-sized flow electrodes having suspended carbon materials. Through this method, ionic species can access charged surface easily, which makes the approach highly energy efficient. On the other hand, a new approach does not require a discharge step to operate, making it advantageous compared to conventional CDI approach. There are several proofs of concept studies that have been successfully performed on a modified CDI technology for medium-scale applications. The efforts are further expected to accelerate further optimization in the system design that may provide a better and more affordable solution for large-scale seawater desalination, which is highly desired for the long-term future.

4.9 Case Studies and Examples of Seawater Desalination

Modern science and technologies have enabled the large-scale and rapid extraction of freshwater from seawater sources through various desalination techniques and processes. However, considering the scale and economics involved in the process, desalination plants are generally large in size and close to the source water (seawater and seashore). On the other hand, desalination plants are generally adopted where the water scarcity is extremely high and all other sources of freshwater are either nil or extremely rare. Therefore, desalination plants in various scales and capacities are typically located in arid areas where seawater is in abundance and energy is sufficient to run the high energy-intensive desalination plants. Therefore, the Middle East region is the home to top 10 high-capacity desalination plants. Among many, Dubai's Jebel Ali hosts the world's biggest desalination plant with a production capacity of nearly 2.23 million m^3/d. There are other considerably larger freshwater production units well known for their seawater desalination production such as Ras Al Khair in KSA and Sorek in Israel. Following is the list of major desalination units mentioned in Table 4.2, operating throughout the world.

Case-1: Jebel Ali, UAE: This huge desalination complex earlier known as Jebel Ali Power Plant and Water Desalination Complex consists of nine separate modules extending for nearly three kilometers along the coast of the Persian Gulf. This huge desalination complex produces a staggering 2.227 million m^3 (490 MIGD) of water per day. In addition to being the world's largest desalination complex, Jebel Ali Complex also houses a power station that helps in running whole desalination units running sustainably.

Table 4.2 List of top 10 desalination plants across the world generating freshwater from seawater resources.

	Location	Capacity (m^3/d)	Technology
Case-1	Jebel Ali, UAE	2,227,587	Reverse osmosis and multi-stage flash distillation (MSF)
Case-2	Ras Al Khair, Saudi Arabia	1,036,000	17-reverse osmosis units and 8-MSF
Case-3	Taweelah, UAE	909,201	100% reverse osmosis
Case-4	Shuaiba, Saudi Arabia	880,000	100% MSF
Case-5	Umm Al Quwain, UAE	682,000	100% reverse osmosis
Case-6	Sorek, Israel	624,000	100% reverse osmosis
Case-7	Jubail, Saudi Arabia	600,000	100% reverse osmosis
Case-8	Rabigh 3, Saudi Arabia	600,000	100% reverse osmosis
Case-9	Fujairah 2, UAE	586,500	MSF and reverse osmosis
Case-10	Sorek 2, Israel	548,000	Reverse osmosis

Case-2: Ras Al Khair, Saudi Arabia: The Ras Al Khair desalination complex is probably the second largest freshwater production facility having a generation capacity of 1,036,000 m^3/d, which is equivalent to 414 Olympic-sized swimming pools every day. Uniqueness of this desalination unit comes as extremely impressive as the Ras Al Khair Complex supplies freshwater through an astounding 535-km pipeline to Riyadh city, which serves the needs of 2.9 million people. Another major advantage of facilities similar to the Jebel Ali Facility is the Ras Al Khair desalination complex also houses a gas-fired power station with a capacity of 2400 MW of power. The major desalination technology involved in this complex is 17 RO modules and eight MSF modules. In this complex, the MSF modules produce nearly 70% of the daily output with the RO units producing the remaining 30%.

Case-3: Taweelah, UAE: This desalination complex is located in Taweelah, Abu Dhabi having freshwater output capacity of 909,201 m^3/d (200 MIGD). Taweelah is one of the desalination complexes ranked as the largest 100% RO desalination plant in the world. On the other hand, both Jebel Ali and Ras Al Khair desalination complexes run on with MSF as the predominant freshwater production process.

Case-4: Shuaiba, Saudi Arabia: The Shuaiba desalination complex is located 120 km to the south of Jeddah in Saudi Arabia, with a capacity of 880,000 m^3/d

(194 MIGD). Currently, this plant uses 100% multi-stage flash (MSF) to produce freshwater through desalination of seawater. Interestingly, this unit has adopted green initiatives to replace high power consumption technique energy-efficient RO units through the process of replacement initiative with 600,000 m^3/d by 2025. This upgrade is expected to reduce CO_2 footprint of the facility by 45 million tons/year.

Case-5: Umm Al Quwain, UAE: This unit is one of the top 5 desalination complexes located in the Emirate of Umm Al Quwain, Even though The Umm Al Quwain desalination complex is in constant expansion. However, once finished, this desalination complex could be the top three by its capacity. However, at present, commercially operating desalination complex has a freshwater production capacity of 682,000 m^3 (150 MIGD)/d. Umm Al Quwain runs on 100% RO technology to desalinate seawater from the Persian Gulf.

Case-6: Sorek, Israel: The first and largest desalination unit commissioned in Israel is located in Sorek. The Sorek desalination complex facility is Israel's largest desalination complex, which utilizes 100% RO technology to produce a maximum output of 624,000 m^3/d (137 MIGD), The Sorek's desalination is behind only Umm Al Quwain and Taweelah in terms of RO technology capacity. This desalination complex was commissioned in 2013 and provides freshwater for nearly 20% of Israel's 9.2 million population. Nonetheless, the Sorek desalination complex was the first in the world to utilize large-scale desalination technologies for seawater desalination leaving a large technology footprint and considered a pioneer in the refinement of large-scale RO technology.

Case-7: Jubail, Saudi Arabia: Jubail desalination, formally known as Jubail 3A Independent Water Plant, is located in Jubail, Saudi Arabia. Jubail desalination complex runs on a 100% RO technology with freshwater output capacity of 600,000 m^3/d (132 MIGD). This desalination complex started operations in 2022 to improve water security and sanitization in the surrounding area of Saudi Arabia.

Case-8: Rabigh 3, Saudi Arabia: Rabigh 3 desalination unit is technically having the similar production capacity to that of Jubail 3A Water facility. So, both are tied at the 7th place. However, these two desalination facilities are located on the opposite side of the country, while the Rabigh 3 desalination complex located in Western Saudi Arabia produces an impressive 600,000 m^3/d (132 MIGD) of freshwater. Rabigh 3 was commissioned in 2021 with the 100% reverse-osmosis (RO) technology.

Case-9: Fujairah 2, UAE: Fujairah 2 desalination complex is located between Khor Fakkan and Fujairah, Dubai. The Fujairah 2 desalination complex has a freshwater production capacity of 586,500 m^3/d (129 MIGD). The F2 Desakubatuib plant operates using a dual process (MSF/RO). Around 450,000 m^3/d production of water comes from MSF and the remaining 136,500 m^3/d product water is produced through RO technology. As the overall complex requires sustainable and uninterrupted power supply, the Fujairah 2 water facility also houses a thermal power station on-site with a production capacity of 2000 MW.

Case-10: Sorek 2, Israel: The second desalination of Israel is expected to complete its commissioning and start operations by 2025. Once commissioned, the Sorek 2 desalination complex started operations with 100% RO technology to produce an estimated 548,000 m^3/d (121 MIGD). Interestingly, this facility is partially funded by the European Investment Bank (EIB). In addition to Sorek 1 and Sorek 2, Israel also has several medium- to large-scale desalination facilities to sustain country's water security. These medium-to-large desalination plants are located in Ashkelon, Palmachim, Ashdod, Hadera, and its older brother, Sorek.

4.10 Future of Desalination Technologies

In this looming water scarcity and overexploitation of water resources, advanced desalination technologies have provided a breathing space. Relatively harsh water stress situations currently exist in various countries, regions, and many densely populated areas of the planet. The situation is expected to worsen shortly. In this condition, several advanced technologies have emerged as commercially viable and affordable processes for the large-scale production of freshwater from seawater sources. Over the last few decades, increased demand for freshwater forced policy makers, investors, innovators, and technologists to develop optimized large-scale water production plants. In this direction, majorly two technologies have emerged as frontrunners, namely thermal desalination systems and membrane-based RO technology. Even though both of these technologies possess several-step methods, optimal use of thermal desalination systems in the form of MSF and RO have been used successfully at large-scale seawater desalination. On the other hand, several other alternative thermal methods are being developed that have shown great potential in large-scale experimental studies. Nonetheless, technologies like HDH, MD, ED, NF, FO, and SC have been oriented toward developing hybridization to create energy-efficient and economically affordable technologies such as MSFD/RO, RM/MD, FO/MD, ED, ED/NF, HDH, MD, AD, and more such technologies. In all cases, reduction of energy consumption and the reduction or replacement of consumable parts to minimize energy inputs is the main focus. As an alternative approach, it is deemed that to make the overall desalination process sustainable and cost-effective, energy, which is supplied in the form of electricity or by combustion sources needs to be replaced by renewable energy alternatives such as solar and wind sources. All these efforts need to be rigorously pursued as the need to reduce overall equipment costs used in seawater desalination technologies is limited and automation is only the near future challenge other than consumption of energy. Nevertheless, recent advances in research in the development of commercial membrane technologies and thermal technologies revolve around drastically reducing production costs, minimizing environmental impact and use of renewable materials, thereby reducing greenhouse gas emissions. Therefore, the focus is on increasing seawater desalination efficiencies, recovering applied or used heat residual and energy, by designing hybrid alternatives, and most importantly, integrating existing and emerging technologies through novel approaches and renewable energy sources. In this way, the near future will see extensive greener approaches and researchers and technologists will focus on the development of seawater desalination technologies that adopt inexpensive equipment, easy-to-use design and modules that operate on a low energy consumption and with high product water generation performance and efficiency.

References

1 https://www.noaa.gov/jetstream/ocean/sea-water#:~:text=In%20sea%20water%2C%20there%20is,of%20high%20and%20low%20salinity.

2 Byrne, R.H., Duxbury, A.C., and Mackenzie, F.T. (2023). Seawater. *Encyclopedia Britannica*. https://www.britannica.com/science/seawater (accessed 17 July 2023).

3 Gale, T (2024). Ocean chemical processes. http://www.waterencyclopedia.com/Mi-Oc/Ocean-Chemical-Processes.html (accessed 2 December 2006).

4 https://www.unep.org/news-and-stories/press-release/half-world-face-severe-water-stress-2030-unless-water-use-decoupled, Paris/Nairobi (21 March 2016).

5 Boretti, A. and Rosa, L. (2019). Reassessing the projections of the world water development report. *NPJ Clean Water* 2019, 2 (1): 1–6.

6 Mekonnen, M.M. and Hoekstra, A.Y. (2016). Four billion people facing severe water scarcity. *Sci. Adv.* 2: e1500323.

7 Kathleen, D. Financing a water secure future. https://www.oecd.org/water/.

8 Do Thi, H.T., Pasztor, T., Fozer, D. et al. (2021). Comparison of desalination technologies using renewable energy sources with life cycle, PESTLE, and multi-criteria decision analyses. *Water* 13: 3023.

9 Alsayed, A.F. and Ashraf, M.A. (2021). Modified nanofiltration membrane treatment of saline water. In: *Water Engineering Modeling and Mathematic Tools* (ed. P. Samui, H. Bonakdari, and R. Ravinesh Deo), 25–44. Elsevier.

10 Toth, A.J. (2020). Modelling and optimisation of multi-stage flash distillation and reverse osmosis for desalination of saline process wastewater sources. *Membranes* 10 (10): 265.

11 Jeandel, C. and Derek, V. New tools, new discoveries in marine geochemistry. *Elements*, In press, Explore the oceans with Geochemistry, ff10.2138/gselements.14.6.xxff. ffhal-02408060f 20: 1–13.

12 ECD (2023). *Endocrine Disrupting Chemicals in Freshwater: Monitoring and Regulating Water Quality*. Paris: OECD Studies on Water, OECD Publishing https://doi.org/10.1787/5696d960-en.

13 Vera, M.L., Torres, W.R., Galli, C.I. et al. (2023). Environmental impact of direct lithium extraction from brines. *Nat. Rev. Earth Environ.* 4 (3): 149–165.

14 Halaj, E. and Kepinska, B. (2019). Conjunctive uses of the geothermal water resources from lower cretaceous formations in the Mogilno–Lodz trough, Poland. *Sustain. Water Resour. Manag.* 5: 1479–1494.

15 Khordagui, H.K. (1999). Desalination. In: *Environmental Geology. Encyclopedia of Earth Science* (ed. W.F. Charles), 124–125. Dordrecht: Springer.

16 Kalogirou, S.A. (2005). Seawater desalination using renewable energy sources. *Prog. Energy Combust. Sci.* 31 (3): 242–281.

17 Ahmed, F.E., Khalil, A., and Hilal, N. (2021). Emerging desalination technologies: current status, challenges and future trends. *Desalination* 517: 115183.

18 Birkett, J.D. (2000). The history of desalination before large-scale use. In: *Encyclopedia of Desalination and Water Resources*, vol. I (ed. C. Wichman and D. Wietelman), 9. History, Development and Management of Water Resources.

19 Al-Mutaz, I.S. (2020). Msf challenges and survivals. *Desalin. Water Treat.* 177: 14–22.

20 Angelakis, A.N., Valipour, M., Choo, K. et al. (2021). Desalination: from ancient to present and future. *Water* 13 (16): 2222.

21 Asadollahi, M., Bastani, D., and Musavi, S.A. (2017). Enhancement of surface properties and performance of reverse osmosis membranes after surface modification: a review. *Desalination* 420: 330–383.

22 Semiat, R. (2008). Energy issues in desalination processes. *Environ. Sci. Technol. 42* (22): 8193–8201.

23 Australian NWC [National Water Commission] (2008). Emerging trends in desalination: a review. Waterlines Report Series No. 9. Turner, Australian Capital Territory.

24 Salman, H.H. and Aswad, Z.S. (2022). Salination methods and their role in solving and managing the problem of salinization in Iraqi water resources. *IOP Conf. Ser.: Earth Environ. Sci.* 1002: 012004.

25 Fritzmann, C., Löwenberg, J., Wintgens, T., and Melin, T. (2007). State-of-the-art of reverse osmosis desalination. *Desalination* 216 (1–3): 1–76.

26 Yang, S. (2022). Water crisis and possible CRs in South/West Asia. In: *Coastal Reservoir Technology and Applications*, 1e (ed. S.-Q. Uamg), 377–428.

27 Antrim, B., Lesan, R., Liu, B., and von Gottberg, A. (2005). Worlds largest spiral element – history and development. *Desalination* 178 (1–3): 313–324.

28 Zubair, M.M., Saleem, H., and Zaidi, S.J. (2023). Recent progress in reverse osmosis modeling: an overview. *Desalination* 564: 116705.

29 https://idadesal.org/ida-2022-world-congress-charting-resilient-water-solutions-opening-ceremony-featured-dignitaries-and-vips-from-around-the-world/. Sydney, Australia, October 13, 2022 (accessed 15 February 2024).

30 Aladwani, S., Al-Obaidi, M., and Mujtaba, I. (2021). Performance of reverse osmosis based desalination process using spiral wound membrane: sensitivity study of operating parameters under variable seawater conditions. *Cleaner Eng. Technol.* 5: 100284.

31 https://www.waterworld.com/drinking-water/treatment/press-release/14274947/worlds-largest-reverse-osmosis-desalination-facility-opens, 15 April 2022.

32 Wilf, M. (2015). Membrane-based desalination processes: challenges and solutions. In: *Mineral Scales and Deposits* (ed. Z. Amjad and K.D. Demadis), 477–497.

33 Shahid, M.K., Mainali, B., Rout, P.R. et al. (2023). A review of membrane-based desalination systems powered by renewable energy sources. *Water* 15 (3): 534.

34 Abid, M.B., Wahab, R.A., Salam, M.A. et al. (2023). Desalination technologies, membrane distillation, and electrospinning, an overview. *Heliyon* 9 (2): https://doi.org/10.1016/j.heliyon.2023.e12810.

35 Toran, M.S., Labrador, P.F., Ciriza, J.F. et al. (2021). Membrane-based processes to obtain high-quality water from brewery wastewater. *Front. Chem. Eng.* 3: 734233.

36 Matchette, R.B. et al. (1995). *Guide to Federal Records in the National Archives of the United States*. Washington, DC: National Archives and Records Administration. 3 volumes, 2428 pages. https://www.archives.gov/research/guide-fed-records/groups/380.html#:~:text=History%3A%20Desalinization%20program%20authorized%20in,2710%2C%20December%2015%2C%201952.

37 Mahanta, P., Munna, M., and Coutu, R.A. (2018). Performance comparison of phase change materials and metal-insulator transition materials for direct current and radio frequency switching applications. *Technologies* 6 (2): 48.

38 Douvi, E., Pagkalos, C., Dogkas, G. et al. (2021). Phase change materials in solar domestic hot water systems: a review. *Int. J. Thermofluids* 10: 100075.

39 Freeman, T.B., Foster, E.O., Troxler, C.J. et al. (2023). Advanced materials and additive manufacturing for phase change thermal energy storage and management: a review. *Adv. Energy Mater.* 13 (24): 2204208.

40 Liu, L., Zhang, Y., Zhang, S., and Tang, B. Advanced phase change materials from natural perspectives: structural design and functional applications. *Adv. Sci.* 25: 2207652.

41 Cescon, A. and Jiang, J. (2020). Filtration process and alternative filter media material in water treatment. *Water* 12 (12): 3377.

42 Baig, H., Antar, M.A., and Zubair, S.M. (2010). Performance characteristics of a once-through multi-stage flash distillation process. *Desalin. Water Treat.* 13 (1–3): 174–185.

43 Chimanlal, I., Nthunya, L.N., and Richards, H. (2022). Membrane distillation crystallization for water and mineral recovery: the occurrence of fouling and its control during wastewater treatment. *Front. Chem. Eng.* 4: 1066027.

44 Naseer, M.N., Zaidi, A.A., Khan, H. et al. (2022). Desalination technology for energy-efficient and low-cost water production: a bibliometric analysis. *Green Processes Synth.* 11 (1): 306–315.

45 Al-hotmani, O., Al-Obaidi, M., John, Y. et al. (2021). An integrated system of multi effect distillation and wind power system – evaluation of total energy saving. *Comput. Aided Chem. Eng.* 50: 81–86.

46 Morris, R. (1993). The development of the multi-stage flash distillation process: a designer's viewpoint. *Desalination* 93 (1–3): 57–68. doi:10.1016/S0011-9164(93)80096-6. Availble at http://www.sciencedirect.com/science/article/pii/0011916493800966.

47 Al-Gobaisi, D. (1994). A quarter-century of seawater desalination by large multistage flash plants in Abu Dhabi (plant performance analysis, assessment, present efforts toward enhancement and future hopes). *Desalination* 99 (2–3): 509–512. doi: 10.1016/S0011-9164(94)00196-0.

48 Youcai, Z. (2018). Physical and chemical treatment processes for leachate. In: *Pollution Control Technology for Leachate from Municipal Solid Waste*, 1e (ed. Z. Youcai), 31–183. Elsevier.

49 Ghazouani, N., El-Bary, A.A., Hassan, G.E. et al. (2022). *Water* 14: 3424.

50 Modla, G. and Lang, P. (2015). Vapor compression for batch distillation: comparison of different working fluids. *Ind. Eng. Chem. Res. 54* (3): 1081–1092.

51 Sharon, H. and Reddy, K.S. (2015). A review of solar energy driven desalination technologies. *Renew. Sust. Energ. Rev.* 41: 1080–1118.

52 Moustafa, S.M.A., Jarrar, D.I., and El-Mansy, H.I. (1985). Performance of a self-regulating solar multistage flash desalination system. *Sol. Energy* 35 (4): 333–340.

53 Hanafi, A. (1991). Design and performance of solar MSF desalination systems. *Desalination* 82: 175–185.

54 Rahimi-Ahar, Z., Hatamipour, M.S., and Ahar, L.R. (2020). Air humidification-dehumidification process for desalination: a review. *Prog. Energy Combust. Sci.* 80: 100850.

55 Abdelmoez, W., Mahmoud, M.S., and Farrag, T.E. (2014). Water desalination using humidification/dehumidification (HDH) technique powered by solar energy: a detailed review. *Desalin. Water Treat.* 52 (25–27): 4622–4640.

56 Mohamed, A.S.A., Ahmed, M.S., Maghrabie, H.M., and Shahdy, A.G. (2021). Desalination process using humidification–dehumidification technique: a detailed review. *Int. J. Energy Res.* 45 (3): 3698–3749.

57 Wang, Q., Zhang, G., Wu, Q. et al. (2022). A combined wall and roof solar chimney in one building. *Energy* 240: 122480.

58 Ming, T., Liu, W., Wu, Y. et al. (2016). Introduction. In: *Solar Chimney Power Plant Generating Technology*, 1e (ed. T. Ming), 1–46. Elsevier Inc.

59 Abraham, S.B. and Ming, T.Z. (2019). Building's solar chimney: The performance of width and inlet to in tropical country. *IOP Conf. Ser.: Earth Environ. Sci.* 354: 012046.

60 Baticados, E.J., Capareda, S.C., Liu, S., and Akbulut, M. (2020). Advanced solar still development: improving distilled water recovery and purity via graphene-enhanced surface modifiers. *Front. Environ. Sci.* 8: 531049.

61 Gabrielli, P. and Mazzotti, M. (2019). Solar-driven humidification–dehumidification process for water desalination analyzed and optimized via equilibrium theory. *Ind. Eng. Chem. Res. 58* (33): 15244–15261.

62 Brunini, A., Gracia, M.C., Melgarejo, A.A., and Rodriguez, R.G. (2021). Humidification–dehumidification desalination process using green hydrogen and heat recovery. *Environ. Res.: Infrastruct. Sustain.* 1: 035005.

63 Xu, L., Chen, Y., Wu, P., and Huang, B. (2020). Humidification–dehumidification (HDH) desalination system with air-cooling condenser and cellulose evaporative pad. *Water* 12 (1): 142.

64 Lawal, D., Antar, M., Khalifa, A. et al. (2018). Humidification-dehumidification desalination system operated by a heat pump. *Energy Convers. Manag.* 161: 128–140.

65 Koronakis, P.S. (1992). Solar chimney dynamic performance under typical mediterranean summer conditions. *Int. J. Sol. Energy* 13 (2): 73–84.

66 Zhou, X., Wang, F., and Ochieng, R.M. (2010). A review of solar chimney power technology. *Renewable Sustainable Energy Rev.* Elsevier 14 (8): 2315–2338.

67 Saad, M., Ahmed, N., Giovannini, L. et al. (2022). Evaluation of possible integration of solar chimney power plants with high-rise buildings: a numerical analysis. *J. Build. Eng.* 60: 105188.

68 Aende, A., Gardy, J., and Hassanpour, A. (2020). Seawater desalination: a review of forward osmosis technique, its challenges, and future prospects. *Processes* 8 (8): 901.

69 Wafi, M.K., Hussain, N., El-Sharief Abdalla, O. et al. (2019). Nanofiltration as a cost-saving desalination process. *SN Appl. Sci.* 1: 751.

70 Aljumaily, M.M., Alshami, A.W., Ismael, B.H. et al. (2022). A review on membrane desalination process in water treatment. *IOP Conf. Ser.: Earth Environ. Sci.* 1120: 012035.

71 Sherwood, T.K., Brian, P.L.T., and Fisher, R.E. (1967). Desalination by reverse osmosis. *Ind. Eng. Chem. Fundam. 6* (1): 2–12.

72 Davenport, D.M., Deshmukh, A., Werber, J.R., and Elimelech, M. (2018). High-pressure reverse osmosis for energy-efficient hypersaline brine desalination: current status, design considerations, and research needs. *Environ. Sci. Technol. Lett. 5* (8): 467–475.

73 Chian, E.S.K., Chen, J.P., Sheng, P.X. et al. (2007). Reverse osmosis technology for desalination. In: *Advanced Physicochemical Treatment Technologies. Handbook of Environmental Engineering*, vol. 5 (ed. L.K. Wang, Y.T. Hung, and N.K. Shammas), 329–366. Humana Press.

74 Greenlee, L.F., Lawler, D.F., Freeman, B.D. et al. (2009). Reverse osmosis desalination: water sources, technology, and today's challenges. *Water Res.* 43 (9): 2317–2348.

75 Cohen, Y., Semiat, R., and Rahardianto, A. (2017). A perspective on reverse osmosis water desalination: quest for sustainability. *AICHE J.* 63 (6): 1771–1784.

76 Yao, Y., Zhang, P., Jiang, C. et al. (2021). High performance polyester reverse osmosis desalination membrane with chlorine resistance. *Nat. Sustainability* 4 (2): 138–146.

77 Yao, Y. et al. (2018). A novel sulfonated reverse osmosis membrane for seawater desalination: experimental and molecular dynamics studies. *J. Membr. Sci.* 550: 470–479.

78 Stevens, D.M., Shu, J.Y., Reichert, M., and Roy, A. (2017). Next-generation nanoporous materials: progress and prospects for reverse osmosis and nanofiltration. *Ind. Eng. Chem. Res.* 56: 10526–10551.

79 Do, V.T., Tang, C.Y., Reinhard, M., and Leckie, J.O. (2012). Degradation of polyamide nanofiltration and reverse osmosis membranes by hypochlorite. *Environ. Sci. Technol.* 46: 852–859.

80 Wang, Z., Ma, C., Xu, C. et al. (2021). Graphene oxide nanofiltration membranes for desalination under realistic conditions. *Nat. Sustainability* 4: 402–408.

81 Wang, Z., Wang, Z., Lin, S. et al. (2018). Nanoparticle-templated nanofiltration membranes for ultrahigh performance desalination. *Nat. Commun.* 9: 2004.

82 Im, S., Jeong, S., and Jang, A. (2021). Forward osmosis (FO)-reverse osmosis (RO) hybrid process incorporated with hollow fiber FO. *NPJ Clean Water* 4 (1): 1–10.

83 Abounahia, N., Ibrar, I., Kazwini, T. et al. (2023). Desalination by the forward osmosis: advancement and challenges. *Sci. Total Environ.* 886: 163901.

84 Hsu, C.-H., Ma, C., Bui, N. et al. (2019). Enhanced forward osmosis desalination with a hybrid ionic liquid/hydrogel thermoresponsive draw agent system. *ACS Omega 4* (2): 4296–4303.

85 Maltby, J., Day, L., McCutcheon, L.E. et al. (2006). Extreme celebrity worship, fantasy proneness and dissociation: developing the measurement and understanding of celebrity worship within a clinical personality context. *Personal. Individ. Differ.* 40 (2): 273–283.

86 Li, L., Shi, W., and Yu, S. (2020). Research on forward osmosis membrane technology still needs improvement in water recovery and wastewater treatment. *Water* 12 (1): 107.

87 Lee, W.J., Goh, P.S., and Ismail, A.F. (2021). Forward osmosis-based hybrid processes for water and wastewater treatment. In: *Integrated and Hybrid Process Technology for Water and Wastewater Treatment*, 121–144.

88 Haupt, A. and Lerch, A. (2018). Forward osmosis application in manufacturing industries: a short review. *Membranes* 8 (3): 47.

89 Criscuoli, A. (2021). Membrane distillation process. *Membranes* 11 (2): 144.

90 Mahmoudi, F. and Akbarzadeh, A. (2018). Sustainable desalination by permeate gap membrane distillation technology. In: *Emerging Technologies for Sustainable Desalination Handbook* (ed. G. Gude), 157–204. Elsevier.

91 Kiss, A.A. and Kattan Readi, O.M. (2018). An industrial perspective on membrane distillation processes. *J. Chem. Technol. Biotechnol.* 93 (8): 2047–2055.

92 Mahmoudi, H. (2015). Water desalination in electrodialysis applications. In: *Encyclopedia of Membranes* (ed. E. Drioli and L. Giorno), 1986–1987. Berlin, Heidelberg: Springer.

93 Al-Amshawee, S., Yunus, M.Y.B.M., Azoddein, A.A.M. et al. (2020). Electrodialysis desalination for water and wastewater: a review. *Chem. Eng. J.* 380: 122231.

94 Nosova, E., Achoh, A., Zabolotsky, V., and Melnikov, S. (2022). Electrodialysis desalination with simultaneous pH adjustment using bilayer and bipolar membranes, modeling and experiment. *Membranes* 12 (11): 1102.

95 Choi, H., Lee, J., and Moon, H. Electrodialysis desalination. In: *Desalination:Water from Water* (ed. J. Kucera), 245–285. Wiley.

96 Far, H.A., Hassani, A.H., Taghavi, L. et al. (2022). Electro dialysis reversal (EDR) performance for reject brine treatment of reverse osmosis desalination system. *PLoS One* 17 (8): e0273240.

97 Katz, W.E. (1979). The electrodialysis reversal (EDR) process. *Desalination* 28 (1): 31–40.

98 Xu, X., He, Q., Ma, G. et al. (2021). Pilot demonstration of reclaiming municipal wastewater for irrigation using electrodialysis reversal: effect of operational parameters on water quality. *Membranes* 11 (5): 333.

99 Hashemifard, S. and Azin, R. (2004). New experimental aspects of the carrier gas process (CGP). *Desalination* 164 (2): 125–133.

100 Aybar, H.S. (2002). Analysis of a mechanical vapor compression desalination system. *Desalination* 142 (2): 181–186.

101 Lucas, M. and Tabourier, B. (1985). The mechanical vapour compression process applied to seawater desalination: a 1,500 ton/day unit installed in the nuclear power plant of Flamanville, France. *Desalination* 52 (2): 123–133.

102 Fergani, Z., Triki, Z., Menasri, R. et al. (2023). Analysis of desalination performance with a thermal vapor compression system. *Water* 15 (6): 1225.

103 Janajreh, I., Zhang, H., El Kadi, K., and Ghaffour, N. (2023). Freeze desalination: current research development and future prospects. *Water Res.* 229: 119389.

104 Zhang, H., Janajreh, I., Hassan Ali, M.I., and Askar, K. (2021). Freezing desalination: heat and mass validated modeling and experimental parametric analyses. *Case Stud. Therm. Eng.* 26: 101189.

105 Jayakody, H. and Mahmoud, S. (2020). Cryogenic energy for indirect freeze desalination–numerical and experimental investigation. *Processes* 8 (1): 19.

106 Montazeri, S.M. and Kolliopoulos, G. (2022). Hydrate based desalination for sustainable water treatment: a review. *Desalination* 537: 115855.

107 Zhang, J., Chen, S., Mao, N. et al. (2022). Progress and prospect of hydrate-based desalination technology. *Front. Energy* 16: 445–459.

108 Babu, P., Nambiar, A., He, T. et al. (2018). A review of clathrate hydrate based desalination to strengthen energy–water nexus. *ACS Sustain. Chem. Eng.* 6 (7): 8093–8107.

109 Kunz, G. (1967). Ion exchange pre-treatment for sea water desalination plants. *Desalination* 3 (3): 363–372.

110 Dong, H., Shepsko, C.S., German, M., and SenGupta, A.K. (2018). Hybrid ion exchange desalination (HIX-Desal) of impaired brackish water using pressurized carbon dioxide (CO2) as the source of energy and regenerant. *Environ. Sci. Technol. Lett.* 5 (11): 701–706.

111 Weiss, D. (1966). The role of ion-exchange desalination in municipal water supplies. *Desalination* 1 (2): 107–128.

112 Alabi, A., AlHajaj, A., Cseri, L. et al. (2018). Review of nanomaterials-assisted ion exchange membranes for electromembrane desalination. *NPJ Clean Water* 1 (1): 1–22.

113 Gebreeyessus, G.D. (2019). Status of hybrid membrane–ion-exchange systems for desalination: a comprehensive review. *Appl. Water Sci.* 9: 135.

114 Klaysom, C., Marschall, R., Moon, S.-H. et al. (2011). Preparation of porous composite ion-exchange membranes for desalination application. *J. Mater. Chem.* 21: 7401–7409.

115 Gohil, J. and Ray, P. (2021). Preparation of heterogeneous cation- and anion-exchange membranes by eco-friendly method: electrochemical characterization and desalination performance. *J. Membr. Sci. Res.* 7 (2): 102–110.

5

Functional Materials in Desalination and Wastewater Treatment

5.1 Polymer-Based Functional Materials such as Water Filters

Material science offers interdisciplinary research options that include fields like material science having interrelationships between compositional studies, structural understanding, process design, and performance evaluation of various materials. Advanced technologies now demand a widely integrated approach with other disciplines to form many interdisciplinary subjects. In reality, modern and advanced technologies often require specially designed materials stock or the involvement of materials with advanced properties. Therefore, the modern technology fabrication process often involves structural materials and functional materials that are specifically designed to obtain desired end results. On the other hand, recent environmental awareness, sustainability, and scarcity of resources have forced researchers and innovators to explore new stock of materials. In this direction, the exploration and application of environmentally benign functional materials, both of natural and synthetic origin, are becoming increasingly popular in water purification and wastewater remediation [1–4].

On the other hand, rapid industrialization, excessive consumption of water resources for developmental activities, and accelerated urbanization result in the production of highly deteriorated contaminated water. Environmental pollution due to the release of wastewater into the ecosystem is getting worse. Moreover, conventional treatment technologies often fail to satisfy the growing demand for the recovery and recycling of high-quality reusable water. Therefore, exploring novel stock of materials, and designing, and developing efficient and economically feasible technologies for large-scale water and wastewater treatment is necessary. In this direction, the arrangement, creating an appropriate combination of functional materials that are categorized as green and sustainable sources, is

Materials and Methods for Industrial Wastewater and Groundwater Treatment,
First Edition. S. K. Nataraj.

expected to greatly improve the efficiency of traditional water treatment processes, thereby facilitating improved water recovery and treatment parameters. This new stock of sustainable materials includes natural clay/minerals, templated materials such as photocatalysts, biochar, ferric hydroxide, synthetic fiber mats, activated carbon, activated alumina, and their composites. Further, the use of these environmentally benign materials for water purification and remediation protocols can be divided into four categories: (i) ion exchange material, (ii) adsorbent, (iii) catalytic oxidation material, and (iv) stabilizing agents [5–12].

Among many functional materials that are largely used in water treatment and wastewater treatment, applications include conventional natural polymer origin [13]. With the rapid development in the advanced and greener materials sourcing, extraction, and exploration efforts, the technologies have also changed their approaches to adopt new stock of materials. Among many, functional polymers have attracted renewed interest in both energy and environmental sectors. Polymers from both natural and synthetic origins have been widely explored for water treatment applications. It is also important that targeted functionalities on the polymer backbone define the degree of success in their end-use. Functional polymers offer several advantages over conventional polymers that cannot serve the important requirement, thereby promoting the design and development of tailor-made polymers that carry specific functionalities. Such polymers can be classified as advanced functional polymers that have potential applications in water treatment, in particular, water purification, waste remediation, oil/water separation or gas separation, desalination, and so on [14–17].

Since the early 1960s, polymer science has gained wider interest and due to increased demand, researchers and innovators have explored novel versions of polymer to cater to interdisciplinary fields encompassing materials science, nanoscience, polymer chemistry, polymer physics, bio-medical, energy, environmental remediation, and water treatment. As polymers offer superior mechanical, thermal, and physico-chemical properties over conventional materials, polymeric materials, advanced functional polymeric materials can be adopted as biocompatible, selective separation, and easily adaptable material stock. Therefore, polymers have lately turned into an important cornerstone of modern applications, industrial livestock, and modern technologies. Nevertheless, advanced materials and functional polymers follow their own characteristics and laws. Among many, the first is the uniqueness of selecting monomers for the synthesized polymer of desired molecular weight and physico-chemical properties; secondly, designing appropriate polymerization reactions, catalysts/initiators, methods, mechanisms, and reaction conditions that needed to be developed by inducing appropriate functionality on the backbone. The third most important part of making a polymer with multiple properties on the polymer is to stabilize the properties of the overall material to withstand the end-use parameters [18–20].

Therefore, it is necessary to combine the preparation, synthesis, processing, fabrication, blending, compositing, and all the morphologies involved in this process to make an advanced material or functional polymer with desired end functionalities. In addition to this, recent developments in both material and technology simulations have enabled innovations in fundamental and cutting-edge development of material and polymer science disciplines that can be utilized to create new materials stock with unique properties. These advanced techniques have also been utilized in improving the performance of the existing materials due to the inclusion of tuneable and desired functionalities that serve the end-use purpose in increasingly popular interdisciplinary sectors. Interestingly, almost all advanced materials and functional polymers such as ceramics, and metals, either in the form of conductive systems, micelles or vesicles, gels, films, nanomaterials, filters, polymer membranes, composites of these precursors with metal-organic frameworks (MOFs), covalent organic framework (COF) layered-double hydroxides (LDHs) and so on have been used to further enhance the properties of the material. Therefore, this chapter has attempted to summarize all the recent advances in advanced and functional materials and polymers. Further, the chapter will provide detailed water treatment potentials of the advanced and functional materials with a focus on understanding functional materials in the context of water treatment in a comprehensive manner [21–25].

5.2 Classification and Recent Advancements

Owing to their multi-functional properties, advanced functional materials can be classified into major four classes, namely (i) metal-based, (ii) ceramic-based, (iii) polymer-based, and (iv) composites of all three materials. Further, advanced functional materials can be broadly selected based on the materials, structural, and physicochemical properties and their end uses as depicted in Figure 5.1. In addition to broader and inherent material characteristics, advanced functional materials also possess unique hierarchical structures, responsive electronic structures, hybrid functionalities, surface characteristics, and active and tenable functional moieties on their parent molecules making these new classes of advanced materials a choice of materials for novel applications and technology development. This chapter, therefore, gives detailed material stocks, their extraction protocols, and methodologies for obtaining advanced functional materials for water desalination and wastewater treatment applications [1].

Another major paradigm shift in material development is the revolutionary transformation of physical nature from the bulky to the nanomaterials, which has been caused by the great advancements of nanotechnology. Nanotechnology enabled in the development of nanomaterials of various bulky material stocks with

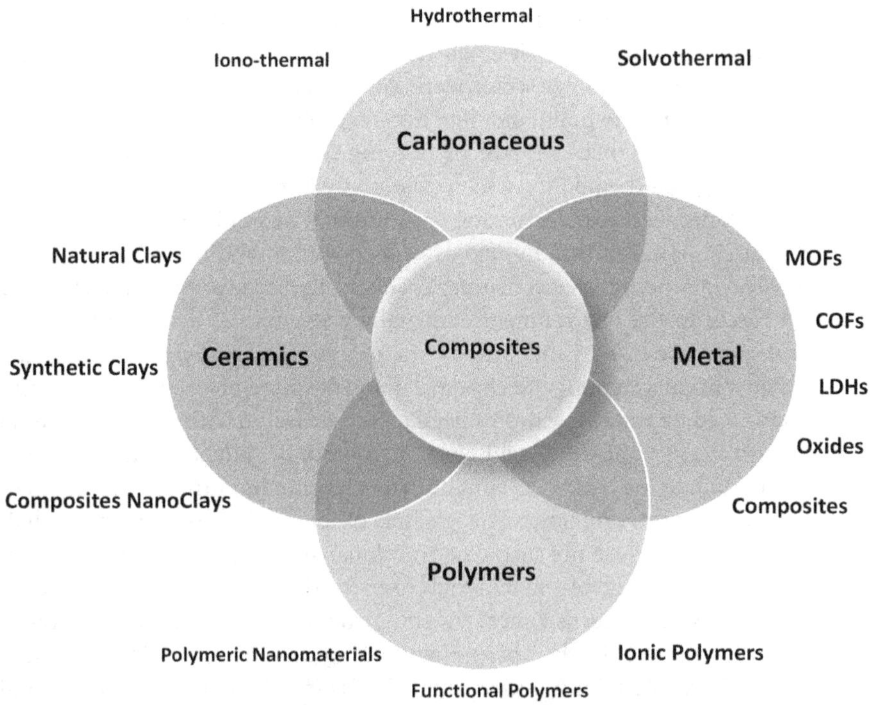

Figure 5.1 Broader classification and materials properties of advanced functional materials that influence their end applications.

their nanoscale sizes has revolutionized various application sectors solving several industrial and environmental issues in their applications such as water treatment and pollutants removal. Therefore, research focuses on multiplying materials properties using nanotechnology principles from the physicochemical properties perspective of the materials that cannot be obtained from the bulk phase of the same material. These developments have shown many advantages to their usage in many fields of sciences, particularly in water treatment and wastewater remediation. In this direction, researchers and innovators have produced several innovative protocols, products, and novel applications. Many biomedical and environmental applications have proven to be disadvantageous due to their dangerous effects associated with the nanotoxicity on human health and their prolonged effects still to be studied. This is particularly alarming in the water treatment applications as the water remains consumed by humans and living beings, which may pose serious challenges especially when the treated water is contaminated with secondary pollutants like nanomaterial that potentially leach from treatment media during the unit operations [26–29].

Nevertheless, nanomaterials have enabled a series of efficient and effective water treatment technologies. First, major set of nanomaterials that were widely introduced in water treatment sector were the photocatalysts and adsorbents for the treatment of organic pollutants like dyes and pharmaceutical contaminants that significantly enhance the risk of altering aesthetic aspects of the water sources and degrade the quality, which potentially harms the aquatics and plant ecosystems. Moreover, ever-increasing consumption of millions of personal and pharmaceutical products that human society consumes and now it is struggling with massive volumes of used consumer products discarded as pollutants that are posing a threat to the environment and many ecosystems. On the other hand, biological contaminants in the various water resources have also increasingly resisted conventional treatment technologies and cleaning protocols. In such situations, advanced treatment strategies need to be discovered with suitable material stock and protocols in place. Nonetheless, in absence of suitable treatment materials, elimination of the biological contaminants layer or bio-film on various water samples, treatment equipment surfaces, and other containers remains a challenge. This is also because the cost of conventional antifouling coatings and treatment techniques are high or the materials used for the purpose are ineffective.

Therefore, recent concerns concentrated on solving both conventional as well as emerging pollutants in the environment using nanomaterials in stable substrate or carrier while taking into account the issues associated with safety and stability of treatment materials. Fundamentally, both conventional as well as emerging pollutants have penetrated freshwater sources, groundwater, and surface water as well as are present in industrial wastewater, which could be an environmental challenge, which requires multi-step expensive treatment protocols for their elimination. Moreover, conventional treatment materials and techniques are proving to be expensive, and ineffective in removing complex pollutants like organic liganded heavy metals, drug molecules, and microplastics. Among these microplastics and organic dye are the main problems that seem to pose a serious threat to already struggling conventional treatment techniques and they need serious and urgent attention from material scientists. Therefore, developing a greener and sustainable approach that solves the potential toxicity of nano-scaled material and also eliminates the pollutants effectively has been the target of researchers all over the world. Nevertheless, sustainable nanomaterials prepared or placed on the stabilized substrate or carriers offer numerous benefits not only in addressing water pollutants but also in providing faster removal rates compared to conventional methods. All these efforts should be implemented with the overall and comprehensive green abundant sources with relatively greater stability at affordable cost.

Basically, green sources are the key to preparing nanomaterials that are intended to use in water treatment applications such as photocatalysts originating from plant or plant extract, and/or natural polymers locust bean gum-based hydrogels that can be used as adsorbate and photocatalytic properties of biomaterial-based systems make them safer. Also, methods of sustainable photocatalyst or absorbate preparation in most cases must be simple and robust. In this direction, several experiments have shown the effect of the right combination of materials and their formulation, degree of dispersion, or degree of loading of nanomaterials in suitable carriers can limit their toxicity. In this direction, recent years have seen many efforts in preparation and development of nanomaterials-based devices in solid-safe configurations. However, stable nanomaterial-based treatment systems also pose several challenges such as formation of aggregates and issues associated with nanomaterial dispersion in substrates or carriers. Nevertheless, advanced approaches recently adopted have minimized the negative effects of nanotoxicity in water treatment. In one such approach, researchers have tried to overcome the agglomeration issue and degree of dispersion and stabilization as parameters, which control and determine their physicochemical properties and, in particular, their nano and cytotoxicity intensities. In this direction, every step involves nanomaterial preparation also to be taken care of to attain sustainability to the overall process. Most of the nanomaterials also escape from the monitoring during either concentration, purification, or separation step [30–34].

Therefore, recycling or reclaiming nano-material-based treatment materials such as nano-cleaners or nano-adsorbents from treatment system post-treatment step is vital. In this direction, several approaches have been developed to extract post-treated adsorbents from cleaning containers such as magnetic nanomaterials separated using externally controlled electromagnet or permanent magnet. Also, iron oxide-based nanomaterials and their derivatives, which can be prepared either through solvo/hydrothermal or by microwave ultrasonic are reasonably considered cleaner techniques. In addition to this other advanced chemical synthesis, protocols can be followed for the preparation of nanomaterials such as reverse microemulsion, sol-gel, and coprecipitation. Nonetheless, keeping the water treatment process as an end application, several protocols are being developed to overcome low-efficiency preparations, reduce high energy and cost inputs, and make the overall process simple and robust [35].

Among many advanced methods, the precipitation and coprecipitation processes can be performed and adopted as potentially large scalable productions. Nano and functional materials prepared from these routes can be used for the removal of heavy metals such as chromium from wastewater sources. Recent awareness of adopting greener and sustainable material stocks has also led to the exploration of the natural environment for new classes of materials to cater to the ecosystem for both preparation and application aspects. Sustainable materials have shown greater potential in encountering modern pollutants like heavy metals and emerging pollutants both in groundwater as well as industrial discharges such as dyes, toxic drug molecules, and harmful effects of emerging pollutants. Among these, the heavy metals, which mainly consist of Cd, Zn, Pb, Ni, Cu, Hg, and Cr ions vastly detected in various kinds of water sources including groundwater, surface water, and wastewater as well as soil sources that their presence thoroughly relates to a variety of human and industrial activities [36–39].

Many studies for years have implicated organic pollutants like dyes as other types of toxic contaminants, which have proven extreme biotoxicity, mutagenic effects, carcinogenic, and threatening the life of various types of aquatic ecosystems. Therefore, adsorptive removal techniques for selective absorption or removal of heavy metals, organic moieties, drug molecules, and dyes not only are urgent but the situation of various aquatic ecosystems can be altered to attain a cleaner system or deviated to retain pristine nature. Literature shows several studies reported on the preparation of carbon-based nanocomposite such as iron or nickel-doped graphene oxides (GOs) and their derivatives having high adsorption capabilities in removing heavy metals (Cr, Pb, As, and so on) from the contaminated water sources. Carbon-based nanomaterials are another set of new adsorbents and photocatalysts, which showed higher efficiencies of water treatment potentials due to their high specific surface area, tunable surface functionalities, and various morphologies. Thus, both magnetic materials and carbon-based nanomaterials have shown greater abilities in absorption process, high potentials of scalability, and reliability in providing desired removal efficiencies [40, 41].

Therefore, this chapter provides a detailed discussion of types of functional materials and their interfacial activation or modification effect on specific parameters that play a key role in removal, abetment, absorption, and degradation features of active materials and focuses on their preparation and functionalization techniques involved. Water treatment materials are generally influenced by various factors that could affect the removal, absorption, degradation, and photodegradation mechanisms. Also, capacity and stability of advanced water treatment materials require optimal working conditions to perform their absorption, removal, and degradation process. Nevertheless, among many material stocks, polymeric systems have played vital roles in several applications including water treatment applications. Figure 5.2 schematically represents the family of functional materials that have been discussed in detail in this chapter.

5.3 Polyelectrolyte Complexes in Wastewater Treatment and Desalination

Water is a unique resource, which cannot be created for the purpose of human activities. However, this abundant source can be reclaimed, recycled, and retained from reaching waste sources. Nonetheless, water is an overexploited natural

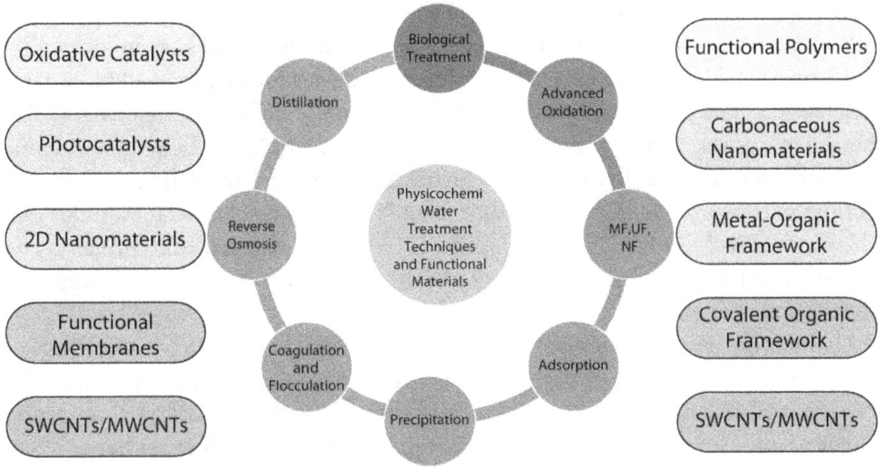

Figure 5.2 Schematic description of various types of functional materials that are being used in water treatment applications.

resource and discharge of freshwater sources is a major concern. The control and effective management of natural water resources and monitoring of discharged contaminated water is the need of the hour. Therefore, before being discharged, contaminated water or wastewater must be treated or purified. In this direction, the main goal should be to prevent discharge of contaminated water characterized by chemical–physical and biological contaminants that are potentially detrimental to aquatic ecosystems and the environment at large. By doing so, water can be prevented from undergoing further degradation as reservoirs or discharge tanks acquire various contaminants at once [42, 43].

Conventional wastewater treatment methods consist of utilizing well-known materials stock to prepare treatment products or applying them through a known technology to enhance or upgrade the quality of treated water from wastewater. Among many conventional techniques, physico-chemical methods play a vital role in treating wastewater. Many physico-chemical methods are used both as individual or standalone techniques and in combination with other techniques like mechanical processes, chemical processes, and biological methods. Due to their simple separation and purification approaches, physico-chemical techniques are being used increasingly as the preliminary step in wastewater treatment before they are being further sent to either chemical treatment technique steps or biochemical purification. Nevertheless, multi-step treatment techniques are being developed in advanced treatment protocols due to the implementation of stricter water treatment guidelines. In this regard, pre-determination of process parameters and their product quality at the end of the treatment is well-defined to achieve optimal treatment quality. Therefore, the degree of purification of wastewater after immediate discharge needs to be monitored in which all suspended pollutants, organic admixtures, and larger solutes are desired to be removed before it is discharged. The physico-chemical techniques largely help separate coarsely dispersed particles. Further, mechanical treatment techniques help in removing finely dispersed particles and microparticles. However, it is difficult to remove colloid particles present in wastewater from both physico-chemical methods and mechanical techniques. Therefore, removal of finely dispersed contaminants and colloidal pollutants is generally treated following advanced chemical processes like coagulation methods.

Colloids are generally defined as particulate matter that falls in the size range of 2–1000 nm. It is estimated that approximately 15–25% of the organic matter or organic pollutants found in municipal wastewater, food-processing effluents and surface water sources are colloidal in nature. Therefore, to achieve higher efficiencies (more than 90%) of organic contaminants' elimination from various wastewater sources like municipal wastewater, treatment facilities should focus on removing major parts of colloidal material. It is also important to note that over a period of time, particles in a colloid range agglomerate to form larger particle aggregates that can be easily settled out and separated either manually or mechanically from the wastewater effluents. This separation process follows several systematic steps. One among these steps of coagulation is flocculation step in which the small suspended particulates form a loose accumulation of aggregates known as flocs. These flocs can be further settled out easily under the influence of colloidal breaking substances called flocculants. The addition of flocculants increases the density of aggregates and with time, the solidification of the flocs formed decreases the consumption of coagulants while simultaneously increasing the colloidal pollutant removal reliability in the wastewater treatment plant. Nevertheless, the rate of flocculation and effectiveness of the process greatly depend on the composition and concentration of the wastewater, the rate of mixing, treatment temperature, and the order in which coagulants and flocculants are introduced into the wastewater. Interestingly, when flocculants are dissolved in wastewater, aggregating agencies may either form a unionized or ionized state. In ionized state, flocculants form soluble polyelectrolytes. Depending on the composition of the polar groups, flocculants can be anionic, cationic, nonionogenic, and amphoteric polymers. Among many flocculants, polymeric flocculants have been widely studied. Specially designed polymeric agents have been used as affordable and effective, and alternative flocculants as innovative flocculants to conventional systems. In this, conventional flocculants initiate the flocculation process through polymerization reactions under the influence of heat-sensitive peroxide, redox systems, etc., that can be used to ionize radiation, like gamma photons or accelerated electrons. This method offers several significant advantages ascribed to the polymerization route in which both removal of colloidal pollutants can be achieved leaving behind relatively high-quality water from wastewater reservoirs. These flocculants can also be used at different rates of separation efficiencies under the influence of external stimuli such as concentration, temperature, stirring, and change in the treatment operations [44–46].

Similarly, several polymeric systems have been used as effective flocculants in advanced wastewater treatment processes. One such type of active material can be obtained through polymerization of acrylamide and acrylic acid [47]. These polymeric compositions are further functionalized to act as organic polyelectrolytes for the treatment of contaminated potable water dewatering wastewater treatment plant sludges through coagulation and flocculation. Following coagulation and flocculation, contaminations separate from the polluted water followed by sedimentation and separation of product water. With a suitable separation mechanism of sediment upon the flocculation of contaminated water, post-filtration can be omitted. In absence of appropriate sedimentation, flotation is used as an effective alternative to separate contaminants such as algae-laden waters. Further, value-added products can also be recovered from the sludge and additional water from the various separation processes. This also helps in considerably reducing the water content in sludge that results in relatively dry mass, which considerably minimizes the transportation costs. In all these cases, polymer-based polyelectrolyte complexes play a significant role in sludge conditioning [48].

Nevertheless, polymers have been widely adopted in various forms and designs to act as coagulants or flocculants in sedimentation processes for water purification applications. Polymers offer several advantages in comparison with alum in water treatment. Some of them can be listed as: (i) conventional coagulants need to be used in large doses; however, polymer coagulants efficiently work at low doses, (ii) polymer coagulants generate smaller volumes of sludge, (iii) polymer coagulants are effective with a smaller increase in the ionic load for the treated water, (iv) polymers can reduce the level of aluminum in treated water, (v) polymeric coagulants cost less making them affordable up to 25–30%. Therefore, polymers-based functional materials are especially beneficial in harsh conditions of wastewater coping with the issues of slow-settling flocs in low-temperature coagulation. Polymeric coagulants also effectively perform in treating soft-colored wastewaters, where they improve settleability and concentration of flocs. These effects can vary based on wastewater treatment plant capacity and polymeric systems can facilitate the formation of larger and stronger flocs, which further influence the rate of solid and water phase separation. With all these advantages, polymeric coagulants can be effective over conventional coagulants in terms of dosage. Further, the range of wastewater that can be treated becomes broader. Nevertheless, polymeric coagulants also possess some disadvantages such as: (i) polymeric coagulants are relatively expensive, (ii) the selection of a suitable system to suit the nature of the wastewater environment that factors being the main concern, (iii) defining appropriate dosage against the sensitivity of the coagulant with respect to particular wastewater is a difficult task, and (iv) polymeric coagulants are less efficient against high turbidity and natural organic contaminants. Nonetheless, polymer structure greatly influences the drinking water treatment process including coagulation/flocculation, the rates of both precipitation and sedimentation, product water quality, and the extent of solid content of the final sludge. Specially designed polymeric coagulants/flocculants can be effective against raw water having suspended solids and colloidal pollutants through the physicochemical procedures and the adsorption of soluble material on polymeric substrates such as metal hydroxide flocs [49–52].

5.4 Polyelectrolyte Complexes and Zwitterionic Polymers in Desalination

Target-specific water-soluble polymers with tuneable functionality have been widely used in wastewater treatment applications largely to remove suspended solids and colloidal solid contaminants from the water. Polymer-based treatment materials are now used regularly in municipal, small-scale industrial plants, and stormwater treatment purifications. However, the extent of their use and the material understanding is still limited. Polymers offer advanced solutions as coagulants and flocculants with exceptional treatment efficiencies. Polymer-based coagulants and flocculants are robust and work spontaneously in settling solids out of colloidal and suspended contaminants in real-time. Conventional coagulants take days or even months to separate the solidified suspended or colloidal contaminants to settle under gravity, which can typically be achieved in minutes or seconds using polymer-based coagulants that are designed and prepared through appropriate functionalization process [53].

Polymers are long-chain molecules made up of unit molecules called monomers. Molecular weight of the polymers ranges from a few hundred Dalton (Da) to several millions of monomer molecules in one polymer. Oligomers is the term used for a low polymer made up of few monomers resulting in molecular weight repeat units. It is important to note that the more the number of monomers, the longer the polymer chain, and physical properties are significantly dependent on the length of the chain. Based on the nature of monomers and functionality, polymers can be divided into three main categories, namely anionic, cationic, and zwitterionic polymers. Both polymers having anionic and cationic functionality can influence the settling of liquid-suspended solids or contaminants. As the name suggests, anions and polymers possess a negative charge and positive charge as a result of active functional moieties on the polymer backbone, respectively [54].

Generally, polymeric materials possess low functional group density, relatively low water-holding capacity, and inferior ionic conductivity, which limits their widespread applications in different applications. Even though polymeric materials are robust and rigid and sustain long life, however, limitations associated with a lack of functional moieties on the backbone present serious bottlenecks in the development of polymers in water treatment. Due to lack of suitable functionality on the backbone, polymers are generally categorized as hydrophobic; however, they are sufficiently stable. Also with appropriate functional groups. modification undergoes controllable transitions to exhibit highly sensitive conductivity, which is highly anticipated in water treatment applications. Nevertheless, zwitterionic polymers that possess a pair of oppositely charged functionality groups in their repeating units satisfy the hydrophilicity, ionic conductivity, and water-holding capacity useful for interacting with contaminants and water molecules [55–58].

Therefore, the zwitterionic polymeric structural molecular unit generally carries a pair of cationic and anionic groups. Table 5.1 gives a detailed summary of cationic, anionic, and zwitterionic polymers and charge distribution on polymer backbone. Here, the cationic groups include protonated amino, pyridine, and quaternary ammonium units, while anionic groups include sulfonate, carboxylate, and phosphate. On the other hand, based on the charge distribution on polymeric backbone, zwitterionic polymers can be sub-divided into two different structures, namely polymer with a positive and a negative charge on the same side chain, and in another case, the charges are localized to opposite side distributed alternatively across the polymer chain. Nevertheless, in some cases, distribution of charges on the polymer backbone is complex and random, which does not fall into any of the previously defined categories [59–69].

Conventionally, anionic polymers are widely used in commercially important sectors like sand and gravel production, mining, drilling, and other mineral-based liquid-solid separation industrial applications. On the other hand, cationic polymers have a positive charge and are generally used for sedimentation settling of organic solids such as animal waste or vegetation. Cationic polymers are also used in commercial applications like dredging, agriculture, food processing, municipal wastewater treatment plants, and dairy applications.

In general, zwitterionic polymers can be prepared by optimal mixture of monomer-containing anionic and cationic functional groups. For instance, the initial preparation of polymerizable betaine monomers can be used in appropriate molar ratios followed by copolymerization with other functional monomers, which is a simple approach to synthesize zwitterionic polymers. Also, a simple process of grafting desired zwitterionic functional groups onto existing target polymer backbone chains is also a widely followed approach. There are several factors playing important roles in the synthesis of optimal zwitterionic polymer such as electrostatic attraction and repulsion interaction forces, pH, and salt ion in the

Table 5.1 Common cationic and anionic groups in zwitterionic polymers. Structural characteristics of zwitterionic polymers.

Type of Charge Structures of Charged Groups	Cationic	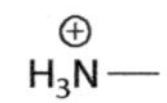 Amino	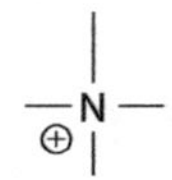 Quaternary Ammonium	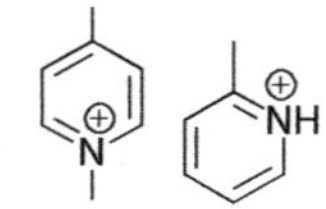 Pyridine
	Anionic	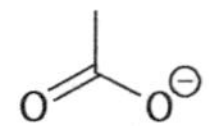 Carboxylate	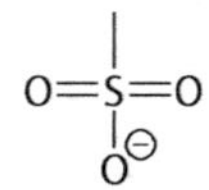 Sulfonate	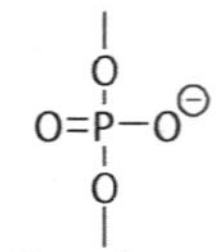 Phosphate
Structure of Charged Group and Typical Structure Description	Distribution of Negative and Positive Charges Zwitterionic Polymer Examples	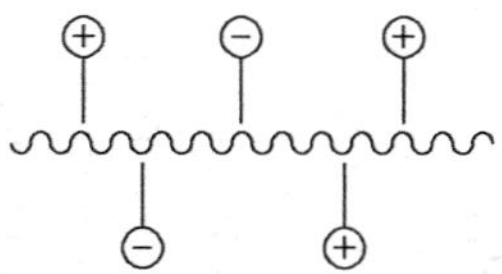	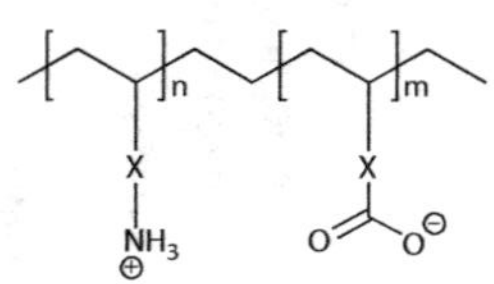 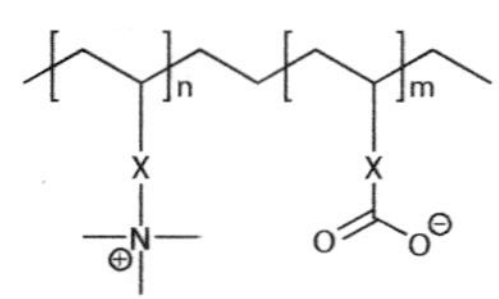 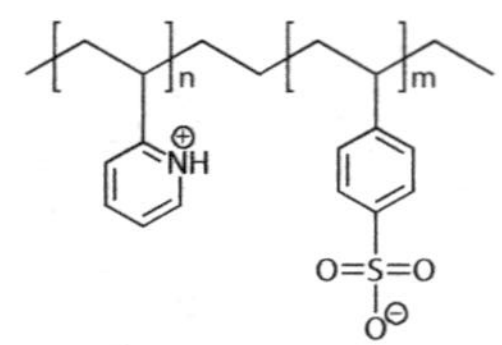	
		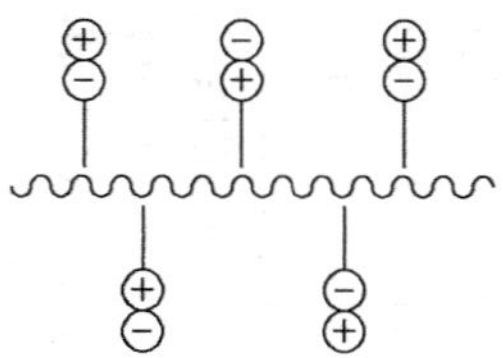	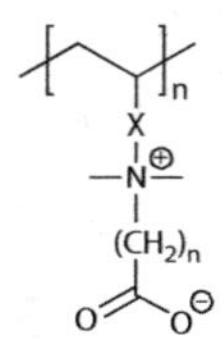 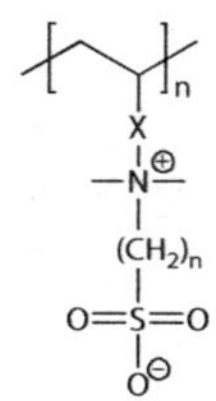 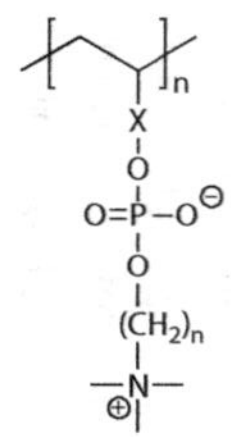 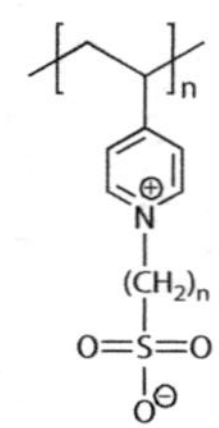 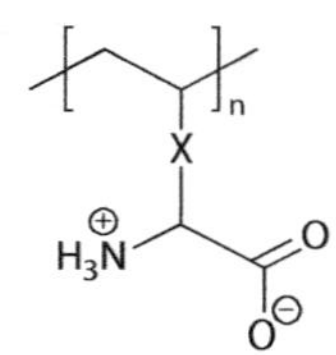	

preparation environment. More significantly, sulfonic acid betaine-type zwitterionic polymers have relatively good chemical stability. However, their properties and structure are relatively pH-sensitive. The zwitterionic polymer long-chains adopt a folded or coil-type conformation through electrostatic attraction that is initially curly and then becomes spherical, whereas the zwitterionic polymer long chains expand through electrostatic repulsion. Interestingly, these attractive and repulsive interactions give unique and robust mechanical characteristics to zwitterionic polymers. At the same time, the structural configuration and altering characteristics of the rich cationic and anionic groups on the polymer chain impart much-needed superior hydrophilicity to the zwitterionic polymers. Interestingly, when both positive and negative charges are located on the same side polymer chain, the zwitterionic polymers tend to show a greater hydrophilic nature due to the accumulation of larger charge density of the cationic and anionic groups. All these characteristics in zwitterionic polymer provide specific impact on the hydrophilicity, charge density, and ionic strength that are extremely useful in water treatment applications.

Even though membrane-based technology has been widely used for years now as a commercially viable and robust desalination process that mainly uses polymer-based film or membrane as separation media. Membranes that are prepared from either organic, inorganic, or both composite origins have been widely used in both porous and dense formats for separation and purification of wide range of contaminants. In particular, dense or nonporous membranes have been used in diverse applications for simultaneous water treatment and alternative energy storage and conversion applications. These applications include electrodialysis (ED), reverse electrodialysis (RED), reverse osmosis (RO), forward osmosis (FO), pressure-retarded osmosis (PRO), and fuel cells. In majority of these cases, the ion/salt transport properties mainly depend on the polymer membrane or films that determine the separation efficiencies and energy conversion/storage application performances. Zwitterionic polymers contain both a negatively and a positively charged moiety on the same pendant group handing on parent backbone while maintaining overall charge neutrality. Zwitterionic polymers have been now recognized as a new class of materials that play a significantly important role in treating charged contaminants or ionic species [70].

In this direction, initially, zwitterionic polymers have been used as a new generation of antibiofouling desalination materials due to favorable resistance to charged solutes or contaminants like protein, colloidal suspensions and biological moieties (bacteria) attachment. For instance, researchers in their initial exploration used interfacial polymerization technique to immobilization of zwitterions via crosslinking on an existing commercial nanofiltration (NF) filtration membrane to achieve antifouling properties. This approach not only improves surface characteristics of conventional NF membrane but also improves desalination performance in long-term operations. Further studies have directly used zwitterionic monomers to prepare novel zwitterionic NF membranes through interfacial polymerization. Membrane performance in seawater desalination was greatly achieved simultaneously retaining antifouling properties of the membrane. However, zwitterionic polymeric moieties may suffer a serious breakdown in treating complex contaminants in which oppositely charged pollutants may influence the separation, deposition, and fouling process in competition with each other. Therefore, even though zwitterionic polymers have shown great potential as antifouling coating layers on membrane surfaces, the salt separation and transportation properties of zwitterionic polymers need further studies to understand overall separation and transportation mechanism.

Generally, water and salt or ionic species transportation through dense or nonporous polymer membranes of films is understood following solution-diffusion theory. In dense polymer membrane case, polymer structure and their film-forming properties play a vital role in the water uptake characteristics, which consequently strongly influence salt and water wettability, sorption, diffusion, and permeation, which lead to further separation properties. On the other hand, several charged polymers used as membranes for desalination purposes contain ionizable or charged functional groups on the polymer long chain. Therefore, it is significant to evaluate and understand the impact of presence and extent of charge on transport properties' membranes in separation and purification applications.

Nevertheless, polymers with charged and uncharged long chains exhibit significantly different salt/water sorption, diffusion, and separation properties. Interestingly, uncharged polymers that do not contain functional groups but swell in water can exhibit water/salt transport properties following a simple partitioning mechanism. On the contrary, charged polymeric network containing fixed charge groups that are covalently interconnected to the polymer backbone can ionize when the polymeric network structure is swollen with water. Therefore, charged polymers exhibit significantly different ion sorption and diffusion behavior compared to that of uncharged polymers. Initially, the ion sorption in an uncharged polymeric network proceeds with a simple partitioning mechanism and majorly depends on the ion diffusion properties and salt concentration, which can be mainly correlated to osmotic de-swelling phenomenon. Whereas, ion sorption in a charged polymer proceeds with both ion exchange phenomenon and simple partitioning mechanisms and ion diffusion, on the other hand, is strongly influenced by electrostatic interactions within the polymer network structure.

Zwitterionic polymeric membranes carry high charge density of both negative and positive charge groups. However, they are overall charge-neutral. Even though it has been established that water uptake properties are greatly influenced by the presence of charges, however, it is not clear from limited studies that the water and/or salt transport properties in zwitterionic polymers are similar in charged or uncharged polymers [45, 71–75].

Several attempts have been made in the recent past to prepare zwitterionic polymer membranes having both antifouling characteristics as well as separation performance. Among many, zwitterionic polymer membranes used for desalination purposes can be divided into three generations, namely (i) 2-hydroxyethyl methacrylate (HEMA)-based, (ii) PEGylated-based, and (iii) zwitterionic-based materials. HEMA-based materials are rich in hydroxyl (—OH) moieties, which help in generating a tight hydration layer, which considerably helps in reducing the possibilities of organic or hydrophobic foulants reaching the membrane surface layer. However, antifouling properties of these hydroxyl groups may reduce or vanish once they interact with complex foulant or solute media like plasma, human blood serum, and other biological moieties. These types of complex physicochemical interactions on the membrane surface between the protein and the functional groups of the surface greatly affect their fouling behavior in long term. The second generation of zwitterionic polymer compositions originated from PEG-based or oligo (ethylene glycol) (OEG) material. This class of zwitterionic systems has overcome many of the limitations of first-generation systems demonstrating an excellent antifouling behavior even again biomacromolecules like protein and cell deposition and adhesion onto the membrane surface interface due to the close hydration layer formed around OEG chain. However, the major limitation arises with chemical scalability with PEG-based materials when oxygen and transition metal ions are found in biochemical environment. Both HEMA and PEG zwitterionic materials have shown a moderate antifouling characteristic performance. Interestingly, in both HEMA- and PEG-based zwitterionic systems, once the positively charged solute or contaminant came in contact, the attractive forces led to the fouling enhancement. Meanwhile, a good fouling resistance was achieved for the negatively charged solutes or pollutants due to the existence of electrostatic repulsion, thus making zwitterionic-based material an interesting stock of materials in the development of third-generation antifouling membrane material [76–78].

Zwitterionic materials can be categorized into third-generation antifouling materials. Most common use of zwitterionic materials has been in membrane surface modification purposes in which charged moieties have shown promising results in enhancing surface hydrophilicity. This helps in enhancing both antiorganic and antibacterial fouling properties of the membrane as a result of formation of a strong hydration layer through electrostatic and hydrogen bond interactions. The typical zwitterionic material systems are sulfobetaine methacrylate (SBMA), carboxybetaine methacrylate (CBMA), and phosphobetaine methacrylate (PBMA) as shown in Figure 5.3. These zwitterionic systems contain both cations and anions on the same side polymeric chain, which helps in

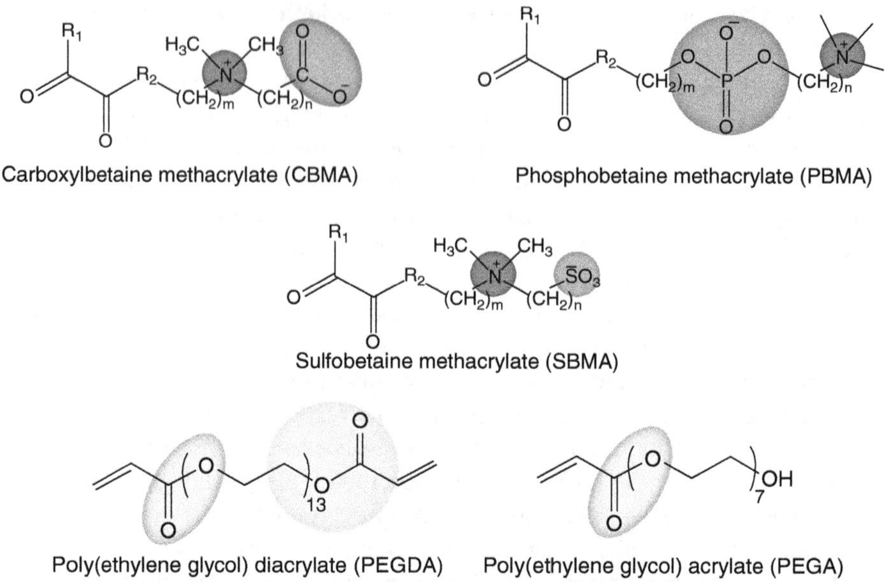

Figure 5.3 Chemical structures of important antifouling zwitterionic material used in the preparation of antifouling surfaces namely, PBMA, SBMA, and CBMA systems and list of co-monomers and crosslinker used in the preparation of UV crosslinked polymer films or membranes for antifouling applications.

preventing surface adhesion of either negatively or positively charged solutes or foulants [79–82].

On the other hand, impact of zwitterionic groups on salt and water transport properties has been studied using a series of zwitterionic materials. In such an attempt, UV crosslinkable membranes containing sulfobetaine and carboxybetaine groups have been used along with poly (ethylene glycol) diacrylate (PEGDA) to attain crosslinked zwitterionic systems. Further, zwitterionic systems were compared with a crosslinked poly(ethylene glycol) acrylate (PEGA) membrane as a control. The chemical structures of the two widely used zwitterionic co-monomers (carboxybetaine methacrylate, CBMA and sulfobetaine methacrylate, SBMA) and the neutral co-monomer, poly(ethylene glycol) acrylate (PEGA) have also been given in Figure 5.3. Through the process of simultaneous polymerization and crosslinking two zwitterionic polymers, poly(carboxybetaine methacrylate) (PCBMA), and poly (sulfobetaine methacrylate) (PSBMA) and a neutral polymer, poly(ethylene glycol) acrylate (PEGA) were prepared following UV photopolymerization. This simple route of UV-photo-polymerization enabled obtaining optimal crosslinking density of the zwitterionic polymers via controlling the ratio of co-monomer to PEGDA in the product mixture [83–86].

In all these co-crosslinked zwitterionic systems, studies revealed that the water uptake has substantially increased with higher concentrations of co-monomer content due to the incorporation of charged moieties within the polymer and optimal crosslinking density also retained much-needed polymer chain flexibility in the membrane that led to enhanced polymer swelling, thereby enhancing the water transportation. The influence of type of co-monomer on water uptake in the membrane was also evident that was prepared using similar crosslink density and co-monomer concentrations. While PSBMA membranes exhibited the highest water uptake, whereas PCBMA and PEGA followed the trend. Interestingly, the different water-binding capacities of the monomers are the major characteristic behavior of the overall system. Here, the PEG ether groups formed hydrogen bond interactions with water via the lone pair electrons on the oxygen atoms. Whereas, the hydrogen bond strength is less than that for the polymers that contain highly polar charged groups. As evident that the sulfonate group is more polar than the carboxylate group, so hydrogen bonding between water and the sulfonate group is the strongest of the three materials. As a result, water uptake is highest in PSBMA and smallest in PEGA.

In all these cases, the hydraulic water permeability of the three cross-linked membranes was measured as a function of co-monomer concentration and water content. Results reveal that the hydrated film showed the thickness range from 110 to 170 μm, which followed the trend of water permeability as a function of the water sorption capacity and co-monomer concentration. Interestingly, water permeability also showed increasing trend with the increased concentration of co-monomer content. The relative values of the water permeability with respect to the type of co-monomer also showed a similar trend to the order of water uptake, viz. for a listed co-monomer content, water permeability decreases in the order SBMA > CBMA > PEGA [87].

Similarly, zwitterionic polymer composite is also designed in membrane-based separations through a novel synthesis method. Here, the amphiphilic copolymer poly(arylene ether sulfone-co-sulfobetaine arylene ether sulfone) (PAES-co-SBAES) was prepared, which was then blended with native polysulfone (PSf) to prepare free-standing membranes. Further, composites of PSf/PAES-co-SBAES blend membranes with different zwitterionic SBAES segment contents and concentrations were fabricated via the nonsolvent-induced phase separation (NIPS) process. As obtained membranes showed enhanced porosity in the support layer and drastically improved surface hydrophilicity due to the incorporation of hydrophilic SBAES zwitterionic segments. In all the cases, the water permeability and antifouling ability of the PSf/PAES-co-SBAES blend membranes were both considerably improved to 2.5 L/m^2/h/bar with 94% flux recovery ratio, respectively, whereas, salt rejection was achieved at a high level more than 98% for feed containing accelerated high concentration of chlorine content. The study gave an interesting outlook into the preparation of commercially successful polysufone-based composite membranes using active zwitterionic species useful for desalination applications with high salt rejection and considerably high permeate flux. These efforts in making suitable desalination membranes for future challenges at a commercially viable scale shows the great potential of zwitterionic polymer systems as effective water and salt separation barriers. The strategy of using commercially successful PSf as basic matrix on which zwitterionic active layers act as performance enhancer is potentially viable solution for achieving efficient desalination at a large scale [88].

Similarly, PSf-based asymmetric thin-film composite forward osmosis (TFC-FO) membranes were fabricated via second interfacial polymerization (SIP) processes involving polyamide (PA) followed by secondary layer coating of zwitterionic moieties, a zwitterionic monomer N-aminoethyl piperazine propane sulfonate (AEPPS), on the PA active layer. Unlike traditional blending, compositing and in-situ modification, a simple SIP approach gives precise control over selective layer coating thickness and physico-chemical property optimization. This method also overcomes the major disadvantage of conventional approach in avoiding the influence of substrate, and uneven distribution of selective layer and its structure. In addition to this, secondary coating structure acts as a buffer layer during the wastewater treatment application making it an exceptionally good antifouling and antibacterial barrier. Here, electrostatic interaction between zwitterion layer on the PA surfaces and water molecules remarkably improves the surface wettability by lowering the water contact angle from 80° to 15° and oil contact angle from 110° to 160°, which, in turn, confirms the introduction of superhydrophilicity or superoleophobicity, respectively. Also, having minimal barrier coating on PA surface significantly reduces the resistance to the permeability of water in which zwitterionic thickness was optimized from 1.06 to 0.66 mm, resulting in substantially improved water flux and permeabilities. Moreover, the presence of zwitterionic layer on the PA selective layer substantially reduced the affinity toward negatively charged bovine serum albumin (BSA) and the positively charged lysozyme (LYZ). Further accelerated fouling study using model organic foulant like sodium alginate (SA) revealed that the water permeability characteristics of the membranes remained stable during long-term FO operations indicating considerable improvement in antifouling characteristics of the PSf/PA FO membranes. Further, biological waste treatment involving bacterial attachment studies on FO-modified membrane revealed ten times lower affinity for E. coli on zwitterionic-modified membrane than the unmodified membrane.

Similarly, transport of salt/water, water, and ionic species through zwitterion-modified CNTs in PA-TFC membranes (Z-CNTs) were studied using both experimental and atomically detailed modeling techniques. In this study, zwitterion groups having the chemical structure of $-COO-(CH_2)_3-N^+(CH_3)_2-(CH_2)_2COO^-$ was used to modify the CNT surface, which was further incorporated into PA layer of composite Z-CNTs/PA membrane. Desalination studies using Z-CNTs/PA showed a flux of 20,000 GFD with a pressure drop of 530 psi. This was 3 orders of magnitude greater than our previously reported similar approaches [89].

In a completely different approach, commercially sourced technologically important ultrafiltration (UF) polyethersulfone (PES) membranes were surface-modified with zwitterionic polymer 3-(3,4-Dihydroxyphenyl)-l-alanine (l-DOPA) layer to significantly improve membrane antifouling characteristics for the treatment of shale oil wastewater. Commercial PES-UF membranes were surface modified by dipping in l-DOPA solution following a dip coating technique. This study gave detailed membrane characterization both for physicochemical characteristics as well as membrane separation performances. Membrane testing experiments revealed a substantially improved permeate flux obtained due to presence of zwitterionic layer in comparison to uncoated commercial membrane. Also, antifouling behavior confirms the fouling resistance phenomena existed at the interface of PES and influent solute/pollutants. Further, experiments were conducted to understand the thickness of the zwitterionic layer on the commercial PES membrane surface in which varying coating times from one day to three days were conducted. Amongst the three-day polymer, 3-(3,4-Dihydroxyphenyl)-l-alanine (l-DOPA)-coated PES-UF membrane showed improved flux performance and greater fouling resistance. However, flux decline was noticed for uncoated PES membrane by 38.4%, whereas flux decline was around 16% for the three-day coated membrane after the 5-hour wastewater feed run. This was attributed to a negatively charged surface developed on the FO membranes after the zwitterionic coating [90].

5.5 Polymer-Based Coagulants and Flocculants

Generally, flocculation and coagulation (flocculants and coagulants) agents play a vital role in the typical wastewater treatment and groundwater purification process. The process follows four important steps, namely (i) Step- 1: in the first step, coagulants or flocculants are added to the contaminated water system. The contaminated water and coagulation agent rapidly mix until coagulation occurs. (ii) Step-2: slow mixing of agents takes place in which the aggregated particles initiate flocs formation. Upon uniform mixing, overall mixture is transferred to a sedimentation tank. (iii) Step-3: at this point, flocs start settling down to the bottom of the sedimentation tank and the clarified water is floated to the top. In the next (iv) Step-4: a filtering system facilitates the separation of clarified or purified water from flocs or other sediment particles. Finally, in the last (v) Step-5: purified water is disinfected and discharged for immediate use or stored for later use. Overall, flocculation and coagulation processes facilitate the optimal conditions to treat and eliminate undesired suspended and colloidal pollutants from groundwater and wastewater sources [91].

Polymers are unique class of macromolecules available in diverse physicochemical forms and states. Polymer-based films, membranes, nanomaterials, functional moieties, and so on have been widely used in recent years for water treatment applications. Polymers separate dissolved, dispersed, or suspended solids from liquid waste systems through a process called flocculation. Even though majority of polymers have been classified as hydrophobic plastics, however, biopolymers and specialty polymers show considerably high hydrophilicity, which makes them suitable for water treatment applications. For water treatment purposes, it is highly desired to have water-soluble characteristics in selected polymers that help in close interaction with water molecules and dissolved suspended pollutants. Therefore, polymer-based flocculants generally are water-soluble polymers referenced polymer flocculants. Flocculation using polymers is the ability of a polymer to break the colloidal characteristics of wastewater resulting in aggregate formation of dissolved or suspended solids leading to sedimentation process, which is central to their role in wastewater treatment. Polymer-based flocculants can be effective alternatives on their own and can be efficient and highly impactful in converting wastewater into highly concentrated sludge and clear reusable water when combined with an appropriate coagulant. Most of the water-consuming industrial processes generate wastewater that is highly characteristic of colloidal waste and suspended particles carry negative electric charge on their surfaces. Colloidal systems with evenly distributed negative charge on their surface make them stable under colloidal dispersion conditions. Similarly, positively charged colloidal systems afloat in the stable suspension form for days, months, and in some cases, even years, making them some of the worst sources of contamination in the environment.

For instance, simple muddy water can take several days to form aggregates of sand and gravel. Whereas, industrial wastewater such as leather industry, distillery industry, and sugar industry wastewater requires years to settle their colloidal pollutants. Therefore, addition of flocculant and coagulant to muddy water or industrial wastewater initiates a coagulation, followed by flocculation resulting in sedimentation of highly concentrated sludge. Here, in this case, polymers play an important role in neutralizing surface charge or unbalancing the repulsive forces resulting in aggerate formation as shown schematically in Figure 5.4. Once neutralized or unbalanced, suspended foulant particles come together to form larger aggregates of particles called micro-flocs or pin flocs, which further grow to experience the gravitational forces. In the end, macro-floc formation results in sedimentation of overall colloidal system leaving behind clear water to be extracted as reusable water [92, 93].

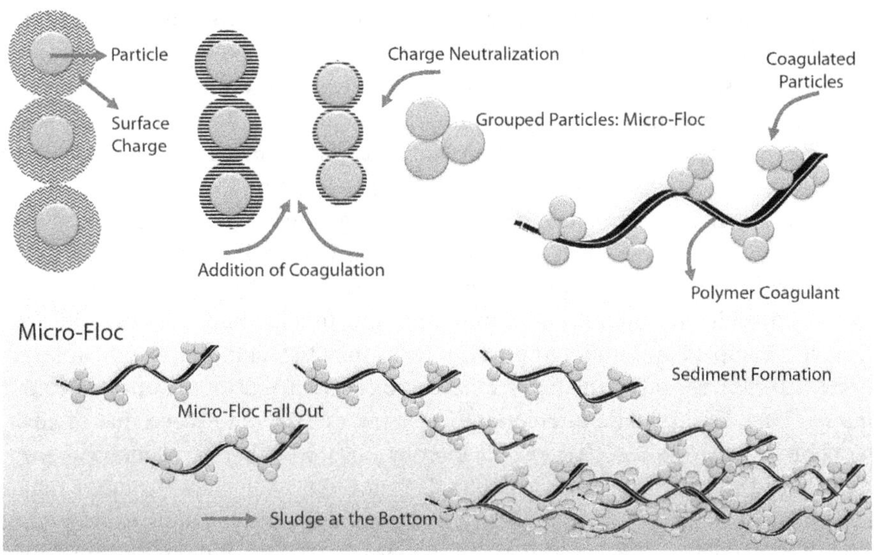

Figure 5.4 Schematical representation of polymer used wastewater treatment following flocculation, coagulation, and sedimentation leading to clarification of reusable quality water and highly concentrated sludge formation.

Therefore, polymers with long chains, high charge density, and multi-functionality characteristics play a vital role in wastewater treatment. Polymers also help in thickening of sludge upon sedimentation process in addition to separation of solids via floc and coagulation steps. Various polymers have been used for dewatering sludge, which helps in handling and disposal. Change of state from liquid to solid waste has significant impacts on waste management, handling, disposal cost as well as tipping costs. International and industrial norms are stringent in handling and managing industrial waste, which is generally defined by the standards solid or liquid is designated as EPA Method 9095B (SW-846 Test Method 9095B: Paint Filter Liquids Test) or "the paint filter test."[94] In this standard waste test procedure, an industrial waste sample in the form of sludge is placed in a standard 60 mesh conical paint filter for 5 minutes. If liquid seeps through the mesh, it is considered liquid waste, whereas if no liquid passes through the mesh filter, the sludge is deemed solid waste and permitted to dispose of which costs far less than disposing of liquid waste.

This section further elaborates on advances in cationic and anionic polyelectrolyte (CPE, APE) used for treating different types of wastewater. CPEs have played an important role in effectively coagulating and flocculating impurities in wastewater from industrial wastewater like papermill, mining, textile, farming industries, and domestic human usage. The CPEs also have shown great antifouling and antibacterial properties. Most studies on CPEs have focused on designing and applying polymer flocculation-charged coagulants to remove contaminants like soluble dyes, natural organic material (NOM), industrial pollutants in oil sands, tailings ponds, contaminants in papermill wastewater, and so on. Additionally, it is important to understand the physico-chemical dynamics of polymer-based flocculants/coagulants in flocculation and coagulation mechanisms comprising natural anionic and cationic polymers in water treatment [93].

The CPEs are polymers having positively charged functional moieties and anionic counterions present on each repeating unit or a large portion of units. Common functional active groups present in CPEs include polymer backbone decorated with phosphonium, ammonium, and imidazolium groups. CPE systems are complex and undergo physical crosslinks through electrostatic interactions between various functional moieties often referred to as supramolecular interactions. Electrostatic interactions are weak and do not exceed that of covalent bonds, nevertheless, electrostatic interactions impart better mechanical properties and thermal transitions compared to nonionic polymers. Another major advantage of CPEs is their adaptability, and potentially create tailor-made system, tuning molecular weight, and positively charged moiety, inducing optimal anionic counterion or considerably high charge density, which makes them appropriate for various wastewater treatment applications. For instance, CPEs are effective against microbial-contaminated wastewater in which antimicrobial and nonviral gene delivery properties of these materials play a vital role in obtaining high-quality reusable water from wastewater sources. For this purpose, ammonium-, imidazolium-, phosphonium-, and zwitterion-containing polymers have been explored for nonviral gene delivery and to discover structure–property relationships. In this new preparation approach, multiple post-polymerization techniques have been adopted to modify phosphines, amines, and imidazoles with alkyl halides to obtain cationic groups with tailored charge density, hydrophilicity, water uptake property, hydration, and so on. These approaches significantly enhanced CPE's flocculation and coagulation efficiencies when used for water treatment purposes [95–97].

Polymer-based complexes used for wastewater treatment via flocculation and coagulation are desired to be water-soluble and allow interactions with solutes, pollutants, or impurities. However, effect of flocculant/coagulant in wastewater treatment depends on the positively charged moiety and anionic counterions dissociate, which enables the polymer backbone to repel itself, type of wastewater, and pH, that influence adopting a ridged-rod conformation and forms the interactions with other negatively charged moieties. Figure 5.5 depicts the CPE backbone before and after dissociation of counterions in the presence of water. However, the degree of the dissociation of CPEs depends on the pH of the water. When CPEs dissolve in the optimal solution pH, the salts dissociate resulting in the strongest electrolyte possible.

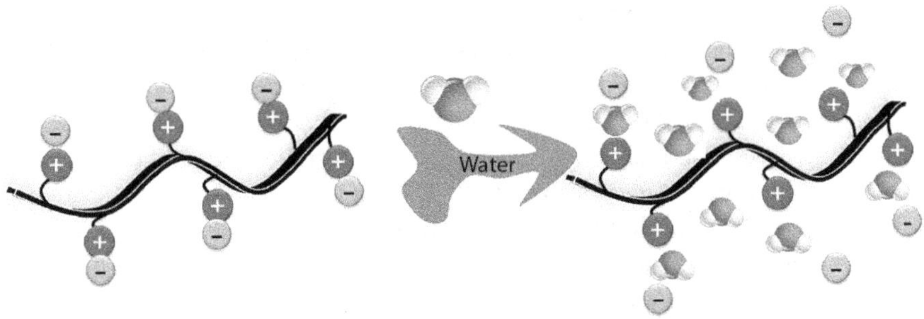

Figure 5.5 Schematical representation of cationic polymer electrolyte (CPE) in its unhydrated state shown left side and CPE in dissociation with the counterion occurs at right side.

Interestingly, synthetic-origin CPEs can be directly used as primary flocculants and also sometimes as substituents for secondary coagulation and these agents occasionally outperform inorganic salts and natural CPEs. Natural CPEs generally used as primary flocculants. CPEs upon flocculation yield stable and filterable aggregated contaminant impurities leaving behind high-quality clear water making it an affordable water treatment technique. In comparison to inorganic salt-based coagulation, polymer-based CPE systems leave fewer dissolved solids in the final product making them highly efficient flocculant and coagulant. This makes them work efficiently with 90% less (polymer) dosage to achieve the same flocculation efficiencies of inorganic systems. On the other hand, natural CPEs offer biodegradable properties in addition to efficient clarification qualities. However, they need to be induced at higher dosages in comparison to synthetic CPEs. Nevertheless, synthetic CPE-based flocculants suffer from several disadvantages agents such as a narrow optimal dosage range, potential toxic precursors including monomer contamination, and are nonbiodegradable in nature. Nonetheless, there are more advantages associated with synthetic CPE flocculants that outweigh natural CPEs. Therefore, the majority of impurities found in either clarified water or concentrated sludge post-flocculation of wastewater generally originate from the utilization of synthetic CPE flocculants [98–100].

5.6 Functional Polymers in Organic Matter and Other Secondary Pollutants' Treatment

Use of freshwater and low-concentration saline water have been essential commodities used in farming activities, industrial manufacturing, and domestic use. Worldwide, on daily basis, billions of gallons of freshwater are being withdrawn from freshwater sources. According to some estimations, withdrawn water constitutes 14% of saline water sources, whereas 86% of used water originates from freshwater reservoirs. At the end of every use, the freshwater or saline water from all sectors including farming, industrial processes, and domestic results in medium to extremely high contamination, which is either directly discarded or partially treated into the environment. It is estimated that the total freshwater withdrawals in the near future will continue to increase and even may exceed by *ca* 20% from now until end of this century to cater for the increasing demands from industrial sectors and at domestic front due to ever expanding population. Therefore, increased consumption will exponentially produce contaminated water that requires more efficient, affordable, and sustainable water treatment solutions across the sectors and the world [101, 102].

The major contaminants in industrial wastewater and some geographically specific drinking water supplies present are NOM and dissolved organic matters (DOMs), respectively. NOMs represent major contaminants on surface and groundwater sources and are found in almost every natural water reservoir like lakes, wells, tanks, and other modern storage and pumping facilities. Nonetheless, all the common storage sites used for distribution of drinking water for urban and rural consumption can be counted for NOM contamination. NOMs originate from various natural sources either from plant or animal decaying. NOMs can also be either aquagenic, i.e. originating from water, or pedogenic, i.e. contaminated from soil. Geological and ecological systems combined with weather patterns, such as floods and droughts, govern the types of NOMs present in water sources. However, typically major sources for NOM contamination in groundwater occur from decaying plants, plant trimmings, animal waste, and manure, which with time leach organic materials such as carbohydrates, biopolymers, and proteins to water sources. Generally, presence of substantial amounts of NOMs in bad odor, color, and taste. Further, NOMs act as a host or substrate for secondary pollutants such as heavy metals that travel with organic contaminants over a period of time resulting in potentially biological contamination growth. The effective technique to remove NOM contamination greatly depends on the nature of organic contamination and quality of NOMs present in contaminated water. Nonetheless, sand filtration, flocculation, coagulation, and sedimentation techniques of purification water offer economically viable options for NOMs and secondary impurities' removal from groundwater sources [103, 104].

Interestingly, black water containing high concentrations of NOMs post flocculation and coagulations can be used directly as a value product in irrigation, and the clarified water can be used for industrial cooling towers and community toilet water purposes. On the other hand, sewage wastewater that is highly contaminated with NOMs of animal and human manure is separated into two main groups: (i) Greywater originates from kitchens, sinks, showers, and laundry applications, whereas (ii) Blackwater originates from flushed toilet water containing urine, feces, and toilet paper. On daily basis, vast population around the globe produces billions of liters of sewage and municipal wastewater and many towns/cities municipal authorities use multiple sewage wastewater treatment infrastructures to treat and recycle high-quality water. However, blackwater portion is relatively in less volume amongst sewage wastewater. However, this category of wastewater poses a serious health risk. Nevertheless, both black and grey wastewater require more comprehensive, complex, and efficient purification techniques such as anerobic treatment and UV light sterilization to either kill, separate, and eliminate pathogenic contamination of various origins. On the other hand, greywater constitutes nearly 70% of municipal and sewage wastewater and provides a great opportunity to recycle and reuse the vast sources through appropriate treatment techniques [105].

On the other hand, wastewater from large freshwater-consuming industrial segments like textile and sugar industries represents some of the most complex categories of wastewater to treat. For instance, the complexity of substances and chemicals present in textile industry wastewater such as dyes, dispersing agents, surfactants, heavy metals, and conditioning agents complicates wastewater treatment methodologies and due to the co-existence of multiple constituents demands a multi-step purification process. Soluble textile dyes contain carcinogenic chemicals and heavy metals and can be very hazardous upon release into environment without treatment to human, animal, aquatic life, and environment at large. Dyes also form colored layer on the freshwater surfaces that disrupts the biosphere by preventing the penetration of light from entering damaging aquatic life and underlying soil. For years now, NF technique has been successfully used for the removal of soluble dyes from various wastewater sources. However, increasingly, NF membranes suffer from fouling phenomenon and lack long-term sustainability. Therefore, to reduce freshwater contamination from organic matter such as dyes and the risk of biological gel layer growth during the treatment process, additives such as anionic, cationic, or natural polymers are recommended in the flocculation process.

Similarly, paper-manufacturing industries in each process unit produce an estimated 20 thousand to 60 thousand gallons of wastewater per ton of paper production that contains pollutants ranging from pulp, inorganic salts, dissolved solids, biomolecules, organic solvents, and organic matter. The paper-making process starts with the pre-treatment step wherein removal of dirt, debris from wood pulp, and other suspended particles will be achieved during subsequent boiling and further washed for bleaching. Further, pulp bleaching is treated with a series of additives that define their final shape, color, thickness, and texture such as organic compounds, inorganic dyes, and curing agents. Wastewater produced from the presence of these series of pollutants during paper-making process causes severe water coloring, which promotes slime growth resulting in toxic effects on aquatic life upon disposal. Here, due to stringent regulations, wastewater from pulp and paper industrial processes is required to meet set standards before wastewater is released into the aquatic reservoirs of environment. In such a case, even though CPEs do not function actively as primary flocculation agents, they are being used as secondary or tertiary coagulants to treat pulp-paper industries by removing organic materials and inorganic dyes from the wastewater.

Similarly, wastewater from mining activities produces humongous amounts of wastewater that pose serious challenges to water purification infrastructure. For instance, Kaolin clay is a major pollutant in mining wastewater, and oil sand tailings are the major wastewater generated during bitumen extraction [106–108]. Kaolin clay is generated from the mineral kaolinite, which produces clay suspensions having turbidities of 360 nephelometric turbidity units (NTUs) that contain mostly negatively charged particles with an average pH of 5. Therefore, CPEs are used as effective flocculants in treating mining industry wastewater. Here, CPEs are used as effective flocculants to separate kaolin clay particles during treatment process. On the other hand, during bitumen extraction, the sand and other oils present in the reserves are millions of gallons of wastewater that is contaminated with fine clay suspensions generated. During the treatment of these oil sand tailings, polyacrylamide (PAM) and cationic polyacrylamide (CPAM) are effectively used for the flocculation separation of fine clay suspensions to recover clarified high-quality water from oil sand tailings [106, 109, 110].

The CPEs are widely used effectively to treat various types of organic contaminated wastewater through flocculation process to make use of product water as repurposed water and, in worse case, discarded safely into environment. Also, CPEs are increasingly being used for harvesting biological substances called microalgae, which is an important source for biodiesel production, biological value-added materials, and commercially important oils. Biodiesel is a new economic commodity and is gaining increased prominence as a renewable energy source that is expected to replace some percent of fossil fuels in the near future. However, conventionally followed current biodiesel harvesting processes are expensive, complex, and commercially impractical. Here, magnesium, ferric, and calcium salts help in the flocculation of microalgae as a conventional chemical flocculation process. However, this yields results when inorganic salt-based flocculants are used in significantly bulk amounts. On the other hand, polymer-based CPEs flocculate microalgae contamination cultured in freshwater with a lower dosage and generate more stable aggregated cells. Interestingly, inorganic salts present in marine microalgae harvesting inhibit CPE flocculation as a result of charge screening and thus do not effectively aggregate the microalgae cells. CPEs flocculate microalgae effectively because the algae have negative surface charges.

5.7 Mechanism of Polymer-Based Flocculants and Coagulants

Flocculation and coagulation follow similar mechanisms in separation of suspended and dissolved pollutants in wastewater. However, both differ in aggregation mechanisms. Flocculation initiates when a polymer simultaneously adsorbs multiple particles and aggregates the pollutant material and the resulting aggregates are known as flocs, whereas coagulation process introduces small, highly charged molecules known as coagulants into wastewater to destabilize the charges on particles, colloids, or oily materials in suspension. In both processes, a change in pH, temperature, and concentration plays a vital role. However, extent of charge in the coagulation enables rate of coagulation where particles aggregate together.

Interestingly, large macromolecules like polymers act as bridging agents for aggregation of large amounts of suspended particles in a quick time. Here, bridging occurs when high molecular weight (>107 g/mol) polymer-based CPEs facilitate the adsorption of several particles resulting in aggregate formation through the binding of particles as a result of electrostatic forces, which further leads to the formation of a floc. Figure 5.6a illustrates examples of polymer-bridging mechanisms through which two CPE chains interact with three particles resulting in aggregation of pollutants. However, for flocculation through bridging to

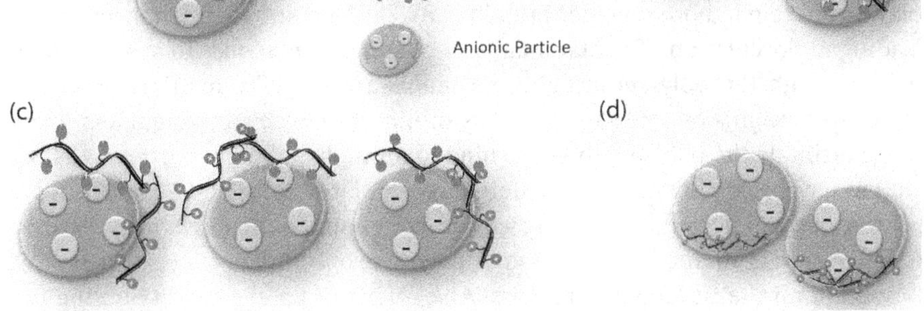

Figure 5.6 The process of coagulation and flocculation involving various types of interaction and separation mechanisms of cationic polyelectrolytes in which (a) polymer bridging; (b) restabilization; (c) charge neutralization; (d) electrostatic patch, gives detailed understanding on the treatment of various types of pollutants.

take place, an optimal dosage of CPEs with specific charge density range is required, which is known as the highest flocculation efficiency (FE). For instance, Figure 5.6b illustrates a charge balancing effect called restabilization in which there is an excess of positively charged species from CPE system over negatively charged particle, which prevents the CPEs from binding multiple particles. Therefore, success of CPE flocculation through polymer bridging mainly depends on molecular weight CPEs and their optimal dosage [111–114].

On the contrary, the alternative coagulation is also explained through the mechanism known as charge neutralization in which the primary mechanism occurs through the involvement of low molecular weight and high charge density of CPEs. Figure 5.6c illustrates the charge neutralization in which CPEs bind with negatively charged pollutant particles, thereby neutralizing the total charge of the media. However, this mechanism restricts the repulsive forces between CPES and negatively charged particles facilitating coagulation over a period of time. Whereas, electrostatic path described in Figure 5.6d is a type of charge neutralization process in which the CPE acts as a patch on the pollutant particle surface, which restricts repulsion phenomenon between particles. However, an electrostatic patch is different from general charge neutralization mechanism as it applies an overall charge neutralization of the target particle, but negatively charged sites remain accessible as CPEs only cover a small surface area. Therefore, the large accessible surface area accessible to media enables other particles to absorb the patches resulting in enhanced FE. For all practical purposes, the measurements of zeta potential for both wastewater and flocculation agent in treatment solution give the degree of charge neutralization. Generally, a zeta potential of ~0 ± 0.1 mV implies a neutral solution that indicates that the cationic groups neutralized the anionic particles, thus enabling the aggregate formation leading to complete coagulation.

Nevertheless, molecular weight of polymeric systems and their charge density play a vital role in both polymer bridging and charge neutralization mechanisms leading to flocculation of contaminants in wastewater systems. To achieve stable flocs through the polymer-bridging mechanism essentially requires minimum molecular weights of 1×106 g/mol. This minimum molecular weight and above range facilitate the polymer chains to bind to multiple pollutant particles. Several studies have used polymers with different molecular weights and architectures to achieve optimal flocculation during wastewater treatment. For instance, researchers have used both linear as well as graft polymers to compare effects of aggregate formation on the flocculation process. Also, effect of charge density on the floc formation has been well recorded in which charge density of the CPEs with a range of 10–30% enables binding simultaneously preventing the particle restabilization. However, charge densities over 30% have been noticed that not only result in restabilization but also potentially influence the repulsion of the particles creating stable suspensions. This particular range (10–30%) is very crucial as the complex constituents of wastewater where most of the suspended particles are neutral and positive and negative charges are equally distributed. Nevertheless, flocculation and coagulation are complex processes and mechanism of pollutant separation depends on many other factors such as the type of wastewater, pH, concentration, and the hydrophobicity/hydrophilicity of the CPEs. Also, as mentioned previously, CPE dissociation in the target waste mixture greatly depends on the pH of the system, which results in either production of a stronger or weaker electrolyte solution. Thus, stronger electrolytes possess a high potential to bind pollutant particles resulting in highly denser and more stable flocs. Other operational and physical factors that influence the flocculation process are the order and speed of addition of the CPEs into the wastewater mixture and physical characteristics of suspended particles such as the particle size of impurities.

5.8 Functionalized Magnetic Nanosorbents

Polymer-based flocculants and coagulants offer sustainable solutions for effective treatment of wastewater. Polymer-based materials also offer relatively low toxicity and abundant functionality and are available in different molecular weights and architecture, which can be tailor-made to suit their affinity toward target pollutants. The functional characteristics and multi-physico-chemical properties of polymers and their composite materials have shown great potential in water treatment applications in the form of adsorbents, coagulants, or flocculants as discussed in the previous sections [115, 116]. However, one of the major disadvantages associated with the polymer-based adsorption and sedimentation treatment process is the recovery of spent sorbents and agents. Presence of effective recovery methods would make them affordable and cost-effective [117]. Therefore, combining polymers-based functional materials with magnetic iron oxide nanocrystals results in formation of magnetic composites that are easy to recover and reuse using external magnetic field or stimulation. This gives overall process a sustainable, cost-effective, and clean separation for water treatment through adsorption or sedimentation process [118, 119]. Particularly, large-scale treatment process using adsorption technique turns into a realistic. Nonetheless, a rational modification of surface of magnetic materials, particularly at the nanoscale with the functional polymer, is essential to attain high adsorption capacity, pollutant specificity, and reusability. For instance, covalent immobilization of the chitosan or blend of chitosan/SA onto the surface of the magnetic nanoparticles is vital to ensure the successful recycling and reuse of the biosorbents.

Similarly, grafting a biopolymer onto a siliceous network that acts as a surface coating of Fe_3O_4 nanoparticles through an encapsulation process results in an organic/inorganic hybrid shell architecture. This encapsulation process involves tetraethyl orthosilicate (TEOS) as SiO_2 precursor in a hydrolysis reaction followed by condensation of a mixture resulting in a silicon alkoxide derivative of the biopolymer in the presence of magnetic particles that are dispersed as colloidal particles. Here, the preparation of the magnetically active biosorbents using the biopolymer κ-carrageenan with a silane coupling agent functionalized with isocyanate groups (3-isocyanatopropyl triethoxysilane-ICPTES) and then followed by a coating of the Fe_3O_4 nanoparticles. This method results in robust coatings and a higher degree of functionalization when compared to the traditional techniques of noncovalent coating and surface-grafting approach. This method is advantageous compared to conventional techniques to obtain reusable magnetically active biosorbents with enhanced adsorption capabilities. This robust preparation technique has paved the way for a series of magnetically active materials using various types of synthetic as well as bio-polymers from polysaccharides with unique ionic characteristics and versatile chemical functionalities that are efficient in the uptake emerging pollutants like active pharmaceutical ingredients (APIs), organic dyes, pesticides, and nonpolar organic solvents from water [120–123].

Interestingly, abundant biopolymer resources allow the preparation of magnetic nanosorbents having various chemical functionalities and tuneable surface charge and densities. This crucial methodology has opened new avenues to obtain easily recyclable adsorbents of high adsorption capacity that possess electrostatic interactions. Here, the interaction between oppositely charged groups of the pollutant and biopolymer play a vital role in their adsorptive interaction and subsequent removal. In addition to this, other prominent interactions such as H-bonding and hydrophobic interactions may also be counted for superior adsorption capacities in this new class of materials [124, 125].

Similarly, to magnetically active materials, carbon-based functional 2D materials have also attracted greater attention among technologists. Carbon has been identified as one of the abundant elements in the environment and human body has humongous material source. Carbon is a unique element specifically having an extraordinary ability to interact and bind to other elements. Carbon has extensive allotropy. However, until recent decades, a few carbon allotropic forms were known, namely graphite, diamond, and fullerene. In recent decades, many other allotropes of carbon have been rediscovered and adapted for technology development with a deeper understanding of fundamental aspects. Recent advances in 2D carbon materials have opened new avenues in exploring functional carbon-based materials in energy and environmental applications including water purification and wastewater treatment. These lists include sheets of graphene, GO, reduced GO, and ball-shaped buckminsterfullerene having distinct structural and functional dimensionalities.

On the other hand, 2D-dimensional nanomaterials are referred to as thinnest nanostructures among the various dimensional groups. Contrary to bulk materials, 2D materials possess a high surface area-to-volume ratio. In the 2D nanostructures, the relative number of external to internal atoms is higher, and their different functional versatility leads to a change in the physico-chemical behavior of 2D nanomaterials in comparison to the other dimensional groups. This imparts them a variety of special characteristics including physical, chemical, physico-chemical, and electrical properties. In this direction, graphene is a 2D material with a hexagonal structure composed of sp2-hybridized C atoms in which carbon atom is covalently bound with each other in the same plane [126]. Due to its relatively high hydrophobicity, graphene experiences poor water compatibility when used for water treatment applications. The pristine graphene experiences serious nonwettability and interaction issues due to poor solubility and agglomeration tendency. However, to overcome these limitations associated with 2D carbon nanosheets, researchers and technologists have devised several protocols and techniques to synthesize new class nanomaterials that are structurally similar like graphite-like nanoplatelets (GNPs) and GO that can be pared from various carbon sources like polymers, biomass, discarded agriculture waste and so on following simple but well established top-down methods [127, 128].

In this direction, one interesting study attempts to explore GNP for adsorption capabilities and ability to improve the adsorption capacity in various composite forms aiming to remove toxic metal ions from groundwater or recover valuable elements from wastewater sources. Graphene-based composite adsorbents have been used for recovering as technologically precious elements such as chromium and nickel from various waste sources. Therefore, there has been increasing interest in a new class of 2D-carbon-based materials due to emerging key technologies that are intrinsically dependent on a limited number of elements and their resources often termed by technology critical elements (TCE), which includes all lanthanoids and the platinum group elements. On the other hand, GNPs are simple, viable, and inexpensive 2D nanomaterials that can be used as robust and highly stable substrates for the preparation of magnetic nanocomposites useful for water treatment applications via adsorption mechanism. This approach allows GNP-based nanocomposite use in adsorptive separation and recovery through magnetic separation technologies following established magnetic separation protocols. Further, the GNPs decorated with spinel ferrites MFe_2O_4 (M = Fe, Mn, Co), nanoparticles have been explored for TCE recovery from contaminated sources. Here, magnetically active spinal ferrites decorated facilitate 2D nanomaterials post-treatment separation via magnetically induced separation upon exposure to an external magnetic gradient [129].

Briefly, in this case, the magnetic nanostructure of GNP composites has been prepared via in situ alkaline hydrolysis or electrostatic assembly of particulates resulting in spinel ferrites, with average particle sizes below 100 nm decorated onto the GNP substrate surface. Advanced analytical tools used for the characterization of novel composite materials have revealed robust magnetic response capabilities when exposed to an external magnetic field. In this case, estimated saturation magnetization tested at 300 K is recorded at ~29 emu/g for the $CoFe_2O_4$/GNP and $MnFe_2O_4$/GNP and ~45 emu/g for Fe_3O_4/GNP composites. However, the saturation magnetization values for all these ferrite nanomaterials are lower than the values of the respective spinel ferrites at RT, viz. Fe_3O_4 92 emu/g, $CoFe_2O_4$ 55 emu/g, and $MnFe_2O_4$ 48 emu/g.

Most importantly, nanocomposites of Fe_3O_4/GNP have been further investigated to recover selected technologically important lanthanoids (La, Eu, and Tb) from different wastewater sources like unary and ternary solutions. All the performances of nanocomposite active materials have been compared with pristine isolated components of Fe_3O_4 and GNPs. Here, uptake studies were conducted for 24 hours' exposure to active materials each having equal amounts of active materials GNP, Fe_3O_4 NPs, and Fe_3O_4/GNPs in a ternary solution containing La(III), Eu(III), and Tb(III) of elemental concentration ~100 μg/L. Post-treatment studies reveal that distinct affinity between active materials and precious elements, in particular, Fe_3O_4/GNPs composite showing exceptional removal capability to target elements from water source compared to other counterparts (La: 4.7% (GNPs), 3.1% (Fe_3O_4 NPs), and 15% (Fe_3O_4/GNPs); 0.2% (Fe_3O_4 NPs), Eu: 1.2% (GNPs), and Tb: 0.2% (GNPs), 37.6% (Fe_3O_4/GNPs); 0.1% (Fe_3O_4 NPs), and 35.0% (Fe_3O_4/GNPs)). These enhanced uptake capabilities of the nanocomposites can be attributed to increased number of active sorption sites available on optimal samples due to the introduction of excessive oxygen moieties into the carbon lattice. Interestingly, trivalent lanthanides are classified as hard Lewis acids with strong chemical affinity toward oxygen donors. More so, the sorption studies showed that adsorption capabilities of optimized GNP composite were very much dependent on pH of the solution with maximum capacity achieved feasible for pH values higher than 5.2. These conditions favor due to the formation of experimental isoelectric point of the GNPs composite. The study was also extended to understand the recyclability aspect of nanocomposites via consecutive sorption/desorption experiments using ternary solutions of TCE (100 μg/L) at optimized experimental conditions of pH 7.5, 250 mg/L and active Fe_3O_4/GNPs using 0.1 M HNO_3 as eluent solution. These sorption/desorption studies showed the recovery efficiency of the target precious elements ranged from 82 to 97% for Tb (III), 82–92% for La (III), and 88–104% for Eu (III). The detailed analysis revealed that the dilute nitric acid solution (0.1 mol/L HNO_3) played a vital role as an efficient eluent for the recovery of elements using the magnetic composite [130].

In another similar study, Fe_3O_4/GNPs, $MnFe_2O_4$/GNPs, and $CoFe_2O_4$/GNPs nanocomposites were utilized for the recovery of mercury (Hg) and arsenic (As) from unary and binary solutions. These toxic elements originated from various natural and anthropogenic origins pose a serious threat to the environment due to their severe toxicity, rigidity, and nondegradable nature, and bio-accumulative character. On the list of priority hazardous substances published in 2019 by the Agency for Substances and Toxic Diseases, As and Hg occupy first and third positions, respectively. This effectively demands devising a protocol and technique to efficiently treat and remove toxic elements from water. Systematic studies were conducted on capturing As and Hg after a 24-hour exposure of equal amounts of MFe_2O_4/GNPs active nanomaterials to unary solutions containing As(III) (1000 μg/L) and Hg(II) (50 μg/L). These results reveal that (i) Fe_3O_4/GNPs were the least effective in removing As(III) or Hg(II), and (ii) As(III) removal was higher for the $CoFe_2O_4$/GNPs at pH 7 (90.4%), whereas $MnFe_2O_4$/GNPs showed higher removal efficiency for Hg(II) at pH 6 (90.6%). However, in binary solutions, $CoFe_2O_4$/GNPs nanocomposites have shown greater removal efficiency to As(III) with removal efficiency reaching nearly 90% in ultra-pure water and ~80% in tap water. Interestingly, $MnFe_2O_4$/GNP's preferably removed Hg(II) up to 85% in both matrices. Even though all the GNP-based nanocomposites were synthesized following similar experimental conditions, however, composites containing Co and Mn ferrites resulted in better than the Fe_3O_4/GNPs and showed greater potential for metal leaching. Nevertheless, the 2D-nanosheet-supported ferrite nanoparticles have displayed varying metal leaching (20–50% of Co, 20–30% of Mn, and <8% of Fe). Interestingly, free Co, Mn, or Fe sites present on the surfaces of the respective active ferrites can be further improved to exchange precious as well as toxic ions in solution through optimization of active sorption sites to achieve by exploiting their different degree of affinity toward the ions [131, 132].

Therefore, the use of 2D functional nanomaterials has shown great potential in removal, recovery, and/or treatment of both technologically important elements as well as toxic elements from various water and wastewater sources. Various studies also highlighted the importance of recycling active coagulants, flocculants, adsorbents, and active magnetic materials post-treatment of wastewater sources to achieve sustainability in the overall process.

5.9 Future Perspective and Conclusions

In summary, functional polymers and nanomaterials have played a significant role in addressing serious issues of groundwater contamination and wastewater treatment. There have been a series of polymers developed such as cationic, anionic, polyelectrolytic, functional, and composite forms to effectively treat various types of water contamination. Another major advantage of new approaches that researchers and innovators have incorporated during the development of advanced materials is the sustainable recovery and reuse criteria. These considerations have led to the development of uniquely designed functional polymers and nanomaterials that have a target-specific end use. In many cases, functional materials have provided a favorable outcome in recycling and reusability parameters. This is a major change among new advanced class of water treatment materials in which technological advantage in recycling through various external mechanisms enables the technologists to design and deliver a treatment plant, which is affordable, cost-effective, environment-friendly, and sustainable in the long run.

References

1 Wu, S., Shi, W., Li, K. et al. (2022). Recent advances on sustainable bio-based materials for water treatment: fabrication, modification and application. *J. Environ. Chem. Eng.* 10 (6): 108921. ISSN 2213-3437.

2 Oprea, M. and Voicu, S.I. (2023). Cellulose acetate-based materials for water treatment in the context of circular economy. *Water* 15 (10): 1860.

3 Ahmed, M., Mavukkandy, M.O., Giwa, A. et al. (2022). Recent developments in hazardous pollutants removal from wastewater and water reuse within a circular economy. *NPJ Clean Water* 5: 12.

4 Eke, J., Yusuf, A., Giwa, A., and Sodiq, A. (2020). The global status of desalination: an assessment of current desalination technologies, plants and capacity. *Desalination* 495: 114633.

5 Chen, Y., Liu, Y., Li, Y. et al. (2020). Synthesis, application and mechanisms of Ferro-Manganese binary oxide in water remediation: a review. *Chem. Eng. J.* 388: 124313.

6 Adusei-Gyamfi, J., Ouddane, B., Rietveld, L. et al. (2019). Natural organic matter-cations complexation and its impact on water treatment: a critical review. *Water Res.* 160: 130–147.

7 Pan, Z., Zhu, X., Satpathy, A. et al. (2019). Cr (VI) adsorption on engineered iron oxide nanoparticles: exploring complexation processes and water chemistry. *Environ. Sci. Technol.* 53: 11913–11921.

8 Zhang, S., Hedtke, T., Zhu, Q. et al. (2021). Membrane-confined iron oxychloride nanocatalysts for highly efficient heterogeneous fenton water treatment. *Environ. Sci. Technol.* 55: 9266–9275.

9 Waqas, S., Bilad, M.R., Man, Z.B. et al. (2021). An energy-efficient membrane rotating biological contactor for wastewater treatment. *J. Clean. Prod.* 282: Article 124544.

10 Ownby, M., Desrosiers, D.-A., and Vaneeckhaute, C. (2021). Phosphorus removal and recovery from wastewater via hybrid ion exchange nanotechnology: a study on sustainable regeneration chemistries. *NPJ Clean Water* 4: 1–8.

11 Yao, Y., Zhang, P., Jiang, C. et al. (2021). High performance polyester reverse osmosis desalination membrane with chlorine resistance. *Nat. Sustain.* 4: 138–146.

12 Yin, R., Blatchley, E.R. III, and Shang, C. (2020). UV photolysis of mono-and dichloramine using UV-LEDs as radiation sources: photodecay rates and radical concentrations. *Environ. Sci. Technol.* 54: 8420–8429.

13 Oprea, M. and Voicu, S.I. (2020). Recent advances in applications of cellulose derivatives-based composite membranes with hydroxyapatite. *Materials* 13: 2481.

14 Dhanraj, N.D., Hatha, A.A.M., and Jisha, M.S. (2022). Biodegradation of petroleum based and bio-based plastics: approaches to increase the rate of biodegradation. *Arch. Microbiol.* 204: 258.

15 Bandehali, S., Sanaeepur, H., Amooghin, A.E. et al. (2021). Biodegradable polymers for membrane separation. *Sep. Purif. Technol.* 269: 118731.

16 Thiagamani, S.M.K., Krishnasamy, S., and Siengchin, S. (2019). Challenges of biodegradable polymers: an environmental perspective. *Appl. Sci. Eng. Prog.* 12: 149.

17 Rai, P., Mehrotra, S., Priya, S. et al. (2021). Recent advances in the sustainable design and applications of biodegradable polymers. *Bioresour. Technol.* 325: 124739.

18 Gohil, J.M. and Ray, P. (2017). A review on semi-aromatic polyamide TFC membranes prepared by interfacial polymerization: potential for water treatment and desalination. *Sep. Purif. Technol.* 181: 159–182.

19 Mamah, S.C., Goh, P.S., Ismail, A.F. et al. (2021). Recent development in modification of polysulfone membrane for water treatment application. *J. Water Process Eng.* 40: 101835.

20 Johari, A., Razmjouei, M., Mansourizadeh, A., and Emadzadeh, D. (2020). Fabrication of blend hydrophilic polyamide imide (Torlon®)-sulfonated poly (ether ether ketone) hollow fiber membranes for oily wastewater treatment. *Polym. Test.* 91: 106733.

21 Wen, Y., Dai, R., Li, X. et al. (2022). Metal-organic framework enables ultraselective polyamide membrane for desalination and water reuse. *Sci. Adv.* 8: eabm4149.

22 Cong, S., Yuan, Y., Wang, J. et al. (2021). Highly water-permeable metal–organic framework mof-303 membranes for desalination. *J. Am. Chem. Soc.* 143: 20055–20058.

23 Zhang, Y., Guo, J., Han, G. et al. (2021). Molecularly soldered covalent organic frameworks for ultrafast precision sieving. *Sci. Adv.* 7: eabe8706.

24 Jian, M., Qiu, R., Xia, Y. et al. (2020). Ultrathin water-stable metal-organic framework membranes for ion separation. *Sci. Adv.* 6: eaay3998.

25 Li, Y., Wu, Q., Guo, X. et al. (2020). Laminated self-standing covalent organic framework membrane with uniformly distributed subnanopores for ionic and molecular sieving. *Nat. Commun.* 11: 1–9.

26 Carpenter, A.W., de Lannoy, C.-F., and Wiesner, M.R. (2015). Cellulose nanomaterials in water treatment technologies. *Environ. Sci. Technol.* 49: 5277–5287.

27 Kang, Y.-L., Zhang, J., Wu, G. et al. (2018). Full-biobased nanofiber membranes toward decontamination of wastewater containing multiple pollutants. *ACS Sustain. Chem. Eng.* 6: 11783–11792.

28 Abouzeid, R.E., Khiari, R., El-Wakil, N., and Dufresne, A. (2018). Current state and new trends in the use of cellulose nanomaterials for wastewater treatment. *Biomacromolecules* 20: 573–597.

29 Mohammed, N., Grishkewich, N., and Tam, K.C. (2018). Cellulose nanomaterials: promising sustainable nanomaterials for application in water/wastewater treatment processes. *Environ. Sci. Nano.* 5: 623–658.

30 Jjagwe, J., Olupot, P.W., Menya, E.H., and Kalibbala, M. (2021). Synthesis and application of granular activated carbon from biomass waste materials for water treatment: a review. *J. Bioresour. Bioprod.* 6: 92–322.

31 (2021). Honeycomb-like activated carbon with microporous nanosheets structure prepared from waste biomass cork for highly efficient dye wastewater treatment. *J. Hazard. Mater.* 416: 125896.

32 De Clercq, R., Dusselier, M., and Sels, B. (2017). Heterogeneous catalysis for bio-based polyester monomers from cellulosic biomass: advances, challenges and prospects. *Green Chem.* 19: 5012–5040.

33 Mosier, N., Wyman, C., Dale, B. et al. (2005). Features of promising technologies for pretreatment of lignocellulosic biomass. *Bioresour. Technol.* 96: 673–686.

34 Machado, G., Santos, F., Lourega, R. et al. (2020). Biopolymers from lignocellulosic biomass: feedstocks, production processes, and applications. *Lignocellul. Bioref. Technol.* 125–158.

35 Manohara, H.M., Aruchamy, K., Chakraborty, S. et al. (2019). Sustainable water purification using an engineered solvothermal carbon based membrane derived from a eutectic system. *ACS Sustain. Chem. Eng.* 7 (11): 10143–10153.

36 Sengupta, A., Jebur, M., Kamaz, M., and Wickramasinghe, S.R. (2022). Removal of emerging contaminants from wastewater streams using membrane bioreactors: a review. *Membranes 12*: 60.

37 Vardali, S.C., Manousi, N., Barczak, M., and Giannakoudakis, D.A. (2020). Novel approaches utilizing metal-organic framework composites for the extraction of organic compounds and metal traces from fish and seafood. *Molecules* 25: 513.

38 O'brien, T.F. (2002). Emergence, spread, and environmental effectof antimicrobial resistance: how useof an antimicrobial anywhere can increaseresistance to any antimicrobial anywhere else. *Clin. Infect. Dis.* 34: S78–S84.

39 Wilkinson, J.L., Hooda, P.S., Barker, J. et al. (2016). Ecotoxic pharmaceuticals, personal careproducts and other emerging contaminants: are view of environmental, receptor-mediated, developmental, and epigenetic toxicity withdiscussion of proposed toxicity to humans. *Crit. Rev. Environ. Sci. Technol.* 46: 336–381.

40 Leite, A.J.B., Sophia, A.C., Thue, P.S. et al. (2017). Activated carbon from avocado seeds for the removal of phenolic compounds from aqueous solutions. *Desalin. Water Treat.* 71: 168–181.

41 Nataraj, S.K. (2022). *Emerging Pollutant Treatment in Wastewater*, 1e. CRC Press https://doi.org/10.1201/9781003214786.

42 Huang, P., Xia, D., Kazlauciunas, A. et al. (2019). Dye-mediated interactions in chitosan-based polyelectrolyte/organoclay hybrids for enhanced adsorption of industrial dyes. *ACS Appl. Mater. Interfaces* 11: 11961–11969.

43 Bediako, J.K., El Ouardi, Y., Mouele, E.S.M. et al. (2023). Polyelectrolyte and polyelectrolyte complex-incorporated adsorbents in water and wastewater remediation – a review of recent advances. *Chemosphere* 325: 138418.

44 Ge, Q., Wang, P., Wan, C., and Chung, T.S. (2012). Polyelectrolyte-promoted forward osmosis-membrane distillation (FO-MD) hybrid process for dye wastewater treatment. *Environ. Sci. Technol.* 46: 6236–6243.

45 Zhao, P., Gao, B., Xu, S. et al. (2015). Polyelectrolyte-promoted forward osmosis process for dye wastewater treatment – exploring the feasibility of using polyacrylamide as draw solute. *Chem. Eng. J.* 264: 32–38.

46 Xiong, S., Han, C., Phommachanh, A. et al. (2021). High-performance loose nanofiltration membrane prepared with assembly of covalently cross-linked polyethyleneimine-based polyelectrolytes for textile wastewater treatment. *Sep. Purif. Technol.* 274: 119105.

47 Betraoui, A., Seddiki, N., Souag, R. et al. (2023). Synthesis of new hydrogels involving acrylic acid and acrylamide grafted agar-agar and their application in the removal of cationic dyes from wastewater. *Gels* 9: 499.

48 Yu, L., Liu, X., Yuan, W. et al. (2015). Confined flocculation of ionic pollutants by poly(L-dopa)-based polyelectrolyte complexes in hydrogel beads for three-dimensional, quantitative, efficient water decontamination. *Langmuir* 31: 6351–6366.

49 Abiola, O.N. (2019). Polymers for coagulation and flocculation in water treatment. In: *Polymeric Materials for Clean Water*, Springer Series on Polymer and Composite Materials (ed. R. Das). Cham: Springer.

50 Mohammed, S.S., Abujazar, S.U., Karaağaç, S.S. et al. (2022). Recent advancement in the application of hybrid coagulants in coagulation-flocculation of wastewater: a review. *J. Clean. Prod.* 345: 131133-131146.

51 Wang, S., Li, E., Li, J. et al. (2020). Preparation and coagulation-flocculation performance of covalently bound organic hybrid coagulant with excellent stability. *Colloids Surf. A. Physiochem. Eng. Asp.* 600: 124966.

52 Annals, N. and Natural, O.F. (2012). Physicochemical treatment of brewery sludge with Naturally Sourced Coagulant (Chitosan and Clay). *Nigerian Anim. Nat. Sci.* 12: 8–16.

53 Amy, G., Ghaffour, N., Li, Z. et al. (2017). Membrane-based seawater desalination: present and future prospects. *Desalination* 401: 16–21.

54 Keyu, Q., Yuan, Z., Wang, Y. et al. (2022). Structures, properties, and applications of zwitterionic polymers. *ChemPhysMater* 1 (4): 294–309.

55 Sharker, K.K., Shigeta, Y., Ozoe, S. et al. (2021). Upper critical solution temperature behavior of pH-responsive amphoteric statistical copolymers in aqueous solutions. *ACS Omega* 6: 9153–9163.

56 Ali, S.A. and Haladu, S.A. (2013). A novel cross-linked polyzwitterion/anion having pH-responsive carboxylate and sulfonate groups for the removal of Sr^{2+} from aqueous solution at low concentrations. *React. Funct. Polym.* 73: 796–804.

57 Liu, L., Wu, J., Ling, Y. et al. (2013). Synthesis of a novel amphoteric chelating polymer flocculant and its performance in Cu^{2+} removal. *J. Appl. Polym. Sci.* 127: 2082–2094.

58 Fu, F., Chen, R., and Xiong, Y. (2006). Application of a novel strategy-coordination polymerization precipitation to the treatment of Cu2+-containing wastewaters. *Sep. Purif. Technol.* 52: 388–393.

59 Wang, Y., Li, L., Li, J. et al. (2016). Stable and pH-responsive polyamidoamine based unimolecular micelles capped with a zwitterionic polymer shell for anticancer drug delivery. *RSC Adv.* 6: 17728–17739.

60 Wei, R., Song, W., Yang, F. et al. (2018). Bidirectionally pH-responsive zwitterionic polymer hydrogels with switchable selective adsorption capacities for anionic and cationic dyes. *Ind. Eng. Chem. Res.* 57: 8209–8219.

61 Beltrán-Osuna, Á.A., Ródenas-Rochina, J., Ribelles, J.L.G., and Perilla, J.E. (2019). Antifouling zwitterionic PSBMA-MSN particles for biomedical applications. *Polym. Adv. Technol.* 30: 688–697.

62 Kaner, P., Dudchenko, A.V., Mauter, M.S., and Asatekin, A. (2019). Zwitterionic copolymer additive architecture affects membrane performance: fouling resistance and surface rearrangement in saline solutions. *J. Mater. Chem. A* 7: 4829–4846.

63 Wang, J., Wang, Z., Wang, J., and Wang, S. (2015). Improving the water flux and bio-fouling resistance of reverse osmosis (RO) membrane through surface modification by zwitterionic polymer. *J. Membr. Sci.* 493: 188–199.

64 Yang, Q. and Ulbricht, M. (2012). Novel membrane adsorbers with grafted zwitterionic polymers synthesized by surface-initiated ATRP and their salt-modulated permeability and protein binding properties. *Chem. Mater.* 24: 2943–2951.

65 Zhao, Y.F., Zhang, P.B., Sun, J. et al. (2016). Electrolyte-responsive polyethersulfone membranes with zwitterionic polyethersulfone-based copolymers as additive. *J. Membr. Sci.* 510: 306–313.

66 Chen, P.C., Lai, J.J., and Huang, C.J. (2021). Bio-inspired amphoteric polymer for triggered-release drug delivery on breast cancer cells based on metal coordination. *ACS Appl. Mater. Interfaces* 13: 25663–25673.

67 Xia, Y., Adibnia, V., Shan, C. et al. (2019). Synergy between zwitterionic polymers and hyaluronic acid enhances antifouling performance. *Langmuir* 35: 15535–15542.

68 Jiang, G., Peng, S., Li, X. et al. (2018). Preparation of amphoteric starch-based flocculants by reactive extrusion for removing useless solids from water-based drilling fluids. *Colloids Surf. A Physicochem. Eng. Asp.* 558: 343–350.

69 Liu, F., Jiang, G., Peng, S. et al. (2016). Amphoteric polymer as an anti-calcium contamination fluid-loss additive in water-based drilling fluids. *Energy Fuel* 30: 7221–7228.

70 Nataraj, S.K. *Advanced Materials and Methods for Water Treatment.* I.K. International Pvt. Ltd ISBN-13, 978-8195966257.

71 Ahmed, M., Kumar, R., Garudachari, B., and Thomas, J.P. (2019). Performance evaluation of a thermoresponsive polyelectrolyte draw solution in a pilot scale forward osmosis seawater desalination system. *Desalination* 452: 132–140.

72 Hamad, M.J.A. and Chirwa, E.M.N. (2019). Forward osmosis for water recovery using polyelectrolyte PolyDADMAC and DADMAC draw solutions as a low pressure energy saving process. *Desalination* 453: 89–101.

73 Kim, J.-j., Kang, H., Choi, Y.-S. et al. (2016). Thermo-responsive oligomeric poly(tetrabutylphosphonium styrenesulfonate)s as draw solutes for forward osmosis (FO) applications. *Desalination* 381: 84–94.

74 Rud, O., Borisov, O., and Košovan, P. (2018). Thermodynamic model for a reversible desalination cycle using weak polyelectrolyte hydrogels. *Desalination* 442: 32–43.

75 Duong, P.H.H., Zuo, J., and Chung, T.-S. (2013). Highly crosslinked layer-by-layer polyelectrolyte FO membranes: understanding effects of salt concentration and deposition time on FO performance. *J. Membr. Sci.* 427: 411–421.

76 Cui, X., Tatsuya Murakami, Y., and Hoshino, Y.M. (2017). Anti-biofouling phosphorylated HEMA and PEGMA block copolymers show high affinity to hydroxyapatite. *Colloids Surf. B: Biointerfaces* 160: 289–296.

77 Batys, P., Zhang, Y., Lutkenhaus, J.L., and Sammalkorpi, M. (2018). Hydration and temperature response of water mobility in poly(diallyldimethylammonium)-poly(sodium 4-styrenesulfonate) complexes. *Macromolecules* 51: 8268–8277.

78 Reurink, D.M., Te Brinke, E., Achterhuis, I. et al. (2019). Nafion-based low-hydration polyelectrolyte multilayer membranes for enhanced water purification. *ACS Appl. Polym. Mater.* 1: 2543–2551.

79 Cheng, G., Mi, L., Cao, Z. et al. (2010). Functionalizable and ultrastable zwitterionic nanogels. *Langmuir* 26 (10): 6883–6886.

80 Ma, M.-Q., Zhang, C., Chen, T.-T. et al. (2019). Bioinspired polydopamine/polyzwitterion coatings for underwater anti-oil and -freezing surfaces. *Langmuir* 35: 1895–1901.

81 Liu, H., Yang, L., Dou, B. et al. (2021). Zwitterionic hydrogel-coated cotton fabrics with underwater superoleophobic, self-healing and anti-fouling performances for oil-water separation. *Sep. Purif. Technol.* 279: Article 119789.

82 Niu, J., Wang, H., Chen, J. et al. (2021). Bio-inspired zwitterionic copolymers for antifouling surface and oil-water separation. *Colloids Surf. A. Physiochem. Eng. Asp.* 626: Article 127016.

83 Li, M., Zhuang, B., and Yu, J. (2020). Functional zwitterionic polymers on surface: structures and applications. *Chem. Asian J.* 15: 2060.

84 R. Lalani, L. Liu, Electrospun Zwitterionic poly(sulfobetaine methacrylate) for nonadherent, superabsorbent, and antimicrobial wound dressing applications *Biomacromolecules* 2012, 13, 1853–1863;

85 Sun, N., Liu, M., Wang, J. et al. (2016). Chitosan nanofibers for specific capture and nondestructive release of CTCs assisted by pCBMA brushes. *Small* 12: 5090–5097.

86 Kobayashi, M., Terayama, Y., Hosaka, N. et al. (2007). Friction behavior of high-density poly(2-methacryloyloxyethyl phosphorylcholine) brush in aqueous media. *Soft Matter* 3: 740–746.

87 Ni, L., Meng, J., Geise, G.M. et al. (2015). Water and salt transport properties of zwitterionic polymers film. *J. Membr. Sci.* 491: 73–81.

88 Yang, Y., Ramos, T.L., Heo, J., and Green, M. (2018). Zwitterionic poly(arylene ether sulfone) copolymer/poly(arylene ether sulfone) blends for fouling-resistant desalination membranes. *J. Membr. Sci. 561*: 69–78.

89 Chan, W.-F., Marand, E., and Martin, S.M. (2016). Novel zwitterion functionalized carbon nanotube nanocomposite membranes for improved RO performance and surface anti-biofouling resistance. *J. Membr. Sci.* 509: 125–137.

90 Nguyen, A., Azari, S., and Zou, L. (2013). Coating zwitterionic amino acid l-DOPA to increase fouling resistance of forward osmosis membrane. *Desalination* 312: 82–87.

91 Teh, C.Y., Budiman, P.M., Shak, K.P.Y., and Wu, T.Y. (2016). Recent advancement of coagulation–flocculation and its application in wastewater treatment. *Ind. Eng. Chem. Res.* 55 (16): 4363–4389.

92 Jiang, X., Li, Y., Tang, X. et al. (2021 Sep). Biopolymer-based flocculants: a review of recent technologies. *Environ. Sci. Pollut. Res. Int.* 28 (34): 46934–46963.

93 Mohamed, E.M., Ali, S.M., Moniem, A. et al. (2022). Innovative polymeric inorganic coagulant-flocculant for wastewater purification with simultaneous microbial reduction in treated effluent and sludge, South African Journal of. *Chem. Eng.* 42: 127–137.

94 SW-846 test method 9095B: paint filter liquids test, https://www.epa.gov/hw-sw846/sw-846-test-method-9095b-paint-filter-liquids-test. last updated on August 17, 2023.

95 Bolto, B. and Gregory, J. (2007). Organic polyelectrolytes in water treatment. *Water Res.* 41 (11): 2301–2324.

96 Gregory, J. and Lee, S. (1990). The effect of charge density and molecular mass of cationic polymers on flocculation kinetics in aqueous solution. *J. Water SRT – Aqua* 39: 265–274.

97 Zeweldi, H.G., Limjuco, L.A., Bendoy, A.P. et al. (2018). The potential of monocationic imidazolium-, phosphonium-, and ammonium-based hydrophilic ionic liquids as draw solutes for forward osmosis. *Desalination* 444: 94–106.

98 Garg, G., Chauhan, G.S., and Ahn, J.H. (2012). Strontium(II) ion uptake on poly(N-vinyl imidazole)-based hydrogels. *J. Appl. Polym. Sci.* 124: 3721–3727.

99 Liu, J., Ma, Y., Xu, T., and Shao, G. (2010). Preparation of zwitterionic hybrid polymer and its application for the removal of heavy metal ions from water. *J. Hazard. Mater.* 178: 1021–1029.

100 Liang, W.J., Wu, C.P., Hsu, C.Y., and Kuo, P.L. (2006). Synthesis, characterization, and proton-conducting properties of organic-inorganic hybrid membranes based on polysiloxane zwitterionomer. *J. Polym. Sci., Part A: Polym. Chem.* 44: 3444–3453.

101 EL-Ghoul, Y., Alminderej, F.M., Alsubaie, F.M. et al. (2021). Recent advances in functional polymer materials for energy, water, and biomedical applications: a review. *Polymers* 13: 4327.

102 Vajihinejad, V., Gumfekar, S.P., Bazoubandi, B. et al. (2019). Water soluble polymer flocculants: synthesis, characterization, and performance assessment. *Macromol. Mater. Eng.* 1800526.

103 Bolto, B., Dixon, D., Eldridge, R. et al. (2002). Removal of natural organic matter by ion exchange. *Water Res.* 36 (20): 5057–5065.

104 Bolto, B., Dixon, D., and Eldridge, R. (2004). Ion exchange for the removal of natural organic matter. *React. Funct. Polym.* 60: 171–182.

105 Paulo, P.L., Azevedo, C., Begosso, L. et al. (2013). Natural systems treating greywater and blackwater on-site: integrating treatment, reuse and landscaping. *Ecol. Eng.* 50: 95–100.

106 Nourani, M., Baghdadi, M., and Javan, M. (2016). Gholamreza Nabi Bidhendi, Production of a biodegradable flocculant from cotton and evaluation of its performance in coagulation-flocculation of kaolin clay suspension: optimization through response surface methodology (RSM). *J. Environ. Chem. Eng.* 4 (2): 1996–2003.

107 Magriotis, Z.M., Leal, P.V.B., de Sales, P.F. et al. (2014). A comparative study for the removal of mining wastewater by kaolinite, activated carbon and beta zeolite. *Appl. Clay Sci.* 91–92: 55–62.

108 Yang, H., Wang, Y., Ding, M. et al. (2012). Water-assisted solvent extraction of bitumen from oil sands. *Ind. Eng. Chem. Res.* 51 (7): 3032–3038.

109 Fu, C., Zhang, Z., Li, Y. et al. (2023). Optimizing the flocculation effect of cationic polyacrylamide prepared with UV-initiated polymerization by response surface methodology. *Water 15*: 1200.

110 Wang, H., Zhang, Y., Wang, G. et al. (2019). A molecular-based design of RGO/ TiO2-PAM composite flocculant with photocatalytic self-degrading characteristics and the application of the oil sand tailings flocculant. *ACS Sustain. Chem. Eng.* 7 (7): 6758–6768.

111 Chen, J., Xu, X., Nie, R. et al. (2020). Chitosan modified cationic polyacrylamide initiated by UV-H2O2 for sludge flocculation and new insight on the floc characteristics study. *Polymers 12*: 2738.

112 Hierrezuelo, J., Sadeghpour, A., Szilagyi, I. et al. (2010). Electrostatic stabilization of charged colloidal particles with adsorbed polyelectrolytes of opposite charge. *Langmuir* 26 (19): 15109–15111.

113 Oyegbile, B., Ay, P., and Narra, S. (2016). Flocculation kinetics and hydrodynamic interactions in natural and engineered flow systems: a review. *Environ. Eng. Res.* 21 (1): 1–14.

114 Guibal, E., Van Vooren, M., Dempsey, B.A., and Roussy, J. (2006). A review of the use of chitosan for the removal of particulate and dissolved contaminants. *Sep. Sci. Technol.* 41 (11): 2487–2514.

115 Nasrollahzadeh, M., Sajjadi, M., Iravani, S., and Varma, R.S. (2021). Starch, cellulose, pectin, gum, alginate, chitin and chitosan derived (nano)materials for sustainable water treatment: a review. *Carbohydr. Polym.* 251: Article 116986.

116 Yaashikaa, P.R., Senthil, K.P., and Karishma, S. (2022). Review on biopolymers and composites – evolving material as adsorbents in removal of environmental pollutants. *Environ. Res.* 212: 113114. https://doi.org/10.1016/j.envres.2022.113114.

117 Baskar, A.V., Bolan, N., Hoang, S.A. et al. (2022). Recovery, regeneration and sustainable management of spent adsorbents from wastewater treatment streams: a review. *Sci. Total Environ.* 822: 153555.

118 Abdel Maksoud, M.I.A., Elgarahy, A.M., Farrell, C. et al. (2020). Insight on water remediation application using magnetic nanomaterials and biosorbents. *Coord. Chem. Rev.* 403: 213096.

119 Soares, S.F., Fernandes, T., Trindade, T., and Daniel-da-Silva, A.L. (2020). Recent advances on magnetic biosorbents and their applications for water treatment. *Environ. Chem. Lett.* 18: 151–164.

120 Soares, S.F., Simões, T.R., António, M. et al. (2016). Hybrid nanoadsorbents for the magnetically assisted removal of metoprolol from water. *Chem. Eng. J.* 302: 560–569.

121 Soares, S.F., Trindade, T., and Daniel-Da-Silva, A.L. (2021b). Enhanced removal of non-steroidal inflammatory drugs from water by quaternary chitosan-based magnetic nanosorbents. *Coatings* 11: 964.

122 Soares, S.F., Nogueira, J., Trindade, T., and Daniel-da-Silva, A.L. (2022). Towards efficient ciprofloxacin adsorption using magnetic hybrid nanoparticles prepared with κ-ι-and λ-carrageenan. *J. Nanostructure Chem.* 1–20: 10.

123 Fernandes, T., Soares, S., Trindade, T., and Daniel-da-Silva, A. (2017). Magnetic hybrid nanosorbents for the uptake of paraquat from water. *Nanomaterials* 7: 68. https://doi.org/10.3390/nano7030068.

124 Crini, G., Lichtfouse, E., Wilson, L.D., and Morin-Crini, N. (2019). Conventional and non-conventional adsorbents for wastewater treatment. *Environ. Chem. Lett.* 17: 195–213.

125 Badsha, M.A.H., Khan, M., Wu, B. et al. (2021). Role of surface functional groups of hydrogels in metal adsorption: from performance to mechanism. *J. Hazard. Mater.* 408: 124463.

126 Kallumottakkal, M., Hussein, M.I., and Iqbal, M.Z. (2021). Recent progress of 2D nanomaterials for application on microwave absorption: a comprehensive study. *Front. Math.* 8: 34.

127 Joseph, A., Sajith, V., and Sarathchandran, C. (2021). Graphene: the magic material. In: *Handbook Carbon-Based Nanomaterials* (ed. S. Thomas, C. Sarathchandran, and J.C. Mareno-Pirajan), 517–549. Elsevier.

128 Rao, N., Singh, R., and Bashambu, L. (2021). Carbon-based nanomaterials: synthesis and prospective applications. *Mater. Today Proc.* 44: 608–614.

129 Pinto, J., Lopes, C.B., Henriques, B. et al. (2021). Platinum-group elements sorption by living macroalgae under different contamination scenarios. *J. Environ. Chem. Eng.* 9: 105100.

130 Afonso, E.L., Carvalho, L., Fateixa, S. et al. (2019). Can contaminated waters or wastewater be alternative sources for technology-critical elements? The case of removal and recovery of lanthanides. *J. Hazard. Mater.* 380: 120845.

131 Lopes, C.B., Coimbra, J., Otero, M. et al. (2008). Uptake of Hg^{2+} from aqueous solutions by microporous titano- and zircono-silicates. *Quim Nova* 31: 321–325.

132 Tavares, D.S., Lopes, C.B., Almeida, J.C. et al. (2020). Spinel-type ferrite nanoparticles for removal of arsenic(V) from water. *Environ. Sci. Pollut. Res.* 27: 22523–22534.

6

Advanced Catalytic and Biogenic Materials for Water and Wastewater Treatment

6.1 Introduction to Catalytic and Biogenic Materials in Water Treatment

Groundwater pollution and generation of large quantities of industrial wastewater are highly critical global problems and a serious challenge for people involving water management activities. This problem is expected to escalate in the near future due to the rapid population growth, increased modern agricultural activities, use of stimulators in agriculture practices, industrial development, and their associated resource consumption. According to sustainability criteria, futuristic solutions that aim to address water-related issues require an alternative materials stock and technologies that are efficient, affordable, and have no-to-low environmental impact on their operations. Nonetheless, some of the important advances in materials development aiming at groundwater and polluted waters have been on catalytic processes. Also, efforts in developing green catalysts that are inexpensive, eco-friendly, and offer reusability and recyclability reduce or eliminate the generation of secondary pollutants or use of hazardous substances for recycling purposes. Therefore, recent interest in the development of green and affordable catalysts for water treatment applications has significantly increased. This chapter gives an overview of the state of the art in the design, development, and use of green catalytic and biogenic materials for the treatment of groundwater as well as wastewater.

A century ago (early 20th century), chemical oxidants such as chlorine and ozone were invented in industrialized countries, and introduced in municipal water treatment systems for disinfection purposes. Initially, disinfectants were directly used for treating surface water before distributing to human and municipal consumption. However, as technology matured, the advanced versions of disinfection approaches came into practice such as drinking water treatment in

Materials and Methods for Industrial Wastewater and Groundwater Treatment, First Edition. S. K. Nataraj.

which a combination of chemical disinfection with sand filtration dramatically reduced waterborne diseases [1–5]. Even though this approach gained prominence and was recognized as a major step forward in drinking water supply chain, however, sooner or later, the detection of the secondary pollutants originating from disinfection by-products (DBPs) raised serious concerns about the technological feasibility and health safety. For instance, chloroform was detected in the mid-1970s in disinfected water, followed by the discovery of hundreds of other halogenated DBPs formed during and post-chlorination detection measurements. This was also potentially linked between disinfection and DBPs to the reports of a new class of diseases such as bladder cancer or miscarriages. These new revelations led to the creation of a revolutionary water policy framework and stringent guidelines for water treatment practices. DBPs are generated when disinfectants (chemical oxidants) are mixed to react with pollutant components such as dissolved organic matter (DOM), iodide, and bromide. On the other hand, during ozonation, DBPs such as bromate and oxygen-rich compounds such as ketones, aldehydes, and carboxylic acids readily form after interaction with bromide or DOM, respectively. Therefore, to overcome serious concerns regarding generation of DBPs, modified versions of disinfectants such as chloramine or chlorine dioxide were used in disinfection and distribution systems. This approach led to a considerable decrease in trihalomethane (THM) formation. However, the use of modified disinfectants further led to the generation of a new class of DBPs such as *N*-nitrosodimethylamine (NDMA) or chlorite/chlorate, respectively [6–8].

With this, focus shifted to trade-off between DBP formation and use of appropriate disinfection chemical agents, which is indirectly referred to as a choice between acute and chronic toxicity caused by the presence of DBPs, which was closely monitored and evaluated under new water quality management guidelines. Since then, due to increased demand for safe water, disinfection has been given the highest priority without compromising the control of DBPs. Nevertheless, considering the potential long-term health risks, many countries have entirely discontinued the use of chlorine and/or chemical disinfection in their disinfection and distribution systems. This move has also been taken to avoid public health including taste and odor problems and environmental consequences that originate from the treatment of chlorine. Such actions have been implemented in many countries to conserve and safeguard water resources, and natural attenuation, to develop multibarrier treatment, and to carefully maintain distribution of safe water production infrastructure [9, 10].

On the other hand, parallel to chemical disinfection implementation, oxidative control of aesthetic parameters such as taste, color, odor, iron(II), and Mn(II), and other inorganic contaminants like sulfide, nitrite, and ammonia, treatment strategies have also been implemented in water treatment [11–13]. On the other hand, innovation in qualitative as well as quantitative detection of various contaminants in various water sources led to a better understanding of oxidative process over a period of time. Then the oxidative treatment technique was widely adopted in the water treatment process for the abatement of synthetic organic compounds, and other organic micropollutants due to their presence and widespread detection in drinking water sources [14, 15].

By the mid-20th century, numerous chemical oxidants had been discovered and applied for the disinfection of both groundwater and surface water sources as standalone or in combination with adsorptive materials like activated carbon (AC) or nanofiltration (NF) and reverse osmosis (RO) membrane processes for effective removal of organic as well as other micropollutants [13, 16–19]. Among several chemical oxidation processes, techniques like ozone (O_3) treatment and the advanced oxidation processes (AOPs), O_3/UV, O_3/H_2O_2, that were used for the enhancement of hydroxyl radical ($^{\bullet}OH$) formation, ultraviolet (UV) light combined with H_2O_2 to generate hydroxyl ($^{\bullet}OH$) radicals, permanganate, bisulfite-activated permanganate, (photo)-process, UV combined with $S_2O_8^-$ or HSO_5^- to produce sulfate radicals and their application for *in situ* chemical oxidation and ferrate(VI). In addition to this, in recent decades, a significant innovation has been made to develop photocatalytic water treatment using a series of nanomaterials and their composite counterparts such as TiO_2-based photocatalysts. Even though water remediation using photocatalytic applications is yet to be explored at large-scale operations, novel materials explored for the photocatalytic water treatment applications have shown great potential in affordable degradation of hazardous pollutants from contaminated sources. Overall, most of the large-scale catalyst-based processes in drinking water treatment practices employ conventional homogeneous ozone- or $^{\bullet}OH$-based processes due to their versatility, simple and relatively low-energy operations, and broader effective action against various types of organic micropollutants and through the controlled operations, formation of DBPs can be minimized [20–26].

6.2 Catalytic Treatment of DOM in Ground and Wastewater

DOM is abundant in ground as well as surface water sources and further get contaminated at the storage points of drinking water supply systems (DWSSs). DOMs in DWSSs broadly occur at either a water source, reservoir, storage site, distribution points, a treatment plant, or distribution network, and have often proven detrimental to delivery and water treatment processes. DOM not only destroys the quality of water but is also responsible for severely damaging water distribution network and infrastructure and is a single major cause of the fouling of treatment equipment such as tanks, membrane modules, and other active materials surfaces (e.g. AC) [27–29]. This harmful effect substantially increases the material replacement frequencies adding to overall treatment costs. DOMs also contribute greatly to the formation of hazardous DBPs in drinking water. Moreover, DOMs act as a food source for bacteria present in water, on treatment materials, membranes, and in the distribution network, which can result from gel layer regrowth leading to severe damage to the system. On the other hand, DBPs over a period may cause serious health risks. For instance, exposure of humans to DBPs through inhalation, ingestion, and dermal intake or absorption has been directly linked to adverse effects on health such as bladder cancer. Without precaution, bacterial growth deteriorates over a period in the distribution network, which directly damages the water quality resulting in altercations of taste and odor. Therefore, close monitoring of DOM accumulation and growth in DWSSs is vital for both water treatment management and public health [30, 31].

Due to the innovation in sensor technologies, a range of new tools have been lately used to monitor DOM in different aquatic environments, including measurements of dissolved oxygen (DO), chemical oxygen demand (COD), total suspended solids (TSS), dissolved organic carbon (DOC), UV, and fluorescence spectra. Even though each of these techniques differs in its mechanism of detection and qualitative and quantitative measuring of DOM and the organic fractions in the analyte, however, a combination of these techniques can be effective in the overall analysis of pollutants. For instance, DOC mainly targets the carbon component of DOM giving often accurate quantification of total DOM. DOC is estimated by high-temperature combustion or chemical oxidation technique followed by the detection of CO_2. Similar to the DOC method, COD measures the oxygen consumed during the oxidation of oxidizable matter by a strong chemical oxidant such as potassium permanganate (potassium permanganate index, COD_{Mn}). In most of the water quality parameters, COD is used as a collective representation of the extent of organic pollution in freshwater sources like lakes, tanks, rivers, and so on. In modern analysis techniques, both COD and DOC analysis protocols are widely used tools for monitoring DOM. There are several official standard methods and protocols for DOM measurement of groundwater, surface water, and wastewater. However, most of the procedures are for quantitative detection and not qualitative assessment. On the other hand, UV and fluorescence spectroscopic techniques are not only used for quantifying DOM levels in various water sources but also give qualitative characteristics of DOM [32–38].

Interestingly, chromophoric DOM is a major constituent of DOM in natural (ground and surface) waters which generally absorbs UV and blue light and exhibit accompanying fluorescence emission. Consequently, chromophoric DOM can be detected and quantified using UV absorption and fluorescence spectroscopic techniques. Nevertheless, chemical composition, existence, and concentration of copollutants greatly influence the intensity and shape of the fluorescence and/or UV absorption spectra, which are the bases of qualitative and quantitative analyses of DOM. Further, fluorescence excitation-emission matrix (EEM) spectroscopy and/or UV absorption spectroscopy have been widely used to determine the concentrations and qualitative identification of organic pollutant sources of DOM in water and to recognize the DOM properties associated with the effectiveness of water treatment, microbial quality, DBP formation, and contamination events. On the whole, each above-mentioned method provides very specialized information on DOM composition. Therefore, the effectiveness of each technique may vary in both qualitative and quantitative measurements in terms of DOM in groundwater, surface water, and wastewater depending on the procedures and protocols followed [39, 40].

On the other hand, in industrial wastewaters, chemical oxidation technology is used to remove odor, color, and organic and inorganic compounds by applying a series of oxidants and catalysts. The most commonly used oxidants in wastewater treatment are ozone, chlorine, peroxide, oxygen, permanganate, and so on. Chemical oxidation process involves the transfer of electrons from an oxidizing reagent (active agent or catalyst) to the chemical species (pollutants) being oxidized. Therefore, in groundwater and wastewater treatment engineering, chemical oxidation serves the purpose of converting putrescible and/or decomposable pollutant matter and substances to less harmful, innocuous, and environmentally benign more stable products. Chemical oxidization follows multistep treatment pathways involving addition of oxidants or generating oxidants in the groundwater or wastewater followed by conversion of the pollutants into innocuous products. A few currently used oxidants include chlorine gas (Cl_2), chlorine dioxide (ClO_2), ozone (O_3), hydrogen peroxide (H_2O_2), peroxy acetic acid ($C_2H_4O_3$), natrium hypochlorite or bleaching liquor (NaOCl), and pure oxygen (O_2). In general practice, gaseous oxidants like chlorine(g) are bubbled into DOM-contaminated water sources that results in hydrochloric acid (HCI) or hypochlorous acid (HOCI) [41–45].

For instance, most mechanistic approaches to contaminant oxidation are recorded in ultrapure water. However, to treat DOM-contaminated water in realistic conditions, upon addition of the oxidant, the largest fraction of chemical oxidants is consumed by DOM opresent in the contaminated water as depicted in Figure 6.1. Due to the extent of majority of DOM, this originates from phenolic moieties (1–2 mmol/g DOC, that is, μM DOM-phenol concentrations) that consume a large fraction of oxidants followed by micropollutant concentrations of ≪0.1 μM in ground- and surface waters and wastewater effluents [46–50].

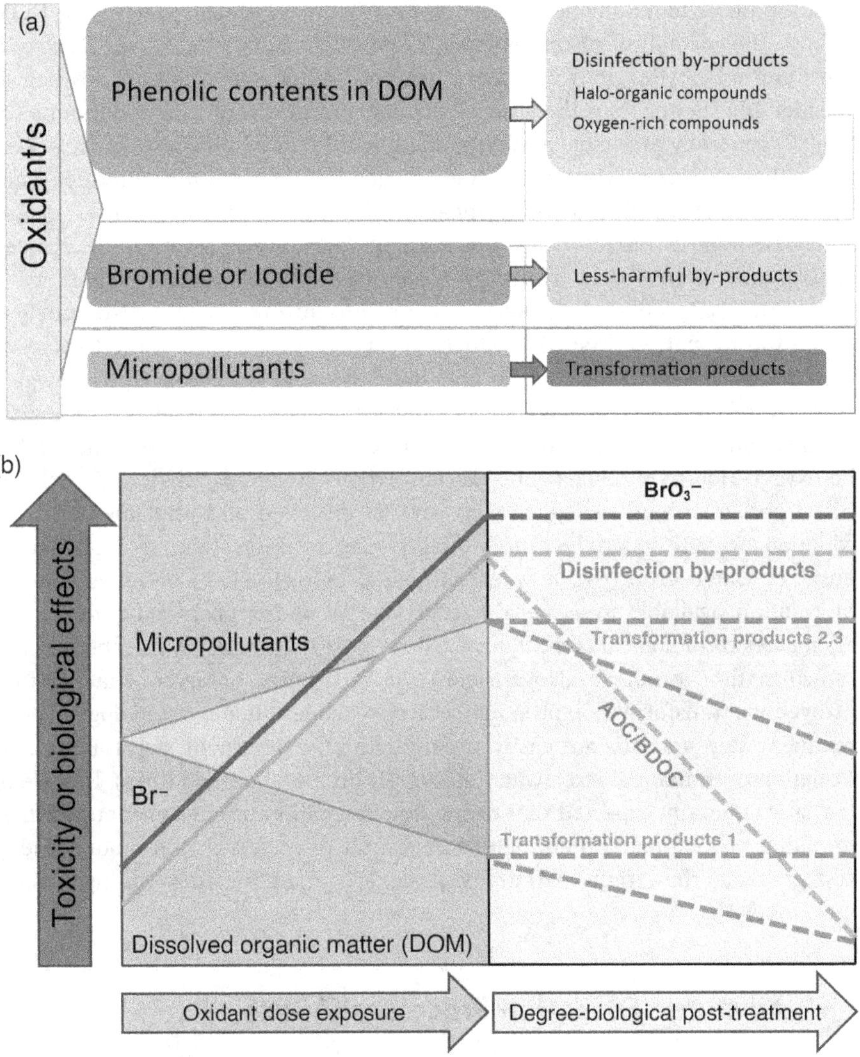

Figure 6.1 (a) Oxidant consumption profile in various water solutes (DOMs) and their extent of consumption shown in a relative thickness of bars. The majority of oxidants are consumed by DOM during water treatment resulting in DBPs, followed by bromide (and, in some cases, iodide) depending on their concentrations. Compared to DOM and bromide, consumption of oxidants by the micropollutants is minimal. Upon reacting with micropollutants, oxidants produce transformation products. (b) schematic represents the relative evolution of toxicity or potential biological effects of different contaminants during oxidative treatment as a function of the oxidant dose and the degree of biological post-treatment. Assimilable organic carbon and biodegradable organic carbon are represented as AOC and BDOC, respectively.

On the other hand, ozone reaction and/or hydroxyl radicle ($^{\bullet}OH$) with DOM leads to the formation of oxygen-rich by-products such as ketones, carboxylic acids and aldehydes in total concentrations often over 100 μg/L, which is considerably higher than the sum of the concentrations of micropollutants (*ca* 25 μg/L) generally estimated and quantified in effluent of a wastewater treatment plant (WWTP) after ozonation. Thus, oxygen-rich oxidation by-products may contribute significantly to the toxicity and/or biological side effects of the treated water than micropollutants (line). For instance, formation of toxic α,β-unsaturated enedials and oxoenals occurs during UV treatment of phenols either in absence or presence of H_2O_2 and quinones or catechols results during ozonation. Nonetheless, during the formation of such products, it has been noticed that the toxicity of certain aquatic organisms considerably increases after ozonation of wastewater. However, often it is observed that an improvement in quality after a biological treatment step of the same effluent takes place (dashed line). It is well known that the oxygen-rich by-products generally undergo easy biodegradation. To achieve this, a post-treatment biological step is often designed and implemented post-oxidation or AOP to produce high-quality safe drinking water or high-quality reusable water from wastewater effluents. Nonetheless, there is limited information available to assess the effectiveness of the post-treatment step or efficiencies of biological techniques used for post-treatment for the abetment of transformation products. Nevertheless, it has been observed that various nitrogenous transformation products such as *N*-oxides often resist biological post-treatment step and are not easily removed in post-treatment step followed by ozonation of municipal wastewater effluent (horizontal dashed lines). In spite of that, it is generally expected that ozone transformation products generated after ozonation process contain aromatic compounds or olefins that are more readily biodegradable than their parent by-product compounds (descending dashed lines) [51–84].

6.3 Advanced Oxidation Processes (AOPs)

In recent years, AOPs have widely used advanced water treatment techniques. AOPs are designed using several conventional as well as advanced catalysts specifically designed to cater to type of effluent or contaminated water. AOPs were first used in the production of high-quality drinking water from contaminated ground and surface water sources in the early 1980s, which are simply referred to as the oxidation processes involving the catalyst that is capable of generating hydroxyl radicals (OH•) in sufficient quantities to influence the oxidation resulting in water purification. Subsequently, concepts of AOPs were extended to the oxidative processes involving sulfate radicals ($SO_4^{\bullet -}$) as reaction initiators.

However, sulfate radicals were different from common oxidants such as ozone and chlorine, which are used for a dual function in water treatment as media for decontamination as well as disinfection purposes. Therefore, AOP technique is generally used primarily for destruction or degrading of organic or inorganic pollutants in groundwater, surface water, and wastewater sources. Interestingly, AOPs inactive against pathogens and pathogenic indicators, therefore, are rarely used for disinfection of biological contaminants. This is because radicals in AOPs possess a short half-life (order of microseconds), which restricts required detention times of pollutants for disinfection and are also ineffective against biological contaminants due to their extremely low radical concentrations. However, when AOPs are used in wastewater treatment, these are often employed as powerful oxidizing agents, which can cause sufficient destruction to wastewater pollutants, resulting in the transformation of toxic pollutants into less harmful and/or even nontoxic products, thereby providing a sustainable and effective solution for wastewater treatment [85–88].

On the other hand, recent advances have adopted photo-Fenton process as an AOP, which makes use of hydroxyl (•OH) radical to disinfect and decontaminate organic contaminants in polluted water. Fenton reaction proceeds with the formation of •OH radicals from H_2O_2 through an electron transfer process between a redox-active metal (often ferrous iron) and H_2O_2 in a homogeneous catalytic media. Its nonselectivity makes it ideal for the removal of a range of organic compounds and microorganisms that are resistant to antimicrobial systems [13, 19, 89–91].

For years, APOs have adopted Fenton and photo-Fenton processes involving H_2O_2 and iron salts to mineralize recalcitrant organic compounds. However, recent advances introducing nano- and bio-based sustainable catalysts have shown substantial improvements in degradation, decomposition, and treatment of wastewater due to development of multifunctional catalysts. In this direction, a catalytic wet peroxide oxidation (CWPO) process uses solid catalysts with redox properties in place of iron salt to produce free radical species from H_2O_2 decomposition [92, 93]. This new approach allows the catalyst to work on a broader pH range than the homogeneous process which is a limited pH range of 3–3.5 and CWPO also enables the reuse of catalyst for further cycles. In addition to this, supported catalytic routes have also been explored to develop advanced catalysts for AOPs. One such approach is the use of iron-functionalized AC as media in CWPO of phenol in synthetic and real water matrices. Here, the functionalization support carbon which was procured from commercial sources was carried out using sulfuric acid, nitric acid (HNO_3), and ethylenediamine that induced the AC surface with functional groups which was further enhanced by the presence of iron. These multifunctional activations enabled advanced catalyst stability and reusability. This also enables the use of low-cost and naturally abundant iron precursors like goethite and magnetite as green catalysts in the CWPO process [94].

The process of oxidation can further be enhanced and intensified based on the choice of catalyst-hazardous cyanotoxin like cylindrospermopsin and a chlorinated organic compound such as 1,2,4-trichlorobenzene with the help of visible LED, respectively. Further, the use of magnetite in combination with H_2O_2 has shown greater efficiencies in removing cyanotoxin through which ~80% pollutant decomposition is achieved. Interestingly, catalytic system showed limited iron leaching from the media, giving it exceptional stability over a period of time. On the other hand, use of goethite in combination with VIS monochromatic LED light (470 nm) proved to be an effective catalyst for the CWPO of chlorinated organic contaminants at neutral pHs. In this case, light source facilitated the reduction of Fe(III) in the parent goethite surface to Fe(II), resulting in hydroxyl radicals faster than Fe(III). Similarly, the heterogeneous photo-Fenton processes in combination with light (CWPO-light) have been successfully used for the abetment of bisphenol A in a contaminated water source. Interestingly, a life-cycle assessment (LCA) protocol was also used for the identification of significant steps involved in using magnetite particles covered with introduction of humic acids (HAs) as effective and green heterogeneous photo-Fenton catalysts for water treatment. Here, the use of humic HAs improved the efficacy and stability of the photo-Fenton catalysts [95, 96].

On the other hand, application of Fe-mineral-based natural materials like pyrite, ilmenite, chalcopyrite, and chromite have been widely used as effective catalysts for the degradation of harsh and rigid contaminants like antibiotic cefazolin. The use of natural minerals in heterogeneous electro-Fenton degradation process yielded significant results with stability and reusability perspectives. Upon reuse of Fe-based chalcopyrite minerals in AOP experiments showed a negligible decline in the catalytic degradation activity of chalcopyrite after five successive recycle and reuse runs. Recyclability and reusability characteristics of Fe-based natural mineral catalysts added significantly to the sustainability and affordability. The empirical assessment also confirmed that natural iron-based mineral catalysts can be effective alternative catalysts for AOP due to their high abundance, ecologically benign nature, high catalytic activity, and low energy intensity for their extraction and processing compared to synthesized catalysts [97–100].

On the other hand, persistent pollutants used for agricultural applications end up reaching water resources. Flood irrigation pushes excessive fertilizers, herbicides, pesticides, and insecticides to surface water. On the other hand, long-term use of excessive chemicals in agricultural activities results in soil contamination which over a period of time reaches surface water through flood irrigation and via soil diffusion and leaching into groundwater sources. Most of the substances used for agricultural activities have been classified as persistent organic pollutants, which can remain in the water bodies for a long period of time. Moreover, most conventional treatment techniques are ineffective against persistent chemicals. Therefore, advanced techniques like AOP are being adopted with suitable catalysts to address the tedious contaminants. In this direction, chelate-modified Fenton process in which organic and inorganic chelating agents have been used to enhance the Fenton process for the remediation of persistent chemicals along with co-pollutants like common organic contaminants. This approach is advantageous as it enables the use of biodegradable and non-toxic chelating in Fenton process. Also, use of renewable sources of energy for photocatalysis also makes it an energy-efficient process.

In this direction, nanocomposite of clinoptilolite with TiO_2 doped in AC has been developed as a sustainable photo-Fenton catalyst for the adsorption/photocatalytic hybrid process of organic pollutant bromophenol blue dye which was exposed to UV-irradiation (400 nm) to achieve complete degradation. Along with experimental assessment, mathematical models were also adopted to predict the degradation pathways of pollutants under the influence of various operational variables, which enabled the determination of the oxidation routes. This multifaceted approach helps in decisive end-process design for the real-time application process. Further, a pharmaceutical drug molecule triclosan used for antifungal and antimicrobial applications is found in contaminated water sources photocatalytically degraded following specific kinetics and monitored for reaction of by-products throughout. Similarly, a systematic model for the degradation kinetic of S-metolachlor was achieved using a heterogeneous photocatalytic system. A detailed study using sustainable catalysts provided a scalable process design having easily adaptable LED–photocatalytic reactors [101, 102].

Novel heterogeneous catalytic-like materials like lanthanum nickel oxide ($LaNiO_3$) were also used for photodegradation of pollutants with the help of support oxidants such as persulfate (PS) serving as adsorbents. Here, lanthanum nickel oxide was very effective as an activator for PS for the photodegradation of micropollutants such as piroxicam, sulfamethoxazole, and methylparaben present in surface water sources. Recycling and reusability studies also showed excellent stability of catalysts under successive experimental runs. On the other hand, natural bio-nanocomposite prepared using eggshell functionalized with Ag-Fe was successfully used as an efficient adsorbent material for the removal of toxic elements like arsenic, lead, and mercury from contaminated water. Therefore, search for sustainable materials stock to realize the large-scale utilization of the photodegradation process with optimal parameters in places requires rigorous scrutiny of the advanced AOP technique. Continued efforts have also shown potential for efficient and sustainable innovation in this direction enabling the pollutant absorption possibilities from gaseous streams and their subsequent abetment. In this direction, the removal of perchloroethylene (C_2Cl_4) from gaseous streams was achieved following two types of absorption tools, namely absorption column and jet absorber which was connected to electrolyzers [103–105].

6.4 Evolution of Oxidative, Magnetic, and Photocatalytic Treatment Composite

Recent developments in water management regulations and stringent policies have led to the forced implementation of advanced water treatment techniques into industrial practices. These advanced wastewater treatment steps have been mainly categorized as tertiary treatment techniques. The tertiary water treatment techniques fulfil the adequate safety of harmless discharges of treated water in both primary and secondary treatment levels. Nanomaterials as catalytic systems in AOPs have now gained significant importance as alternative oxidants in WWTPs. Among many methodologies, CWPO has shown promising characteristics in which highly reactive hydroxyl radicals ($HO^•$) initiate degradation randomly to provoke oxidation of multiple organic pollutants simultaneously. In particular, H_2O_2 in CWPO is used as an oxidizing agent for the generation of $HO^•$ radical (E^0 $HO^•/H_2O_2 = 2.73\,V$). Further, the CWPO is considered a heterogeneous version of the well-known Fenton process, which has mainly emerged as an alternative to conventional Fenton process overcoming limitations of homogeneous treatments like generating secondary pollutants due to iron sludge, or high reagent consumption [106–108].

Therefore, heterogeneous CWPO treatment method has proven an efficient protocol offering effective catalyst recovery/recycling and recovered catalysts can be further reused, which greatly contributes to the reduction of fresh resources and overall treatment cost. However, the immobilization of catalysts (iron or other catalysts) on the support poses compatibility issues, which overall act as detrimental to the catalytic activity of the oxidation process. This is due to the use of high surface area support generally, which attains nonuniform distribution of active catalyst loading which limits it overall performance when dispersed into wastewater sources. Additionally, supports are also designed to provide desired multifunctional active sites for catalysts to interact with pollutants for enhanced adsorption capacity, low catalyst loading, stability in a wide range of operating pH and regeneration of metal species, which results in in situ generation of hydroxyl radical source. For instance, the application of magnetic supports, or use of immobilized magnetic substrate plays the double role of CWPO catalytic site and assists in simple recoverability pathways and the coupling of an electromagnetic field integrated with the catalytic system or utilization of microwave has shown to improve the CWPO performance. In this way, performance of efficient catalysts has been designed for the decomposition of H_2O_2 into $HO^{\bullet}$, which is synchronized with the emerging demand for target oxidation of organic pollutants in the wastewater sources and has been consistently achieved [109–111].

Even though several combinations of organic, inorganic, and composite substrates have been tried and tested as support for catalyst species for the CWPO process. Among all, only carbonaceous materials with suitable surface functionality and surface area have proven effective. Carbonaceous materials not only offer multifunctional characteristics but also can be easily recoverable and remain in relevance as polyvalent support, which can be altered or customized through simple facile modification to induce their surface affinity and functional characteristics. Here, surface chemical functionality, porosity, affinity, and potential physicochemical interaction of carbon substrates play a vital role in the adsorption of contaminants and oxidants, which enable optimal immobilization of metals/catalysts. Further, carbonaceous substrate surface also offers the stability to regenerate catalysts via redox active sites, thereby increasing the efficiency in the formation and regeneration of hydroxyl radicals, durability of catalysts, and recyclability among others. Therefore, in general, surface of carbonaceous materials provides high reactivity, which helps in generating desired functionalization. Also, functionalization is often associated with mesoporosity in high surface area carbon, which favors the accessibility to inner pores, thereby enhancing the density of active sites and accelerating the reaction kinetics through increased surface contact sites [112–116].

Nanomaterials of colloidal nature have demonstrated enormous potential in water treatment applications to substantially improve the water quality through advanced treatment techniques. However, due to their size-related complexities, nanomaterials have been carefully selected for application in water treatment technologies. For instance, nanotoxicity is an emerging concern as majority of nanomaterials used in treatment media have been noticed to leach out into the product water. In such a case, it is made sure that the scale at which nanomaterials are used for functionalizing active water treatment systems differs from nanosorbents in water treatment units and an analytical kit for a laboratory monitoring a target contaminant. Therefore, to avoid unexpected complexities in treatment systems, several biocompatible and eco-friendly substrates are being used as nanomaterial carriers or stabilizers. One such approach is the use of biopolymers as sustainable materials due to their abundance and low toxicity, whose structure possesses functional groups with an affinity toward diverse pollutants. This approach has enabled the assimilation of synergic properties of both biopolymers and their composite constituents useful for water treatment applications, specifically as absorbent techniques. However, one of the major concerns associated with adsorption-based treatment process is the recovery of spent absorbents, which should be easy to recover at a cost-effective procedure and reuse to make it more sustainable. For instance, nanocomposite of biopolymers with magnetic iron oxide nanocrystals yields a stable composite that is easy to recover using external magnetic separation stimuli making them a very practically viable alternative for water treatment through adsorption. Further, adsorption specificity for target pollutants can be achieved by adopting a rational surface modification of magnetic nanocrystals, which can yield a high adsorption capacity, stability, and reusability. For example, covalent immobilization of the biopolymer onto the surface of the particles is critical to ensure the successful recycling and reuse of biosorbents.

Further, various types of magnetically active biopolymers have been prepared as effective water treatment media in the past decade. These magnetically active cleaners can be termed magnetic polymer composites (MPCs) that are mainly composed of two constituents, namely an inorganic metal oxide and an organic system or biopolymer such as chitosan. In general, biopolymers or organic components of the composite do not exhibit magnetic properties; for instance, chitosan is a magnetically inactive constituent of the composite. Therefore, an additional magnetic constituent must be introduced in order to magnetically function and separate the absorbent post-treatment process. The magnetic particles that are typically embedded into chitosan include Fe_2O_3, Fe_3O_4, $CuFe_2O_4$, $CoFe_2O_4$, $NiFe_2O_4$, and $ZnFe_2O_4$. Interestingly, based on the type of preparation method followed, formation of MPCs can attain two different structures, namely magnetic core–chitosan shell and/or magnetic multicores homogeneously dispersed in chitosan as shown in Figure 6.2a and b. Among these two, magnetic core–chitosan shell nanocomposite materials have shown greater interest and proven their potential in treatment of water applications [117–119].

Even though several combinations of magnetic nanocomposites have been prepared and used for various applications including water treatment in the literature, however, iron oxide-based systems like magnetite (Fe_3O_4) or maghemite (γ-Fe_2O_3) are by far widely used cleaners due to ease of synthesis, strong super-paramagnetic behavior, their biocompatibility, and low toxicity. Further, the amine groups ($-NH_2$) functional group in chitosan readily interact with Fe-ions in contaminated aqueous solution through ion-exchange and complexation interactions, which make them versatile and effective chelating agents with a provision to recycle and reuse using external magnetic field. In this composite system, chitosan interacts with negatively charged molecules like hydroxyl (Fe—OH) functional groups present on the surface of magnetite particles. Similarly, various combinations of magnetic nanocomposites have

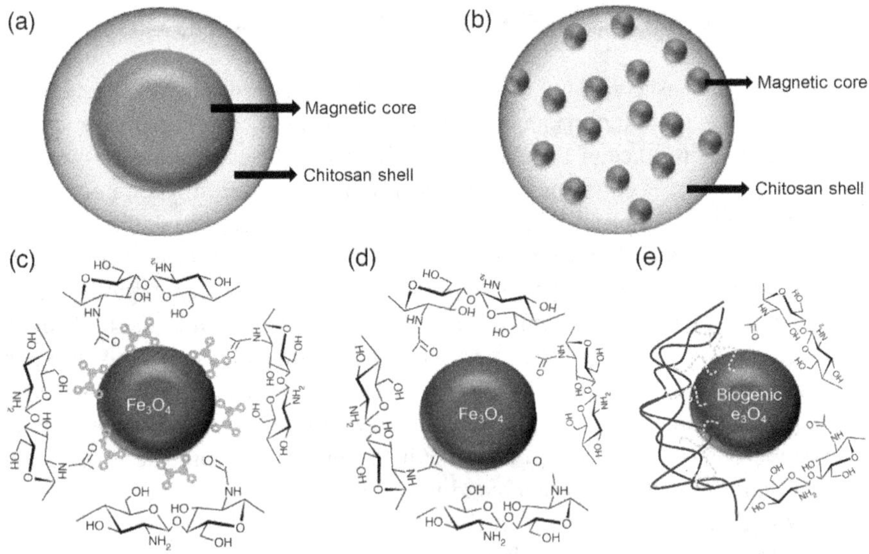

Figure 6.2 Among various preparation strategies, illustration shows the different types of (a and b) morphological arrangements obtained following the method of preparation and (c–e) showing the integrations in the chitosan/Fe_3O_4 magnetic composite materials.

been prepared using chitosan and iron oxide as the core are shown in Figure 6.2c–e [120–122].

Nonetheless, recent developments in advanced oxidation techniques have explored a series of photocatalysts suitable for real wastewater treatment applications. Among tested, visible light-driven photocatalysts have recently gained much attention due to their direct use in ecological decontamination and solar energy renovation in the presence of natural solar light. For visible light photocatalysis, semiconductor photocatalysts are generally preferred as they readily produce hydroxyl and oxidative radicals that are highly reactive leading to degradation of pollutants in the system. Therefore, following sections of this chapter discuss types of photocatalysts, nature of preparation, morphological features, properties, and their influencing factors during photocatalysis and degradation process. Following section also examines the use of doping technique to enhance the photocatalytic activities of the system with metals, bimetallics, and nonmetals, and magnetic, homogeneous, and heterogeneous photocatalysts. Even though recent trends have led to the rediscovery of titania-based composite catalysts and extensive use of zinc oxide-based systems and a similar energy gap of ZnO (3.2 eV) and an energy band pattern. However, titanium(IV) oxide has been the most widely adopted, affordable, and efficient photocatalyst in water treatment sector. This is also largely due to the greater chemical stability of TiO_2 compared to other photocatalysts [94].

6.5 Magnetically Metal Oxide Nanophotocatalysts

Photocatalysis is used in light and a catalyst to accelerate chemical reactions. Therefore, the phenomenon can be defined as "acceleration of a photogenerated electron in the presence of a catalyst," in which neither catalyst undergoes any transformation nor is consumed in the reaction process. Recently, photocatalysis has been adopted as a promising method, which involves chemical, physical, and biological processes. Photocatalysis is a cost-effective, eco-friendly, and highly efficient way to remediate various pollutants such as organic, inorganic, heavy metals, and microbial contaminations in presence of catalysts and light. Photocatalysis also plays a key role in the production of clean fuels through artificial photosynthesis in a green approach. In advanced artificial photosynthesis techniques, nanomaterials have been considered the best photocatalysts due to their properties including high photoactivity, safety, high stability, and low cost. Till now, metal oxide-based semiconductor nanomaterials such as TiO_2, ZnO, Fe_2O_3, WO_3, and $SrTiO_3$, have been considered the most efficient photocatalysts. This is due to their exceptional photocatalytic activities and wide bandgap useful in the visible region applications. In addition to this, additionally, nanocomposites comprising conductive polymers have been considered one of the best photocatalysts. The enhanced photocatalytic activities in these nanocomposites are attributed to their ability to produce electron–hole pairs upon absorption of light. When introduced into a contaminated source, electron–hole pairs in photocatalysts readily interact with the toxic contaminants to generate free radicals, which subsequently react with water to produce various degraded products [123–136].

Nonetheless, TiO_2 has been the most used and efficient photocatalyst for effluent treatment due to its low cost. Photogenerated hole-electrons are effective in producing highly photoactive oxygen species and are photostable over a wide range of concentrations and pH conditions. In particular, colloidal TiO_2 possess several advantages over other photocatalyst systems due to their higher number of delocalized charge carriers over the surface supported by high surface-to-volume ratio, improved charge transportation mechanism, longer shelf-life originated due to their dimensional anisotropy, and effective contribution to the separation of photogenerated holes and electron pairs. These colloidal TiO_2 or ZnO photocatalysts have been successfully used in photodegradation of adsorbable organic halides (AOX) present in pulp and paper mills that originated during the bleaching treatment stage with the help of UV radiation. In this typical process, the metal oxide-based colloidal photocatalysts are dispersed in the effluent at the end of the treatment and subjected to degradation process followed by the separation and recovery of the photoactive materials. Often, separation and recovery of colloidal photocatalyst are cost-intensive tasks and also require careful separation mechanisms to avoid contamination of the recovered systems. Therefore, use of support material for active photocatalysts in wastewater treatment is essential to attain sustainable recovery and reuse strategies.

In this direction, the immobilization of the photocatalyst on active supports is a promising strategy. However, often it is noticed that the conventional immobilized substrates are less efficient in obtaining effective photodegradation due to a drastic decrease in the specific surface area of the catalyst. Thus, to retain effective surface area of the photocatalysts upon immobilization, alternative strategies have been adopted. In one such attempt, coupling TiO_2-semiconducting phase photocatalyst with magnetic nanostructures enabled magnetically induced separation of the hybrid-supported photocatalysts. This strategy eliminates the additional downstream treatment processes, making it sustainable and cost-effective. Nonetheless, these new approaches have been effective against the degradation of smaller organic contaminants such as organic dye pollutants in controlled operating conditions. Further extended studies are underway to understand the effect of supported catalysts on degradation of complex wastewater systems and their recovery and reuse mechanisms. However, magnetic-TiO_2 hybrid colloidal systems have shown great potential in the treatment of real effluents such as photodegradation of AOX from pulp and paper mills [119, 137–140].

In this strategy, TiO_2 photocatalysts were prepared following a solid-state nanomaterial preparation technique in which magnetite (Fe_3O_4) co-precipitated in the presence of TiO_2. The co-precipitation has then resulted in magnetically active water-dispersible TiO_2 nanostructures with a magnetite content varying between 67 and 86 wt% in the reactants ratio. As prepared, hybrid nanostructures were composed of spheroidal magnetite having ~12 nm particle size and TiO_2 nanoparticles ~24 nm with anatase/rutile ratio 80/20 wt% phases. Interestingly, Fe_3O_4/TiO_2 hybrid colloidal photocatalysts showed a favorable absorption edge in the visible region dominated by the Fe_3O_4 whose direct band-gap energy of 2.27 eV. Subsequently, these magnetic Fe_3O_4/TiO_2 hybrid photocatalysts were used for the degradation of AOX compounds present in the Kraft pulp mill effluent generated during the bleaching process with pH = 2 in presence of UV light irradiation and H_2O_2 as the co-catalyst. Detailed studies also revealed that the AOX photodegradation was effective regardless of the ratio of TiO_2 content in the hybrid Fe_3O_4/TiO_2 photocatalyst system. Even though the highest photodegradation of AOX was achieved using pristine TiO_2 in a non-magnetic TiO_2 hybrid, which showed in the range of 59–73% AOX degradation over 60 minutes operation under UV irradiation. However, a greater advantage for sustainable separation and recovery was only achieved with the magnetic composite Fe_3O_4/TiO_2 for real industrial wastewater treatment and their subsequent reuse [141–143].

6.6 POM-Supported Photocatalysts

Polyoxometalates (POMs) are molecular metal-oxide anion nanoclusters that consist of abundant oxygen atoms and a series of early transition metals (TMs) including Mo, V, W, Ta, and Nb in the highest oxidation state. POMs offer multiadvantages featuring a broad range of nanostructures, chemical versatility, and functionalities making them some of the most versatile classes of inorganic molecular materials. POMs have recently attracted widespread attention due to their homogeneous catalytic activities in photocatalysis in which they offer switchability by visible light, the possibility of multielectronic photoreduction, followed by reversible electron exchange with a support substrate, have potentials in heterogenization, and are low-cost materials. Another major feature of POMs is their possibility to undergo heterogenization without losing POM properties, which is the unique feature of POM-based nanomaterials in photocatalysis of effluent treatment applications [144–149].

POMs are generally prepared following a condensation procedure in which metal salts in solution (mostly in water) are condensed at an appropriate pH and temperature. Based on the end use and application, the functionalization of POMs in their pure inorganic form is in situ carried out during preparation steps. This is significant as the functionalization steps followed during the preparation play a vital role in regulating the physical, chemical, and physicochemical properties of POMs. Generally, POMs can be functionalized following three strategies, namely (i) Counter-ion exchange method: In this strategy, the counterions of POMs generally having negative charges are exchanged with organic cations to adjust the solubility of POMs; (ii) Complexation: In this approach, POMs with abundant surface oxygens can be used as inorganic ligands to coordinate with metal ions to create high-dimensional coordinated complexes, (iii) Hybrid formation: In this strategy, POMs can be covalently surface functionalized using organic ligands, which enable the rational design of POM-based inorganic/organic hybrid materials. POMs not only offer exceptional structural diversity and easy synthesis methods but also possess excellent physicochemical properties that are useful in applications like biomedical applications, magnetism, energy storage, and conversion devices as well as photoelectrocatalysis. Therefore, after careful evaluation of physicochemical properties of POMs, researchers have identified semiconductor-like characteristics and potential photochemical properties in various hybrid systems and utilized them for photocatalytic degradation of organic pollutants found in different contaminated waters [150–152].

In addition to proven physicochemical properties, POMS possess several advantages as photocatalysts, namely (i) POMs are rich with TMs constituents such as Mo, Nb, V, W, and Ta, and most importantly, a large number of active functional sites are exposed on the surface, (ii) POMs can be easily modified to adjust the bandgap by changing the heteroatom constituents such as Si and P or POM structures can be altered to obtain the variation in the valence states of metal atoms in their structures to further improve the photocatalytic performance, (iii) POMs can also be functionalized following hybrid strategy in which organic ligands and/or loaded on substrate materials like carbonaceous materials, TiO_2, and other support materials, which enable in achieving synergistic effect between different constituents, and (iv) thorough understanding of POM's molecular structure using single crystal X-ray diffraction can yield specific structural characteristic information, which is beneficial for exploration of structure and function relationship at atomic levels [153].

Therefore, there is renewed interest in POM-based photocatalysts for water treatment applications through the photodegradation process. In particular, recent studies have demonstrated the efficient use of POM-based photocatalysts for outstanding photocatalytic activity and durability during wastewater treatment applications in which organic pollutants like dyes, organic pesticides, and other EPs have been successfully remediated. In this direction, many researchers have also invested their efforts in understanding structural design of POMs, effect of physicochemical characteristics on photodegradation process, and the detailed investigation of photodegradation mechanism that transforms the hazardous substances into less toxic constituents and their transformational kinetics. Further, the influence of POM structures, composition, and use of substrate for stabilization purposes have been discussed in the following section.

6.7 Degradation Mechanism in POM-Based Photocatalysts

POMs are an exceptional class of cluster compounds prepared as a result of coordination between pretransition and/or transition metals (PTMs/TMs) and oxygen atoms, comprising of cations and POM polyanions, in which the metal oxide polyhedra (MO_x, $x = 5, 6$) act as a basic structural unit of POM polyanions, and M generally represents a PTM/TM in a high-valence such as molybdenum (Mo^{VI}), tungsten (W^{VI}), vanadium (V^{V}), tantalum (Ta^{V}), and niobium (Nb^{V}). The POMs possess a semiconductor-like electronic structure, which constitutes an electron-occupied valence band (VB) and an unoccupied conduction band (CB). POMs as ligands have further added advantages due to their inherent stability against oxidation process. However, POMs-based composites and metal complexes are hydrolytically unstable, which tend to leach out metal in solution. Nevertheless, POMs possess the Anderson–Evans (P1), Lindqvist (P2), Keggin (P3) and Dawson (P4) structural classes following well-established protocols for the diversification of POM compositions and their physicochemical properties. POMs are a large group of discrete, mostly anionic polynuclear metal-oxo clusters amenable to a variety of chemical transformations. Interestingly, these complexes, each consisting of an array of corner-sharing and edge-sharing pseudo-octahedral MO_6 (M = Mo, V, Nb, W) units, form an ionic core. Nonetheless, structural representations of the archetypal POM molecules can span from 1 to 6 nm diameter range [154–157].

POMs are abundantly available and inexpensive photocatalysts initially used for the photochemical reduction of CO_2. As photocatalysts, POMs possess following advantages: (i) the redox capability with adoptable stability, (ii) the presence of definite active catalytic sites due to their crystal structures, (iii) the light absorption onto POMs can be altered through the introduction of suitable TMs into the structure, (iv) the reversible specific number of electron donation and acceptance abilities, which makes them remarkable catalysts in various catalytic applications. Therefore, initial interests in using POMs as photocatalytic systems in CO_2 reduction have gradually expanded to other applications involving liquid/liquid and liquid/solid systems such as water treatment and remediation.

In general, the photocatalytic activity in POM catalysts follows a similar trend to that of TiO_2 catalysts in which when the catalyst is exposed to light (energy) equal to or higher than its bandgap, the electrons in its VB are excited to the CB, resulting in photogenerated holes (h^+) and electrons (e^-) as shown in Figure 6.3a. Here, the positive holes will generate hydroxyl radicals ($^{\bullet}OH$) interacting with organic substrates in the presence of water. Further, hydroxyl radicals having strong oxidizing properties act as active species for the oxidation and degradation of organic substrates as depicted in Figure 6.3b. The photocatalytic activity of POMs can be further increased or altered depending on the target application with the introduction of photocatalytically active dopants like semiconductor TiO_2 into the structure. In such a case, POMs generally act as scavengers by capturing photogenerated electrons from the semiconductor moiety and turning into a reduced POM^- species, which delays the process of recombination of h^+ and e^- pairs. This further initiates the production of hydroxyl radicals through h^+ from the semiconductor much more efficiently. At the same time, the POM^--active species transfers an electron to the DO in the solution (wastewater or contaminated water in case of water treatment application) to generate oxygen radicals $O_2{-}^{\bullet}$, which then interreact with water to yield $^{\bullet}OH$ and/or H_2O_2 species. Further, $^{\bullet}OH$ and/or H_2O_2 species lead to the oxidizing of the organic contaminants leading to complete degradation as shown in Figure 6.3c. Among various types of POMs available, Keggin-type is widely used in photocatalytic applications due to their simple and convenient preparation, suitable bandgap, good stability, and efficient degradation of being involved against hazardous organic pollutants. Nonetheless, other types of POMs have also shown great potential in water treatment applications through undergoing structural modifications adjusting the bandgap via reasonable doping and surface modification and loading [158].

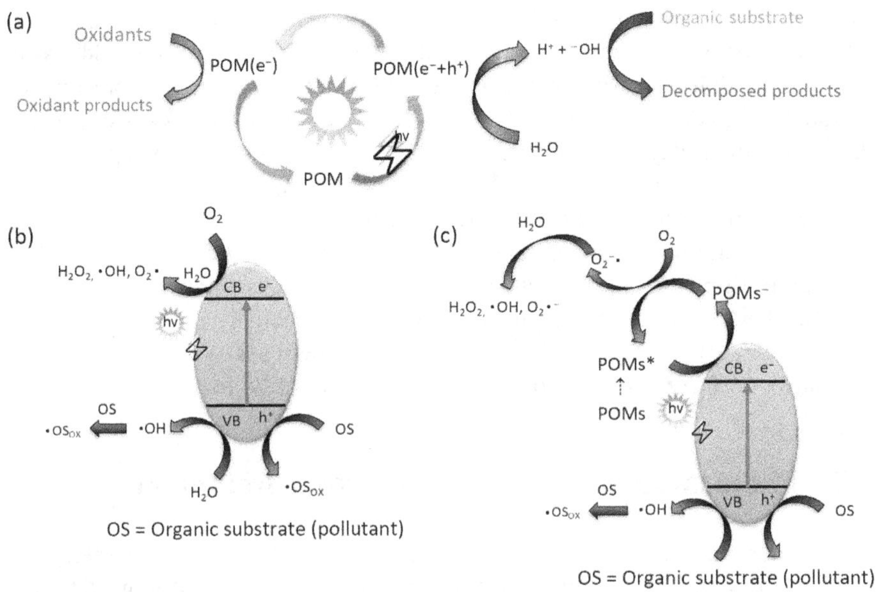

Figure 6.3 (a) Schematic representation of photocatalytic pollutant degradation mechanism in POM-based catalysts; (b) the photodegradation steps involved in pristine POM photocatalysts; (c) the photocatalytic degradation mechanism in supported or composite POM catalysts.

Further, various composites of POMs have been developed following hybrid nanostructured materials preparation techniques. One such approach uses $[PW_{11}CoO_{39}]^{5-}$ POM clusters in hybrid combination with graphene. Here, graphene was supported and further reduced to graphene oxide (rGO). Thus, POM/rGO has been optimized for water treatment applications in which organic pollutants of pharmaceutical origin have been efficiently treated. Pharmaceutical wastes are one of the most dangerous classes of EPs, which require careful remediation to protect environmental health. In this study, the new electron transfer possibilities result in rGO-$PW_{11}Co$ from the synergetic combination of POMs and rGO that have been efficiently explored for photocatalytic degradation of organic pollutants. To further evaluate the photocatalytic capabilities of the POM $[PW_{11}CoO_{39}]^{5-}$ supported in rGO (rGO-$PW_{11}Co$), the experiments involving organic dye like Rhodamine B (RhB) have been tested for its photodegradation under visible light exposure conducted at room temperature. Here, the choice of RhB was made as a model organic dye because it is photostable in the absence of a photocatalyst and possesses strong absorption characteristics in the visible region (maximum at 554 nm, 0 hour). To assess the effect of photoactivity and efficiency of a catalyst in dark conditions, identical experiments were also carried out in a blank experiment without a catalyst, which resulted in no change in solution color or no alteration in the RhB concentration suggesting rigidness of RhB under visible light irradiation. On the other hand, comparative photocatalytic degradation experiments using rGO-$PW_{11}Co$ and the pure solid K_5 [$PW_{11}CoO_{39}$] showed complete degradation of the organic pollutant under similar experimental conditions. Also, rGO-$PW_{11}Co$ showed superior photocatalytic degradation activity compared to pristine POM. Overall, the photodegradation of RhB was ~95% in 600 minutes in the presence of rGO-$PW_{11}Co$ catalyst under visible light irradiation. Similarly, several POM-based heterogeneous photocatalysts (rGO-$PW_{11}Co$) have also shown promising photocatalytic activity under visible light exposure conditions at room temperature making them sustainable and effective alternative for futuristic water treatment applications [159–162].

6.8 Photocatalytic Membranes in Water Treatment

Contaminants of emerging concern (CECs) or EPs are threatening the safety and health of human beings, aquatic ecosystem, and environment at large. CECs or EPs have been known to cause severe hazardous effects due to their complexities and lack of information on effective treatment. Although several methods and materials have been tried and tested against the EPs, however, issues associated with generation of secondary pollutants and their unknown forms have led to added concerns among water management experts. Use of nanomaterials in the form of adsorbents, constituents of carbon composites, and photoactive compounds has already demonstrated effectiveness for the removal of EPs/CECs in water sources and wastewaters. However, most of these efforts have been in either standalone dispersants, suspensions, or blend/composite systems. This approach has led to the generation of secondary pollutants as the nanoparticles tend to leach out into the product water. This limits their wider spread use as standalone treatment materials. Also, nanoparticle systems in a dispersed phase (generally due to colloidal nature of active materials) tend to aggregate over a period of time due to van der Walls interactions, which drastically reduces their active surface area and decreases their overall performance. More significantly, even in case of effective remediation of contaminants, removal of secondary pollutants or the spent adsorbent or active materials requires a tedious post-treatment step for the recovery or the separation of the spent active materials in the form of solid slurry. This multistep approach substantially increases the cost of overall treatment process. Nevertheless, photoactive materials have shown exceptional efficiencies in eliminating hazardous EPs from various sources through photodegradation process. Thus, to make effective use of photoactive materials in water treatment applications, innovative newer strategies need to be adopted. In this direction, immobilization of photocatalytic active nanomaterials has been one of the effective strategies explored following various techniques [163].

Among many such approaches, introduction of active nanomaterials in membrane matrices or polymer gels or ceramics has been explored widely. Membranes used in pressure-driven filtration processes like microfiltration (MF), ultrafiltration (UF), and NF are generally prepared from either polymer such as polysulfone (PS), polyacrylonitrile (PAN), polyether sulfone (PES), polyvinyl fluoride (PVDF), polytetrafluoroethylene (PTFE), and polyaniline (PANI) or ceramic materials which possess high chemical, thermal, and mechanical stability. On the other hand, nanocomposite membranes are prepared to induce specific target functionality to the membrane to achieve target application goals. Composite membranes are generally prepared either by surface coating or entrapping nanoparticles through blending and compositing with the membrane precursor (polymer/ceramic) during the membrane manufacturing process or used as a free-standing catalyst as schematically depicted in Figure 6.4a and b [164].

Besides solving the problem of nanoparticle stability, coupling photoactive nanoparticles with polymer membranes enables easy scale-up potential, and reusability having less environmental risks and offers a fundamental advantage to overall process economics. Moreover, photoactive nanomaterials embedded in the active selective layer of the membranes or uniformly dispersed inside the membrane matrix are of interest. Interestingly, this strategy also reduces the membrane-fouling effect in the long run, which also solves one of the major drawbacks of MF, UF, and NF processes. Following this strategy, the mixed-matrix membrane can act simultaneously as a catalyst support while acting as a barrier

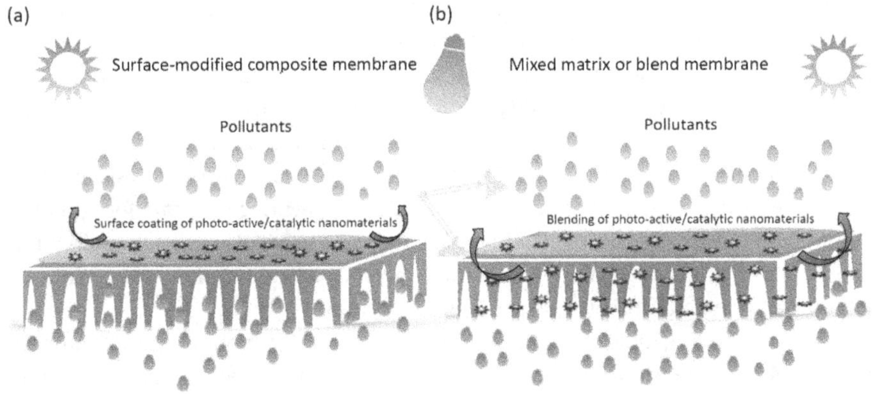

Figure 6.4 Schematic representation of catalytically active nanoparticles immobilized photoactive membranes with (a) surface-coating strategy: active materials on the active surface layer and (b) embedded strategy: nanoparticles dispersed throughout the membrane matrix.

for the pollutants to be degraded. Degraded products are simultaneously retained in the feed side of the filtration setup depending on their physicochemical properties making it a sustainable alternative for water treatment applications.

In this strategy, the introduction of photocatalytic active nanomaterials inside the polymer matrix or on the selective layer promotes changes in morphology membrane, which further induces required hydrophilicity due to presence of polarity. According to the solution–diffusion theory, increase in the surface hydrophilicity of the membrane greatly helps the water molecule absorption, diffusion, and propagation into the bulk of the barrier matrix. Also, controlled architecture of the composite membrane facilitates the controlled mass transport process, thereby improving its permeability. However, immobilization of nanomaterials into the polymer may adversely affect their surface area and in some cases, block their active functional groups present, which hinders the maximum utilization of vital parameters through their exposure to such as light irradiation and mass transfer for the degradation of pollutants into gaseous molecules. Nevertheless, several photoactive metal oxide nanomaterials such as TiO_2, ZnO, ZrO_2, and WO_3 have been explored for photocatalytic remediation of polluted water and wastewater. Also, a series of advanced materials both as standalone and composite forms have also been explored such as carbonaceous nanomaterials such as graphene, graphene oxide (GO), rGO, hydrothermal functional carbons, nanomaterial-induced carbon materials, g-C_3N_4, SWCNTs, and so on [165–169].

To make overfall membrane preparation process sustainable, modern quick techniques like surface-coating via electrospraying, dip-coating, sputter deposition, or gas-phase deposition have been adopted, which generates a specific thickness nanomaterial coating layer on its surface. In this direction, the dip-coating of a commercial ceramic alumina membrane with *N*-doped TiO_2 has been coated as an active functional layer through the sol-gel method. As prepared membranes were used for carbamazepine removal applications, however, presence of a coating layer affected the decreased permeability. Similarly, *in situ* growth of hydrophilic iron oxides (<4 nm) was used to alter the characteristics of ceramic membranes through which porosity, surface morphology, and the intrinsic hydraulic resistance of parent ceramic membranes were retained. However, rejection for organic contaminants like HA has been achieved significantly. Nevertheless, high loading of iron oxide nanomaterials into ceramic matrix obstructed the pore pathways resulting in denser and lower permeability membrane.

On the other hand, mixed matrix membranes prepared by blending nanomaterials in the polymer membrane matrix enable the minimization of active catalytic species leaching from the bulk. In this method, nanomaterials are generally dispersed in the polymer dope solution before mixing casted membrane into a gelation bath in a phase inversion technique. Following this method, photoactive TiO_2 nanoparticles (3–8 wt%) were dispersed in a porous PVDF/NaY-based membrane through a solution casting method. Interestingly, a study revealed a minor loss of (~3%) contamination removal efficiency of membrane compared to the nanomaterial suspension assays. In another study following a similar strategy, researchers immobilized NCQDs/BiOBr/TiO_2 in a PVDF matrix, which showed exceptional antifouling properties under visible light irradiation. The use of blended nanomaterials enhanced the degradation of the hazardous contaminants, which remarkably improved the filtration performance that is directly attributed to lower membrane fouling. However, presence of nanomaterials has been known to block the permeation channels, thereby directly affecting the membrane permeability at high nanomaterials loading. Therefore, immobilization of photocatalytically active nanomaterials in membranes is an effective strategy to overcome the limitations and disadvantages associated with conventional batch-type slurry systems adopted for photocatalytic applications. Nonetheless, suitable surface-coated membranes can remarkably improve hydrophilicity and permeation characteristics and induce antifouling properties. However, nanomaterials leaching from the polymer matrix need to be carefully studied to avoid the nanomaterial contamination in the product water. Thus, the strategy of using photocatalytic nanomaterials in filtration barriers can be ineffective in providing an effective catalytic phenomenon. In this direction, further in-depth studies and optimization of the use of nanomaterials, their performance, dose, porosity, surface dispersion, functionality, high photocatalytic activity, improved mass transfer, and so on are required to design sustainable technology and materials stock to address the broad emerging sustainable market commercialization of self-cleaning membranes for environmental application [170, 171].

6.9 Bio-Based Treatment Systems

Recent advances in sustainable development, green materials stock, and circular economic practices in technology development have led to the surge of interest in exploring bio-based polymer materials for water treatment applications. Over the years, biopolymer-based systems have been used in various personal care and pharmaceutical product development. However, it is only in recent years as the world is adopting sustainable manufacturing practices, research into biopolymer and biopolymer-based composites has gained significant traction. Biopolymers extracted from various terrestrial as well as from seaweed sources offer a cost-effective alternative materials stock and practical solution as they possess well-defined numerous functional active sites that can be effectively utilized for coordinated pollutant recovery, remediation, and/or removal from the contaminated water sources.

On the other hand, synthetic polymers such as polyacrylic acid, polyacrylamide, and polystyrene sulfonic acid, which are readily soluble in water have also been used as effective absorbents, coagulants, and flocculants. Polymers containing important functional groups like $-COO^-$, $-COOH$ groups, $-CONH_2$ groups, $-SO_3H$, and $-SO_3^-$ are responsible for coordination in polyacrylic acid, polyacrylamide, and polystyrene sulfonic acid, respectively. For instance, poly(glutamic acid) (PGA) with outstanding biodegradable characteristics also offers several other advantages like nontoxicity and biocompatibility. PGA is also known as an edible anionic polymer widely used in food industry, personal care products (PCPs), and water purification. On the other hand, biopolymers extracted from terrestrial biomass plants like *Hibiscus/Abelmoschus esculentus* (okra) have demonstrated promising performance for removal of colloidal pollutants even comparable to commercial standard polymer (polyacrylamide) flocculants used for the same purpose. However, biopolymer-based flocculant has shown several advantages such as it can outperform separation and removal activities over synthetic polymers without altering pH and does not require the additional coagulant [172–174].

Currently, synthetic polymer acrylamide-based adsorbents dominate the commercial market for several decades now to their excellent water solubility, high reactivity, and affordability. Polyacrylamide-based composites have also been prepared using different combinations of hybrid materials, which are readily water-soluble and have found applications in various wastewater treatments following coagulation, flocculation, and sedimentation, as well as in membrane forms. Polyacrylamide-based composites have been effectively used for the elimination of nano-/microcontaminants such as bacteria, colloidal solids, and organic pollutants. However, polyacrylamide systems have been lately categorized as toxic substances and nonbiodegradability which hinders their future utilization in water treatment applications. As a consequence, scientists have now diverted their focus on developing biocompatible, biodegradable, and greener polymeric materials stocks that are less toxic, and possess extraordinary contaminant removal capabilities at affordable cost protectives. Therefore, this section discusses the recent advancements in utilizing biodegradable or bio-based polymeric materials for water treatment applications.

Among numerous biopolymers, polysaccharide-based macromolecular systems have shown great potential in water treatment applications. Polysaccharides are made up of repeating units of sugar molecules that can be found in various natural origins such as plants, seaweeds, algae, bacteria, and so on. Historically, it is recorded that polysaccharides exhibit exceptional affinity toward water and help in diffusion, swelling, and transportation of water molecules when used in different forms like gels, films, aerogels, and membranes. A list of biopolymers and their origin generally used for water treatment applications in various forms and compositions have been listed in Table 6.1. Polysaccharide-based macromolecules in the form of hydrogel, solvogel, aerogels, foams, films, membranes, and xyrogels have demonstrated a remarkable ability

Table 6.1 A list of biopolymers originating from various sources and their chemical structures, most common solvents, and gelation mechanisms followed to form gels, hydrogels, solvogels, aerogels, xyrogels, foams, films, in membrane preparation, and further to be used in water treatment applications.

Biopolymer	Chemical structure	Active functional group	Soluble solvent	Water treatment module
Cellulose	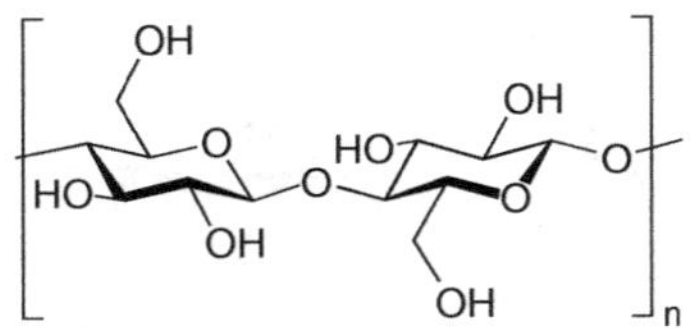	—OH	*N*-Methylmorpholine-*N*-Oxide (NMMO)	Membrane
Chitin	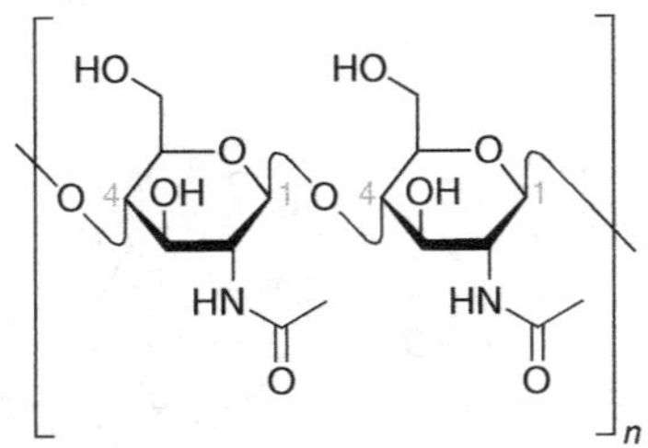	$—NH_2$, —OH	hexafluoroisopropanol (HFIP) acetic acid, formic acid	Adsorbent material, membrane
Chitosan	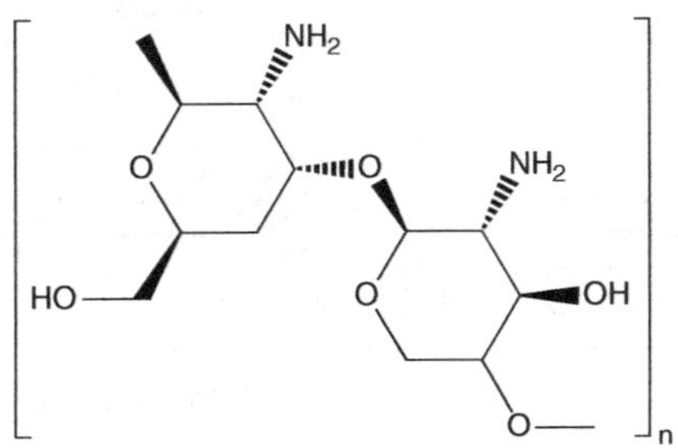	$—NH_2$, —OH	Acetic acid Formic acid Hydrochloric acid Citric acid	Adsorbent material, membrane, gel

(*Continued*)

Table 6.1 (Continued)

Biopolymer	Chemical structure	Active functional group	Soluble solvent	Water treatment module
Gelatin	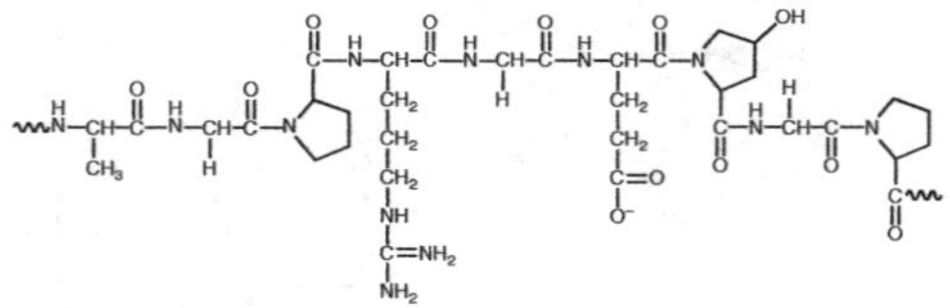	$-NH_2$, $-OH$, $-CH_2$	Water	Gel, membrane
Alginate	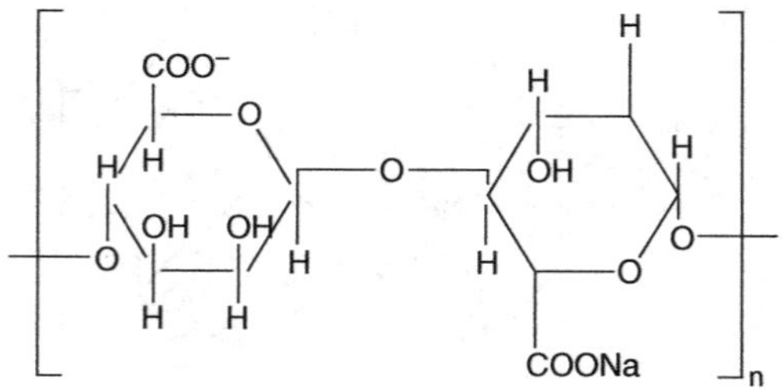	$-COO-$, $-OH$	Water	Gel, bead, membrane
Agar	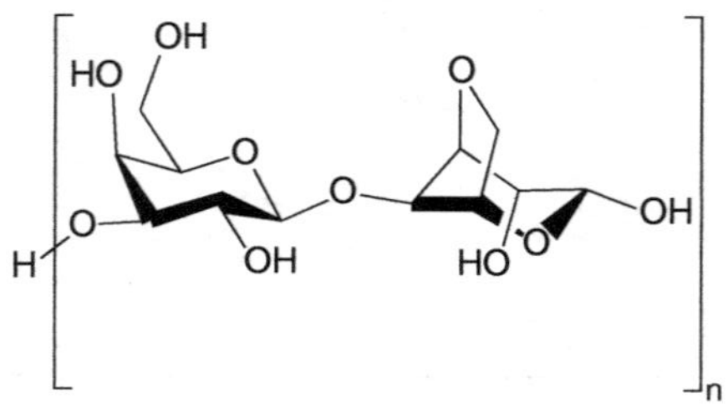	$-OH$	Water	Gel, plate
Hyaluronic acid	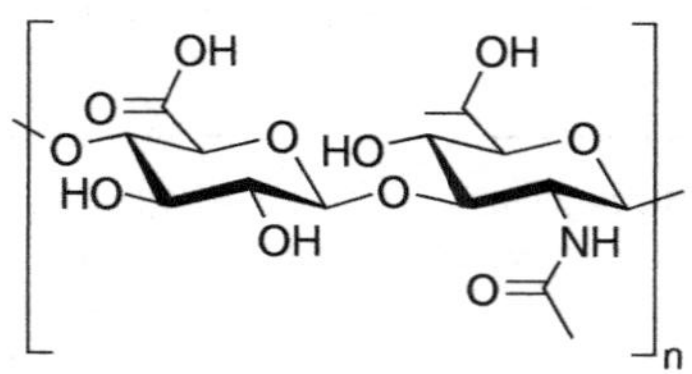	$-NH_2$, $-OH$	Water DMSO	Gel, membrane

Dextran b-cyclodextrin	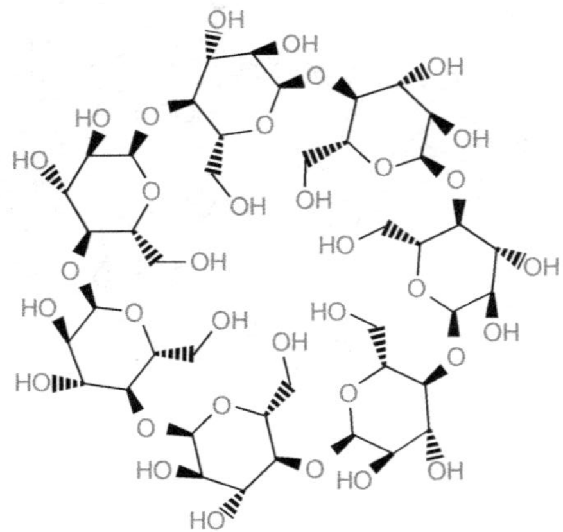	—OH	Water DMSO	Gel, membrane, inclusion complex, bead
Glycogen	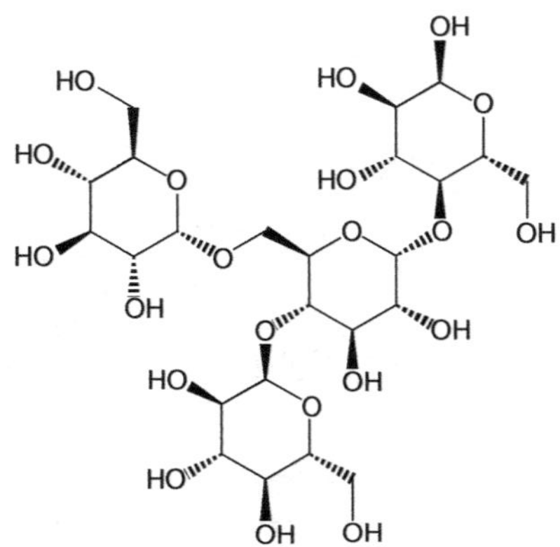	—OH	Water	Granular particles

(*Continued*)

Table 6.1 (Continued)

Biopolymer	Chemical structure	Active functional group	Soluble solvent	Water treatment module
Carboxymethyl cellulose (CMC)	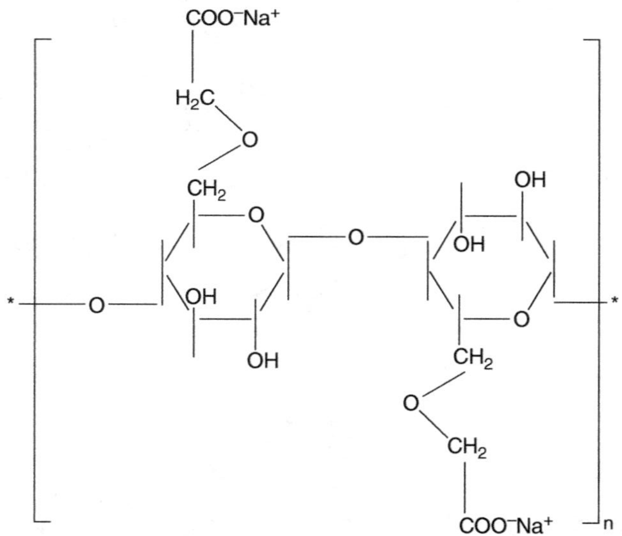	—OH, COO^-	Water, NaOH	Gel, membrane
Polyhydroxyalkanoate.	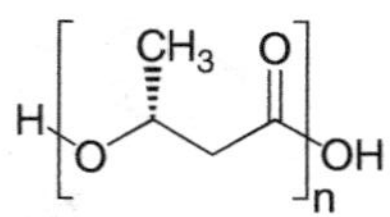	—OH	Chloroform	Film, bed, coating
poly(glutamic acid) (PGA)	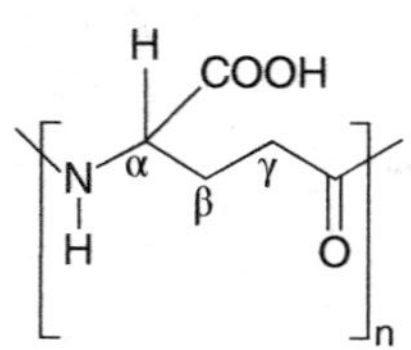	—COO^-NH_2	Water, NaOH	Film, hydrogel

Starch	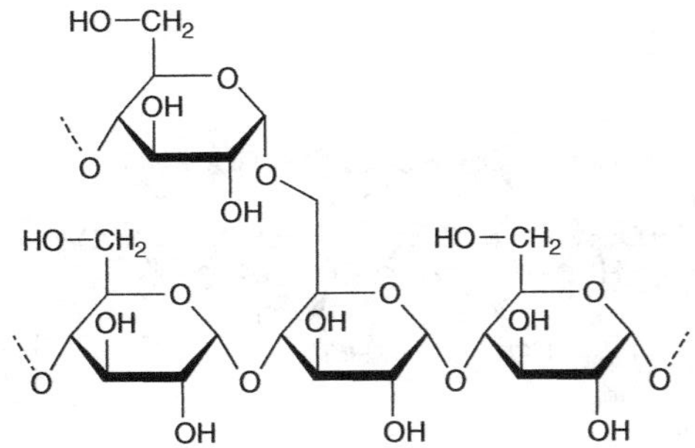	—OH	Water, glycerol	Gel, coating, membrane
Cyclodextrin	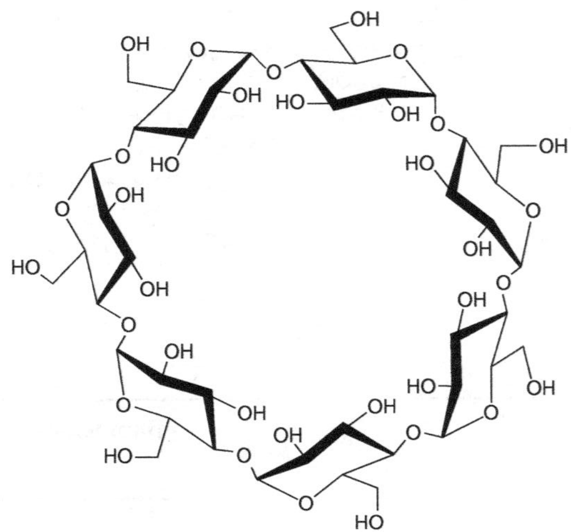	—OH	Water, glycerol, DMSO	Inclusion complex, bead, membrane

(Continued)

Table 6.1 (Continued)

Biopolymer	Chemical structure	Active functional group	Soluble solvent	Water treatment module
Polycaprolactone	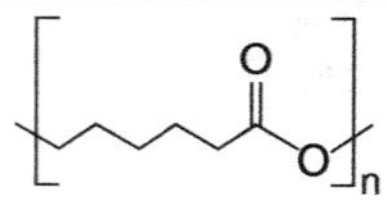	—OH	Chloroform, acetone	Film, scaffold, coating
Guar Gum	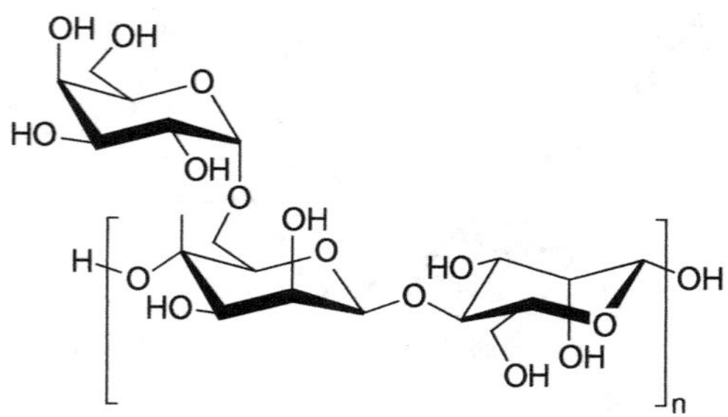	—OH	Water	Gel, coating, membrane
Gellan gum	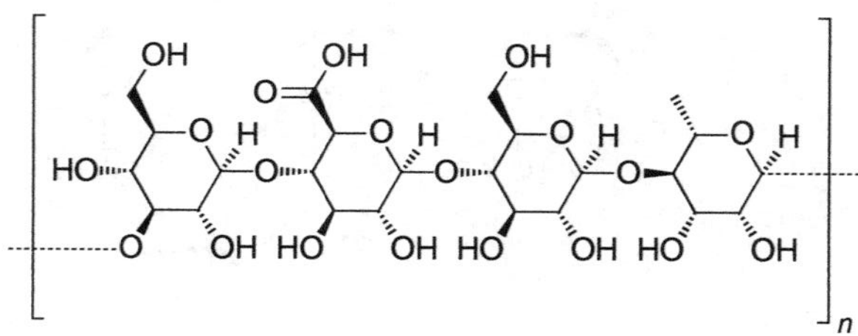	—OH	Water	Gel, coating, film

Polylactic acid	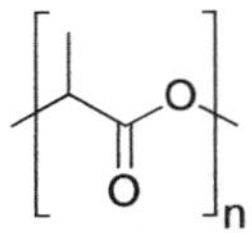	$-OH$	Chloroform, DMSO	Scaffold, coating, film
Carrageenan	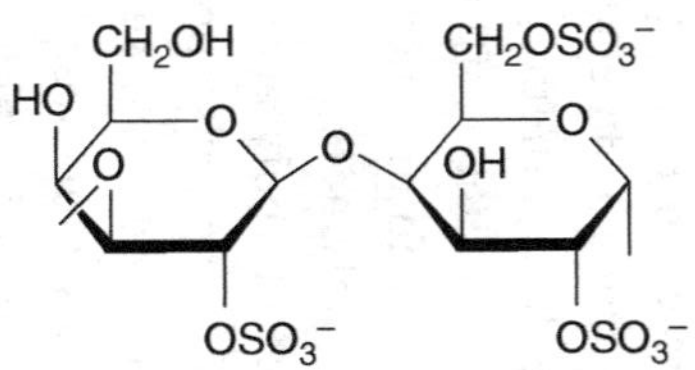	$-SO_3^-$	Water, NaOH	Gel, coating, film
Pectin	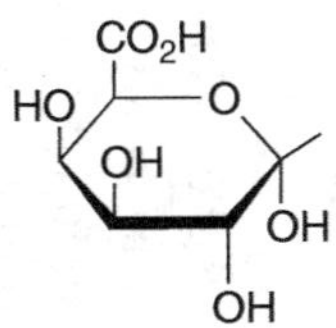	$-OH$	Water	Gel, coating, film

to absorb, interact, and/or eliminate various water pollutants such as organic dyes, heavy metals, pharmaceutical ingredients, surfactants, and other persistent organic pollutants. Notably, chitosan, derived from chitin, alginate, cellulose and its derivatives, gelatine, agar/agarose, gaur gum carrageenans, polylactic acid, dextran, and so on stands out as widely employed polysaccharides in water treatment application with remarkable water uptake abilities, contaminant adsorption capacity, water permeabilities, and, more importantly, antibacterial characteristics. Even though there are issues like solubility, pH sensitivity, and chemical/thermal/mechanical stability that limit direct use of biopolymers in various applications including water treatment. However, these biogenic materials (especially agar, chitin, chitosan, starch, gelatine, guar gum, and cyclodextrin) have been widely used for water treatment applications following suitable cross-linking techniques, blending, composite formation, and device fabrication strategies. On the contrary, chitin/chitosan and alginate have been frequently favored for their unique amino and carboxyl functional groups, respectively. Similarly, agar/agarose carrageenan and others have been explored for their distinctive super hydrophilic and exceptionally high water uptake properties. In all futuristic efforts, studies are focusing on the development of stable, long-lasting, and long-shelf-life biopolymer-based water treatment systems to ensure their effectiveness in real-world, large-scale water treatment applications [175, 176].

Polymer-based flocculants both from natural and synthetic origin have been widely used for water treatment applications in the past. Synthetic polymers in four different categories based on surface charges, namely anionic, cationic, nonionic, and amphoteric. Historically, cationic flocculants have been more effective and found commercial traction in the past than anionic flocculants that are used for the treatment of strong and negatively charged pollutants and vice versa. Even though some of the synthetic polymerics have been widely used as commercially successful flocculant effluent treatment applications. However, bio-based flocculants have found increasing interest as they offer several advantages over synthetic polymers as well as inorganic flocculants. Biopolymers, mostly comprised of polysaccharide origin offer several advantages like biodegradability, fouling resistance, nontoxicity, and abundant sources, which can be easily extracted from seaweed, marine dad species, agricultural waste, or forest resources. Moreover, biopolymer-based flocculants are stable at a range of pH conditions, adaptable for various types of applying conditions, easy to functionalize, eco-friendly, and have high flocculation efficiency at low dosages.

However, over the years, it has been noticed that biopolymers are vulnerable and might undergo quick biodegradation in some operational conditions, which drastically reduces their shelf-life and commercial viability. This requires careful chemical modification of the biopolymers to attain structural and operational stability. Therefore, use of suitable cross-linking agents and stabilizing agents could not only increase the structural stability of biopolymer system but the presence of additional functional and active sites on the backbone can also dramatically increase the flocculation efficiencies. Here, chemical modification techniques like oxidation, cross-linking, graft copolymerization, acetylation, acylation, etherification, esterification, and the Mannich reaction are followed to introduce both stability and multifunctionality to the biopolymers. Among all, grafting and cross-linking techniques are simple, easy, and economically viable techniques to follow.

6.10 Biopolymer-Based Flocculants and Coagulants for Industrial Waste Treatment

Among various biopolymer systems, starch family upon grafting had shown considerable improvement in their flocculation performances, namely starch-grafted-poly(acrylamide), amylopectin-grafted-poly(dimethyl acrylamide-co-acrylic acid), starch-grafted-poly(dimethyl acrylamide-co-acrylic acid), hydroxyethyl starch-grafted-poly(acrylamide), starch-grafted-poly(dimethyl acrylamide), hydroxyethyl starch-grafted-poly(dimethyl acrylamide), amylopectin-grafted-poly(dimethyl acrylamide), amylopectin-grafted-poly(acrylamide), and hydroxyethyl starch-grafted-poly(dimethyl acrylamide-co-acrylic acid). Flocculation performances of the grafted-starch family were then assessed using four synthetic colloidal pollutants including silica (1.0 wt%), coal (1.0 wt%), kaolin clay (1.0 wt%), and iron ore slime (0.25 wt%). Among all, results showed higher flocculation efficiencies for polysaccharide-based PDMA graft copolymers than the polysaccharide-based PAM polymers for silica, coal, and municipal wastewater treatment. Nevertheless, polysaccharide-based poly(DMA-co-AA) showed improved performance than the PDMA-, PAM- and PDMA-based polymers used for the treatment of iron ore slime and kaolin clay suspension. Here, dangling grafted chains of polyacrylamide branches on polysaccharide backbones enable easier access to pollutants, which further enhances the remediation process. Further, it has been noticed that the PAM-grafted biopolymer backbone is rigid due to intramolecular hydrogen bonding between the -NH_2 groups and >C=O of PAM molecules, which restricts the intractability of the grafted PAM chains to suspended pollutants compared to PDMA molecules. On the other hand, the absence of H_2-bonding interactions in PDMA chains enhances the flexibility and approachability of PDMA chain-grafted polysaccharides to various pollutants in the contaminated mixture. Therefore, presence of binary groups such as —COOH and —$CONMe_2$ on the backbone of biopolymers could be beneficial for the adsorption of contaminants compared to the flocculation process. Here, the static repulsion in the drafted polymeric system decreased the flexibility of the polymer chains, thereby decreasing the flocculation characteristics. Interestingly, use of binary monomers and presence of polysaccharide chain types in grafted copolymers severely affect their water treatment abilities. For instance, hydroxyethyl starch-g-PDMA shows improved flocculation efficiencies against colloidal pollutants than starch and amylopectin-based graft copolymers. This is attributed to the branching chains in hydroxyethyl starch that are more flexible and interactable with pollutants than linear and pristine chains in starch and amylopectin. Also, the hydroxyethyl starch-g-PDMA polymer showed remarkable removal efficiencies for suspended pollutants like silica, coal suspension, and iron ore slime from wastewater in comparison to commercially successful flocculants such as Telfloc 2230, Magnafloc 1011, and Percol 181 that are mainly made up of PAM having liner polymer chains [177–184].

On the other hand, acrylic acid (AA) grafting onto the starch backbone showed enhanced removal efficiencies for suspended pollutants in industrial effluents such as kaolin clay. This is enabled due to the presence of electrostatic interaction between aluminum silicate in kaolin clay suspension and carboxylic group (—COOH) in hydroxyethyl starch-g-poly (DMA-co-AA) chains. Similarly, AP-g-PAA graft copolymer showed excellent pollutant treatment efficiencies against suspended pollutants in mining industry wastewater in which turbidity due to suspended contaminants reduced from 510 NTU to 64.7 NTU in treated product water. Similarly, a lignocellulose-acrylamide-carboxymethyl cellulose copolymer was prepared through a microwave-assisted copolymerization technique, which effectively eliminated 92% of the cationic organic pollutant (methylene blue), which is estimated to be 32% more cost-effective compared to commercially used anionic polyacrylamide for same application. In a different approach, a cationic polymer carboxymethyl cellulose-acrylamide-dimethyl diallyl ammonium chloride was prepared following γ-ray co-irradiation grafting method for the removal of kaolin colloidal suspension from wastewater with 86% efficiency. In this process, graft copolymerization was designed following an environmentally benign bagasse extract to prepare a cellulose-based end product, a flocculant, which showed relatively high removal capabilities (>90%) for colloidal pollutants from lake water. Therefore, several other attempts have been made to design graft copolymers for various types of water treatment applications. The main advantage of biopolymer systems comes from their high charge density, which increases with increased chain length conformation in the solution. This long-chain conformation helps in bridging the flocculation effect on the pollutants. Also, biopolymer systems follow three main mechanisms through which flocculation progresses leading to separation of pollutants from the clarified water, namely (i) surface charge neutralization, (ii) polymer bridging, and (iii) electrostatic patch [173].

6.11 Biopolymer-Based Filters

In recent advances, several attempts are being made to directly make use of biopolymers that are derived from renewable sources to design affordable and easy-to-use filtration devices for both groundwater as well as wastewater treatment applications. Polysaccharides are major natural sources and abundantly occurring macromolecular polymers, which are extracted from renewable sources such as plants, animals, microorganisms (fungi and bacteria), and algae (see illustration in Figure 6.5a). Along with other biomolecules such as proteins and nucleotides, polysaccharides also play a vital role in exerting many essential activities in the biological system that include adhesion, cell/cell communication, and molecular recognition in the immune system. Polysaccharides which belong to the third major class of biopolymers, specifically carbohydrate systems, also influence the crucial roles in many different physiological processes providing structural characteristics, adhesion, protection, and stimuli responsiveness and playing additional roles in pathogenesis prevention, protecting immune system, fertilization, blood clotting, and therapeutic efficacy.

Nonetheless, biopolymers possess unique chemical structures, physicochemical characteristics, and functional groups that give specific chemical reactivity such as amino, hydroxyl, and acetamide, or functions in their polymer backbone and extended chain. This contributes greatly to their versatility in adopting biopolymers in various forms, chemical nature, sensitivity, and selectivity. Among many carbohydrate-based biopolymers such as chitosan, alginate, lignin, and carboxymethyl cellulose have gained significant traction in the form of gel (hydrogel/solvogel), aerogel, film, membrane, and powder/beads due to their simple processability, affordability, effectiveness, ease of preparation, and biocompatibility. These versatile biopolymers carry various functional groups on their backbones such as hydroxyl, carboxyl, methoxyl, amine, hydroxide, and phenolic, and most importantly, their high hydrophilic nature makes them suitable for water treatment applications. These functional groups also help in enhancing the remediation of pollutants through various interactions for the removal of heavy metals, chelating hazardous and precious metal ions, organic pollutants, and other emerging contaminants from water. For instance, chitosan having abundant hydroxyl and amino offers excellent adsorption capacity and biodegradability. However, chitosan also suffers from solubility issues due to its limited functionality, which requires acidic media, and is often criticized for relatively poor mechanical properties. These limitations lead to the innovation in composite formations, which adequately address the mechanical properties, stability, additional functionality, controlled swelling in different solution conditions, and adaptability to form desired shapes and sizes.

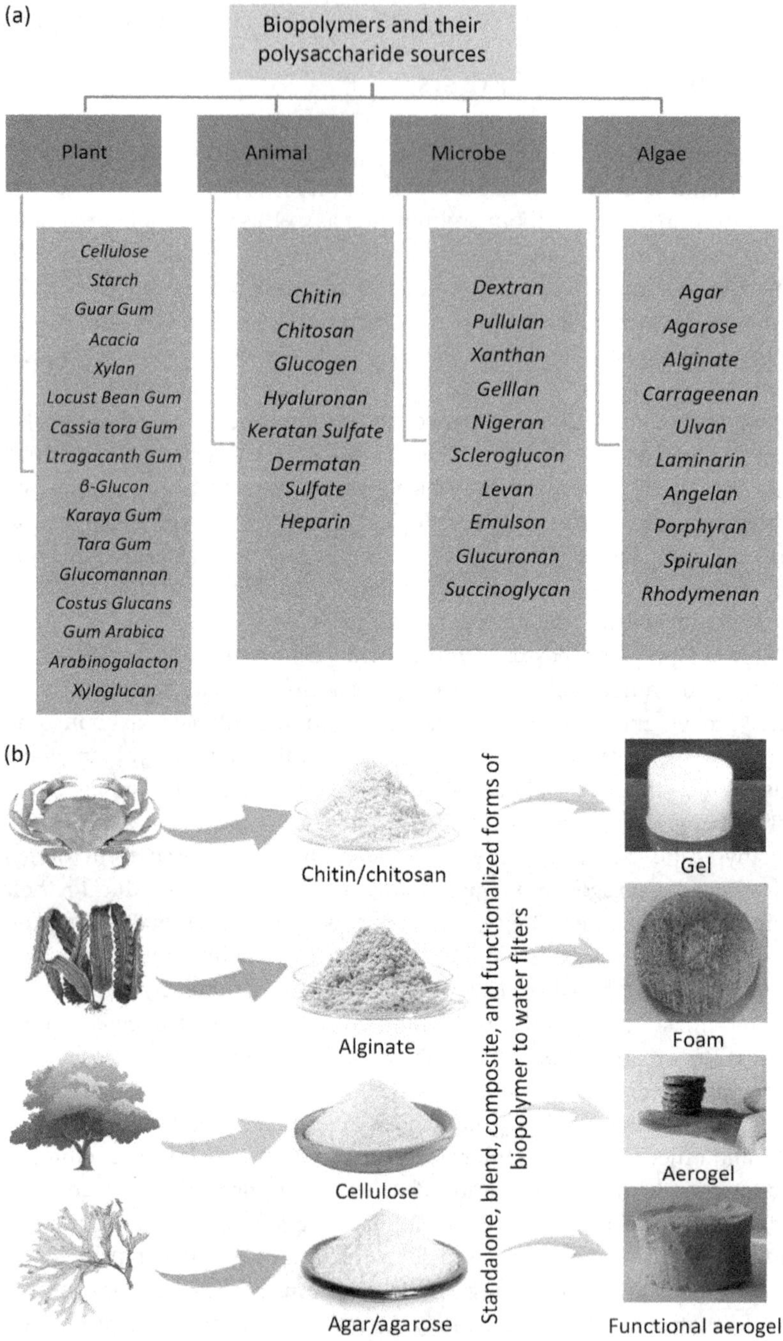

Figure 6.5 (a) List of polysaccharide-based natural polymers and their origin and (b) schematic illustration of sources of biopolymer and their potential engineered forms of water treatment systems. *Source:* Ozgur Coskun/Adobe Stock Photos; jamesjoong/Adobe Stock Photos; Luis Fernández García / Wikimedia Commons / CC BY SA 4.0; with permission of Boreal Bioproducts.

Another major issue with biopolymeric systems is the inadequate specific surface area, which hinders their widespread use in pollutant remediation as an adsorbent. Therefore, to overcome these limitations, biopolymers in general and chitosan in particular are chemically, physically, or biologically modified to create innovative end products of various forms such as gel (hydrogel or solvo-gel), aerogel, film, membrane, nanofibers, nanoparticles, microspheres, scaffolds, and so on (Figure 6.5b). Further, these unique user-friendly product forms can be achieved following simple chemical modification through cross-linking, blending, and compositing with suitable synergic constituents. For instance, in chitosan, chemical modification involves etherification, esterification, *N*-alkylation, and graft copolymerization. On the other hand, surface modification strategies can also be followed for stabilizing biopolymers through various effective strategies to enhance their pollutant interaction, adsorption as well as filtration capabilities which can be achieved through grafting, cross-linking, or functionalization techniques or nanocomposite formation protocols. Porous functional materials can also be prepared using biopolymer precursors through hydrothermal, solvothermal, and ionothermal preparation techniques to yield granular, powdery, as well as spherical functional carbon materials as cost-effective and renewable adsorbents/filters for water purification and the recovery of precious metal ions [185–187].

Further, Lignin is an oxygen-rich common aromatic biopolymer, which is a main constituent of plants along with cellulose. Lignin is the second most abundant material on earth after cellulose. Even though lignin has limited industrial applications, however recently, it has been widely explored as a fuel and an affordable carbon precursor for advanced applications. Lignin-based porous carbon (LPC) has demonstrated great potential in water purification that was contaminated with organic contaminants like dye effluents. However, recycling and reuse of lignin-based carbon is a timeous task and preparation often requires large quantities of pore-forming compounds or activating agents, which increases the overall product cost of the system. However, to overcome such limitations, novel strategies are being adopted to induce high porosity and controlled morphologies, and remarkable microstructure with minimal chemical usage is crucial. In one such attempt, a cost-effective LPC with a microporous hierarchical structure having reasonably high specific surface area (979.3 mg/g) a was prepared, which was further tested against the organic pollutants (2-cationic dyes, BG and CV and 2-anionic dyes, AF and O-II) demonstrating a high adsorption capacity of 700 mg/L. These results are encouraging in designing abundant bio-based resources for water treatment applications of future concern [188, 189].

On the other hand, as the commercial interest in biogenic materials is increasing, technological approaches to make processes more economical are also emerging at a faster pace. Innovative approaches are being adopted to produce high-value products from bio-based polymeric materials targeting water purification and treatment of industrial wastewater as a potential sector. One notable example is PGA, which has been widely used in the development of PCPs and has now been utilized for water treatment applications. PGA, which is an anionic polymer, is biodegradable and nontoxic in nature. It has emerged as a potential candidate for product development targeting industrial wastewater treatment. With the use of PGA, researchers have effectively treated water containing metal ions and organic dye molecules from industrial effluents with efficiencies ranging from 90 to 100%.

Chitosan is another polysaccharide polymer sourced from chitin and its derivatives have also shown great adsorption capabilities for the removal of toxic heavy metals, fluoride, pharmaceuticals, surfactants, and other emerging contaminants. Chitosan is generally extracted from crustacean shells, which possess rich amino and hydroxyl group functionality that can undergo complexation with metal ions. Chitosan has been used in the past for the effective removal of metal ion contaminants through adsorption mechanisms such as chromium, cadmium, lead, copper and so on. Similarly, activated high surface area carbon materials derived from biogenic macromolecules have been widely used for the removal of organic contaminants like dyes, phenols, pesticides, and pharmaceutical contaminants. Therefore, commercial-scale use of biogenic polymeric materials has proven their potential in water treatment offering several advantages over synthetic polymers.

First, biopolymer-based treatment materials possess versatile surface functionality, high surface area, tunable porosity, and affinity for water-soluble pollutants, which help them in effective removal. Secondly, regenerative properties of biogenic materials following simple pH alteration and washing techniques give them commercial traction.

Adsorption technique is widely regarded as a nondestructive technique among many water treatment methods. Adsorbents are generally used and reused during and after the removal of pollutants from water. However, one major drawback with adsorbents is that active materials retain impurities from wastewater or contaminated water sources without degrading the pollutants. However, biomaterial surface recovery and regeneration techniques have been designed to address the issue of secondary pollutants originating from adsorption of impurities. Through various surface regeneration techniques, such as altering pH, solvent washing, altering functional modification and so on, activities of adsorbents are restored, which enables their reusability. Active material regeneration is essential for designing cost-effective technology for large-scale water treatment applications. This is also crucial to address the sustainability aspect of the technology involving waste minimization, waste reduction, resource conservation, and recovering the active functional material.

Desorption of contaminants from spent adsorbents is a common technique followed to regenerate by removing the adsorbed pollutants. The effectiveness of the regeneration, decontamination, recovery, and activity of used adsorbents determines their reusability in subsequent cycles. An efficient absorbent can be reused and recovered several cycles at a commercial scale in industrial applications, which significantly reduces the cost associated with materials, preparation, and activating new adsorbents. However, it is often observed that regenerated adsorbents generally lose significant active sites and exhibit relatively lower adsorption capacity in comparison to fresh adsorbents. Nevertheless, selecting the appropriate regeneration technique/process is vital for enhancing pollutant desorption from the spent adsorbents and their performance. However, several factors need to be considered before designing and selecting a surface regeneration technique such as the type of adsorbent, energy requirements, eco-friendliness, cost-effectiveness, and replacement economics. Nonetheless, historically, several adsorbent regeneration strategies have been designed and investigated at a small scale. However, very limited have been implemented at a large scale. These include thermal regeneration techniques, chemical regeneration, microbial regeneration, microwave-assisted regeneration, ultrasound regeneration, electrochemical regeneration, and pH/thermal stimuli agents. In the literature, these techniques have been extensively studied, discussed, and implemented on at a small scale. Among all, thermal regeneration and electrochemical regeneration methods have shown promising potentials as techniques can eliminate large portions of both organic and inorganic species from their active sites. In particular, an electrochemical method is effective in regenerating porosity compared to thermal or chemical regeneration methods. However, the overall surface regeneration process and rate are slow in most of the cases as the fouling of electrodes in electrochemical system can prolong the process and the regeneration cost. However, it is important to note that there is no single or standalone technique that exists to effectively regenerate a wide range of adsorbents used at present. However, several combinations of techniques have been adopted with a series of optimal regeneration strategies based on the specific adsorbent, nature of active sites, regeneration time, cost of replacement, overall cost of the adsorption–regeneration process, effect of secondary pollutants in case of direct discharge of spent adsorbents and their environmental considerations. The disposing of spent adsorbents which are now categorized as secondary pollutants is also a difficult task and in this situation, recovering adsorbates after separating them from pollutants is a practical challenge.

However, adsorbents are now being used in continuous method to avoid the separating and recycling techniques generally practice over a period of time. Researchers are developing adsorptive membranes, filters, or porous system through which contaminated water is filtered out to produce clean product water. These biogenic porous and functional filters have been designed through various strategies to produce high water flux and effective retention of the pollutants on their surface. The continued efforts have shown in generating porous and functional water filters using biopolymer in the form of gels (hydro/solvo), foams, filters, films, and membranes following advanced material preparation techniques. This not only avoids issues associated with regeneration but also minimizes the equipment requirements during their use and handling.

6.12 Biopolymer-Based Gels, Hydrogels, Solvogels, and Aerogels

Biopolymer-based gels were one of the earliest forms of interpenetrating networks of polymer chains developed in the food industry. However, recent developments in materials science and separation technologies led to the rediscovery of biopolymer-based gel networks in various forms and compositions adopted for water treatment applications. Even though membrane technology has proven commercially successful for desalination applications, it lacks in treating industrial wastewater. On the other hand, several issues such as clogging, fouling, need for pretreatment step, post-treatment conditioning of membrane, and so on associated with the membrane have limited their wider application in wastewater treatment. Current membrane technology broadly follows the spiral wound module technology in its large-scale format to produce large quantities of the product water. Also, currently used membranes have shown several limitations against EPs such as fluoride and pharmaceutical and PCPs (PPCPs) ingredients, pesticide, oil/water, surfactants, microplastics, and persistent organic matters. These rigid pollutants have either penetrated the membrane barrier or have shown affinity toward membrane material upon which they diffuse. This limits their separation efficiency and proves ineffective for the removal. However, recent advances have shown encouraging results in addressing these pollutants using eco-friendly biopolymer-based separation systems. These separation media have either been designed and tested against EPs such as gels, aerogels, foams, films, functional porous matrices, and so on [190–192].

Gels have traditionally been used for selective adsorption of the contaminants in a solvent-rich media and their potential functional groups have been used to attract the hazardous species. Polymers have been historically used as flocculants for wastewater treatment.

However, there have been limited polymers that have shown large-scale operational potential. On the other hand, polymer-based gels have shown specific affinity toward specific pollutants. However, similar to polymer flocculants, gels have limited technological features for large-scale operations. Nevertheless, development of novel approaches has converted traditional polymeric gels into an effective barrier in which use of biopolymers has shown significant potential. More so, innovations in materials design and development in the nanotechnology sector have led to the modification of biopolymer architectures to achieve desired functionality to retain contaminants. Through these processes, it is possible to modulate the performance of these sustainable biopolymers that can lack in some aspects as mechanical properties, and thermal and chemical resistances of synthetic polymers. However, it is only in recent years that researchers, innovators, and technologists are pursuing biopolymer and biopolymer-composite gels and aerogels that have gained significant traction. This surge for simple and ultraporous filters as alternatives to currently used membranes can be attributed to compelling sustainability considerations, the remarkable and adjustable properties of these materials, and their versatility in terms of functionalization.

Among many porous materials, aerogel or foam-type filters have gained prominence as they offer simple and quick filtration potentials. These aerogels or foam filters have shown remarkable promise not only in retaining EPs but also producing large amounts of the purified product water. Filtration of EPs and secondary pollutants has always posed serious technological limitations for traditional membrane barriers. Therefore, advanced materials processing techniques in converting biopolymer stocks have led to the development of a series of standalone, hybrid, and functional filters. Historically, silica-based aerogels have been developed for applications like thermal insulation. However, now these ultraporous materials have been used for emerging fields such as filtration, oil/water separation, CO_2 capture, catalysis, and medicine.

Traditional inorganic-based aerogels have been prepared decades ago (over 90 years, by Kistler) as highly porous ultra-light, and ultra-high surface area possessing exceptional physicochemical properties. These properties encompass astonishingly ultra-lightweight, low density, high specific surface area (1000 m^2/g), ultra-low thermal conductivities as low as 12 mW/m/K under ambient conditions, and flexibility in forming with various combinations of precursors make aerogels versatile materials. However, as modern technologies demand a new combination of precursors to prepare aerogels in various forms and compositions, in such a situation, the definition of "aerogel" undergoes intense debate and transformation. Traditionally, the term "aerogel" was coined to describe materials prepared following supercritical drying (SCD) processes, whereas highly porous foam-type sol-gel materials are created following a combination of evaporative and freeze-drying techniques that result in ambigels, xerogels, or cryogels followed by porous 3D networks [193–195].

In recent times, developments in specialty polymers and macromolecular design as well as advent of non-SCD techniques have opened new avenues to prepare a variety of porous 3D network structures by retaining most of the original gel microstructure, composition, and intended pore network. Therefore, gradually conventional definition of aerogels has altered based on material properties rather than the drying process itself. Accordingly, various polymers of different origins can be used in this revised perspective with varying densities, micro-/mesoporous, that are all grouped into aerogel category. Even though this definition deviates from stringent IUPAC terminology, this confines aerogels to being strictly microporous 3D network structures.

Even though there has been a trend of referring porous materials prepared following sol-gel methods that predominantly contain macropores as aerogels, researchers have also named them general terms like foams, sponge, open-cell foam, and 3D porous networks that might be more appropriate in specific cases. On the other hand, in separation technologies, when porous materials are used, they have been termed foam filters, aerogel filters, ultraporous filters, and so on. In this direction, several attempts have been made to make use of polymers as effective precursors to obtain a 3D network of highly porous structures. Among many macromolecular precursors, biopolymer-based systems have found major attraction in aerogel preparation through two primary pathways: (i) Colloid formation: In this, precursors are derived from natural feedstock materials such as seaweed origin, alginate agarose, agar, pectin, chitosan, and so on. The aerogel properties prepared using biopolymer stocks are mainly influenced by concentration, functional groups, and use of stabilizing agents such as pore-forming reagents or cross-linking agents, the pH of the sol, gelation condition, and degree of gelation and the specific drying protocol employed. (ii) Powder-based precursors: Instead of using bulk polymers or high-molecular-weight precursors, in this approach, biopolymers or inorganic precursors are used to prepare in particle form in which the particle size ranges from micro- to nanoscale, for instance, chitin nanofibrils, cellulose nanocrystals, protein nanofiber aggregates, natural fiber-based fibrils, and nanofibrous cellulose. During aerogel preparation, these particle-/powder-form precursors also undergo systematic transformation into aerogels, and the resulting properties are shaped by factors similar to the colloid formation approach. Interestingly, in both pathways, the specific characteristics and performance of the resulting biopolymer aerogels are primarily dictated by the concentrations, gelation conditions, extent of cross-linking, extent and nature of functional groups on the precursors, the pH conditions during gel formation, the type and concentrations of cross-linking agents utilized, any optional adjustments made to the gels after their initial formation, and the particular drying process chosen for the final material.

In a typical procedure, precursors (biopolymers) are dissolved in solvents like water or organic solvents, and powder-/particle-based precursors involve reducing particles to the nanoscale, leading to aerogels with distinct characteristics. In both cases, the nature of cross-linking is crucial, which creates a 3D network structure following either hydrogen bonding to covalent bonding or weak van der Waals forces influencing the strength and properties of the resulting aerogels. Cross-linking of macromolecular chains can also occur through ionic interactions and physical entanglements. However, to obtain optimal strength gel followed by a porous final product, a balance of several factors is essential. Nevertheless, it is crucial to control the structural integrity in biopolymer-based aerogels as mechanical properties of biogenic materials are inferior to that of synthetic polymers, which influence the delicate skeleton network structure, high porosity, pore size distribution, and nature of resulting porous material. For this, the choice of drying method is vital that has control over temperature-pressure phase diagram parameters of the solvent. Among many (i) evaporative drying is widely used in which solvent drying process directly leads to the liquid-gas phase boundary. However, during the drying process, the capillary forces within small pores could lead to pore collapse and densification of polymer network resulting in a high-density product. Therefore, evaporative drying is less suitable for biopolymer aerogels. (ii) Freeze-drying or lyophilization process circumvents the capillary forces by sublimating the pore fluid. However, prior to lyophilization, gel needs to be frozen to get the optimal porosity, which comes with a drawback and extends the overall processing times, high energy consumption, and the formation of microcrystals. However, recent advances in cryogenic and freeze-drying technologies such as spray freeze-drying have improved microstructure preservation and properties. (iii) SSCD is considered the most efficient technique among all biopolymers drying leading to foam formation or aerogel without compromising the microstructure. Nevertheless, the high critical temperature and pressure required for water degrade biopolymers making direct use of SCD impractical. Interestingly, organic solvents with lower critical pressures offer better pore-forming abilities and their high critical temperatures still can cause thermal degradation. However, advanced low-temperature SCD like the Hunt process can replace the pore fluid with fluids like CO_2 before drying, which makes the most common choice for biopolymer-based aerogels due to cost and safety considerations [196–199].

Therefore, recent advances in aerogel filtration systems have explored biopolymers as a major materials stock. Selection of biopolymers has also possessed added advantages such as hydrophilicity, antifouling nature, and ease of forming gel network (Table 6.1). The majority of aerogels prepared for water treatment applications are derived from polysaccharide-based macromolecules. Among cellulose and its derivatives, chitosan and alginate are frequently favored for their unique amino and carboxyl functional groups, which offer distinctive properties and applications including removal of diverse contaminants from water sources. Some of the important biopolymers are (i) chitin and (ii) chitosan: Chitin having a highly crystalline structure and low solubility in many conventional solvents can be transformed into aerogels by dissolving in specialized organic solvents like LiCl/DMSO or concentrated formic acid. On the other hand, chitosan, a chitin derivative, can be dissolved in acidic aqueous solutions due to its protonation characteristics. Various gelation techniques, including adjustments in pH or temperature, aldehyde cross-linking, and hemicellulose citrate followed by lyophilization are available to create chitosan aerogels with tailored properties.

Further, chitosan-based aerogels are prepared by combining different polymer precursors and functional composite materials that can lead to enhanced properties and novel functionalities compared to pristine polymers. Incorporating inorganic and nanocomponents can enhance density, strength, deformation resistance, and flammability, and reduce the innate hygroscopicity of biopolymer aerogels. In the context of water remediation, several studies have been reported, where hybrid biopolymers have been utilized as selective separate barriers for various filtration processes. In one such attempt, a novel bio-based polycocolloid foam membrane (FM) was prepared using blend solutions of gelatine (Gel) and agarose (Agr). The Gel/Agr polymer gel structures were created using environment-friendly and nontoxic cross-linking agent genipin (G), which was derived from plant extract. With this approach, overall precursor stock originates from bio-origin making them a sustainable source of materials. These foam membranes were tested under different operational conditions involving both gravity and accelerated crossflow membrane modules. The superhydrophilic microporous FM permeates water selectively across the membrane retaining the oily phase in the feed stream. These seaweed-based polymeric network structures have shown a high degree of oil removal efficiencies (500 $L/m^2/h$ with ~98% pure water) at a low operating cost, which does not involve any applied pressure. Therefore, superhydrophilic and microporous FMs possess several advantages when used for tedious oil-contaminated water separation including natural abundance, easy processing, nontoxic materials stock, stability under different testing conditions, and no hustle in post-use spent membrane disposability. Similarly, another study demonstrated a simple method to recover water directly from stable emulsion waste using an aerogel membrane. In this case, the chitosan-based gel was converted into a highly porous aerogel membrane using bio-origin genipin as a cross-linking agent. The researchers also tested the aerogel's ability to recover high-quality water from an oil spill sample taken from a ship-breaking yard. The aerogel was able to recover high-quality water with a flow rate of over 600 $L/m^2/h/bar$, which is more than 99% pure. After repeated use, the aerogel membranes were evaluated for more environmentally friendly disposal options by biodegrading the membrane in soil [106, 107].

Furthermore, functionalized versions of ultra-porous foam and aerogel membranes were developed as sustainable, low-operating-pressure, and easy-to-use filters for various emerging treatment applications. In one such study, researchers have presented an innovative approach for effective water purification. Researchers have developed functional porous specialized aerogel membranes using a series of biopolymers (chitosan, agarose, agar, alginate, and gelatine) and functionalized their surfaces by tailoring polymer macromolecular architecture with Fe-Al-based nanocomposites, Zr-based composites, and hydrothermal carbon-induced fillers. Among these engineered aerogel membranes composed of ethylenediaminetetra-acetate dianhydride cross-linked chitosan and agarose in a 7:3 weight ratio with α-FeOOH- and γ-AlOOH-based nanocomposites exhibited macropores and short-range channels with a remarkably low density of 0.021 g/m^2, a high swelling ratio of 1974%, and outstanding pure water flux of 19,228 $L/m^2/h$. These results were superior, representing more than sixfold increase in flux compared to previously reported aerogel membranes. Further, a combination of agarose, agar, alginate, and gelatine was explored to prepare stable superhydrophilic aerogel membranes employed for the continuous flow purification of various pollutants including dyes; heavy metals (Cr, Pb); complex pollutants like fluoride, EPs, and arsenate; and biological contaminants under the influence of gravity. The optimized aerogel membrane demonstrated high rejection rates of 95–98.6% for both cationic and anionic organic pollutants like dyes with a permeate flux ranging from 1150 to 1375 $L/m^2/h$, whereas for EPs, porous membranes exhibited a rejection of 89–92% for retaining permeate flux of 1098–1165 $L/m^2/h$. When tested for As(V) removal efficiencies, these bio-nanocomposites (BNC-AG-0.1) retained a high adoption capacity of a q_{max} of 102.45 mg/g at pH 6.5, with over 93% rejection at a flow rate of 1000 $L/m^2/h$. More significantly, the membrane shows excellent arsenic retention performance of 92% tested over multiple cycles making it a commercially viable alternative compared to all existing bio-char-based adsorption systems for ensuring arsenic-free potable water. In addition to this, optimized aerogels also exhibited an adsorption capacity of 81.56 mg/g and rejection of ~99% for fluoride (F^-) at pH 6.5 and in continuous mode, membrane showed 99% rejection retaining flux of 250 $L/m^2/h$.

Similarly, zirconium(IV)-induced chitosan@agar (Zr-CA) bio-foam was prepared following a simple two-step reaction using biopolymer precursors chitosan and agar with *in situ* conversion of zirconia (ZrO_2) for removal of various EPs and heavy metals like Diclofenac (Df), Ciprofloxacin (Cf), Fluoride (F^-), and Chromium (Cr(VI)). On the other hand, solvothermal-derived carbon (STC) was induced into bio-foam with unique adsorption characteristics. These STC carbons having chemical functionality, confined morphology, surface acidity, and high surface area were immobilized in highly porous biofoam to prepare multifunctional STC-induced biofoam (STC-BF) filters. These functional biofoams were then converted into a stable filter device following gelation and lyophilization processes useful for wastewater treatment applications. The optimized STC-BFs were used as adsorptive filters for removal of several pollutants in water, such as organic dyes (methyl orange), anions (Chromate (CrO_4^{2-}) and Fluoride (F^-)), and other emerging pharmaceutical pollutants like Paracetamol (PAR), Ciprofloxacin (CIP), Dichlophenac (DIC), and so on. In all these cases, (STC, C-BF, and STC-BF) filters showed the synergetic effect of both STC and bio-foam (STC-BF) exhibited an exceptionally high maximum adsorption capacity for organic pollutant MO dye (896.56 mg/g), CIP (387.56 mg/g), CrO_4^{2-} ion (142.89 mg/g), and F^- ion (31.45 mg/g). Further, foam-adsorptive filters converted into real-time usable filters tested for multiple pollutants simultaneously, which showed an average flux of ~400/L/m^2/h with >90% rejection for MO, CIP, and CrO_4^{2-} ions and around 60% rejection for F^- ions. Overall, use of aerogel and foam technology for developing sustainable membranes and water filters has shown great potential of advanced approaches in robust removal of various contaminants from different sources while allowing for ultrafast pure water permeation. In all cases, recycling and reuse experiments showcased excellent recyclability features of advanced membranes/filters making them a futuristic alternative for existing technologies [108–113].

In similar attempts, a blend of chitosan-gelatin (CG), and GO-based gels were converted to aerogel as adsorbents for water purification. The study involved preparation of two distinct composites involving CG and GO (2% w/w), resulting in a porous matrix. These ultra-porous adsorbents were further used for the selective removal of fluoroquinolone antibiotics and lead (Pb) contaminants. The study revealed that the GO-coated biopolymer aerogels exhibited swifter adsorption for Pb ions. However, when applied for fluoroquinolone antibiotics, both GO-induced as well as GO-surface coated aerogels showed comparable maximum adsorption capacity. Even though adsorption efficiency for Pb was considerably high, it was more resilient in the embedded aerogel composite adsorbents. The adsorption features of the GO-CG aerogel suggest that the novel materials retained over 500 mg/g of Pb within just 240 minutes, which is 10-fold higher over conventional materials (AC) demonstrated through traditional removal methods. Interestingly, the adsorption of Pb onto GO-CG was greatly influenced by the pH of an effluent in which a maximum adsorption was achieved at pH 5. Moreso, the composite aerogel showed exchangeable capacity in which partial replacement of calcium ions (Ca^{2+}) and sodium ions (Na^+) with pollutant Pb revealed a stronger affinity for Pb compared to other metals like contaminants such as lanthanum (La^{3+}) and copper (Cu^{2+}). Finally, to make overall process sustainable and commercially viable, composite aerogel filters were regenerated using a brief washing of adsorbents with HCl or 2% HNO_3 to recover the active material for subsequent reuse. Therefore, advanced approaches in modulating biopolymer-based filters as an alternative to conventional synthetic polymer-based membranes show promising outcome. These novel composite aerogels/

filters/adsorbents through engineered functionality provide them with unique structures, morphologies, a large specific surface area, and affinity for target pollutants helping in effective binding sites [200–208].

In addition to these efforts, biopolymers have also been used for the surface modification of commercially successful synthetic polymer-based membranes to overcome their shortcomings. Currently used synthetic polymer-based membranes suffer from poor surface characteristics such as fouling propensity, poor hydrophilicity, and undesired functionality, which make them suffer from inferior separation and rejection performance. Therefore, to overcome the demerits of synthetic polymers, surface modification has been adopted through the grafting, functionalization, blending, or compositing of biopolymers. These biopolymers have been studied as mentioned earlier, indeed emphasizing the significant potential of biopolymer-based aerogels in addressing water treatment and remediation challenges. These engineered surface modifications have shown reasonably good efficiency in separating various complex and rigid pollutants from water such as oil, dye, arsenic, industrial waste, organic matter, and fluoride. However, it is also important to note that most of these biopolymers following stabilization steps through chemical cross-linking, which avoid the delamination and leaching from the overall system.

6.13 Prospects of Biodegradable Polymeric Materials in Water Treatment

Even though several studies have proven benefits from the use of biopolymer through various procedures including biopolymer surface modification, which exhibited considerable promise. However, there are several challenges to address and technological demerits to tackle to make these procedures practically viable. These challenges are mainly at the compositional level and encompass optimizing the grafting ratios, compatibility issues associated with blending synthetic and biopolymer combinations, coating process, matching appropriate adhesion properties to ensure long-term stability, and obtaining optimal separation as well as purification performances. Therefore, researchers are actively engaged in enhancing the effectiveness of biopolymers in their modified version either as a standalone or in combination with synthetic polymers. For this, various innovative techniques, assistance of nanotechnology protocols, and grafting techniques are being adopted. Among other techniques are incorporating nanomaterials into the bulk of the polymer membrane matrix, designing functional surfaces, and creating brush-like architecture, which prevent deposition of pollutants during the separation performance, creating hydrophilic permeation channels to promote high flux rates. Organic and bio-fouling are other major issues in conventional membranes as a majority of commercially successful membranes use synthetic petroleum-based membranes that are hydrophobic in nature, which attract organic and biological contaminants over a period of time, which further severely damages the performance. Therefore, use of biopolymers is expected to increase in membrane technologies and separation technologies in general.

In this direction, extensive studies have been reported investigating the regeneration of polymeric materials from deposited pollutants that have been characterized to cause damage to the synthetic polymer. However, use of biopolymers has been considered to add biodegradability, antifouling, and antibacterial aspects to the conventional polymer matrix upon and during the use of a membrane. Biodegradation here refers to the degradation of pollutants on the surface of biopolymer active materials that have been used either as separation media or treatment material through the metabolic activities of microorganisms. In cases like biological contaminants found in groundwater, surface water and industrial wastewater such as fungi, bacteria, and algae play important roles in which biogenic materials act as catalysts to trigger the biodegradation process through their antibacterial activities. Nevertheless, unlike synthetic polymers used conventionally, biopolymers or natural polymers upon use can undergo degradation in the environment, leading to the formation of biomass, CO_2, and CH_4 making overall process sustainable and clean to the environment. Several factors influence the biodegradation of polymeric materials, including the stability of functional groups, hydrophilicity, swelling properties, reactivity, molecular weight, and surface morphology. However, biogenic systems have been used in separation and purification technologies. For instance, the enzymatic hydrolysis of polysaccharide-based graft copolymers and pristine polysaccharides using α-amylase was observed in which *α-amylase* is an enzyme capable of cleaving the glycosidic linkage (1,4-α-D-glycosidic) in starch chains, releasing glucose residues. Interestingly, the concentration of glucose can be determined by measuring the optical density (OD) using a photoelectric colorimeter after adding DNS (2,4-dinitrosalicylic acid) at pH 6.7 in a phosphate-buffered solution (PBS). The polysaccharides exhibited a more rapid degradation compared to the graft copolymers. This can be attributed to the modification of the polysaccharide backbones through the grafting technique. Consequently, the polysaccharide contained a higher number of glucose moieties compared to the graft copolymers. As a result, the release of glucose was relatively lower for the graft copolymers in the solution treated with α-amylase. Additionally, chemically modified starch (HES) exhibited a slower degradation compared to starch and amylopectin. Overall, polysaccharide-based graft copolymers demonstrate biodegradability.

Furthermore, a study of the potential of *Fusarium sp.* for degrading partially hydrolyzed AP-g-PAM could lead to the easy disposal of the biogenic materials post-use in water treatment applications. Here, the growth of *Fusarium sp.* on the partially hydrolyzed AP-g-PAM in a sterilized modified Czapek–Dox medium takes place. Moreover, the enzyme *lysozyme* has been used for biogenic macromolecular component degradation such as polysaccharide-based nanocomposite hydrogels known as NHA, revealing the importance of improved removal efficiency of organic pollutants in the wastewater. Therefore, naturally abundant polysaccharide-based polymers exhibit superior biodegradability compared to synthetic polymers that can be widely adapted in various types of contaminated water treatment applications. Natural origin, special properties, and biocompatibility characteristics make them suitable candidates for applications in wastewater purification processes [209, 210].

On the other hand, use of catalytic nanomaterials in water treatment applications has exotically increased due to innovations in the catalysis and design and development of nanomaterials-based catalysts. Therefore, a combination of nanomaterials that are tuned to perform catalytic activities has led to enhanced degradation and photodegradation applications. There have been a series of biopolymers listed in the previous sections have been tried and tested on various synthetic polymer surfaces that have been further used for real-world water and wastewater treatment applications. Therefore, future holds the promise for use of catalytically active materials and biopolymers, which are deemed of continued innovation in this domain, leading to more efficient and sustainable membrane technologies that address critical societal and environmental needs.

6.14 Limitations, Challenges, and Opportunities

Even though a series of studies have reported the use of both biogenic materials and catalytic systems in water and wastewater treatment applications, however, researchers have enlisted several limitations. Most of the new findings have, during careful evaluation of results, highlighted the limitations of biopolymers when used as one of the components in the existing system such as polymer/polymer compatibility and their long-term stability. Therefore, synthetic polymers have been given due credit for their stability, robustness, and for inducing desired separation and purification end results. Nonetheless, in addition to use of several composite materials in combination with the synthetic polymer-based separation media, significant efforts have been undertaken to mitigate the harmful effects of synthetic polymers as standalone systems through various methods. Among many, grafting inorganic and biopolymer active chains onto the surface of polymer backbones has been the widely used technique. However, retaining grafted short-chain biopolymer on the synthetic polymer substrate has been a difficult task. Nonetheless, grafting of low-molecular-weight systems like polyethylene glycol, acrylamide-based polymers, and other oligomers has been more prominent in the membrane modification sector. However, retaining grafted polymer systems has always been challenging.

Therefore, lately, researchers have adopted a unanimous opinion to avoid relatively toxic polymeric systems in water treatment processes such as development of acrylamide-free polymeric materials. Several studies have highlighted the concerns associated with the use of acrylamide-based systems that potentially leave behind carcinogenic residuals through acrylamide monomers or generate toxic degradation by-products. Therefore, researchers should be careful in selecting the use of raw materials or precursors for a treatment media to avoid toxic effects that potentially leach out into the product water. Nonetheless, trend shows that future water treatment systems have considerably developed and used nonacrylamide-based polymers that offer a compelling alternative that fulfils the requirements of nontoxicity, low-cost, high performance, and biodegradability. However, researchers and technologists still need to be cautious in selecting raw materials or precursors in designing future water treatment technologies, media, and products that directly go into the bulk of the contaminated water sources such as nonacrylamide-based polymer flocculants and low-molecular-weight and unstable surface modification agents.

On the other hand, in many cases, membrane-based technologies, advanced oxidation techniques, and a combination of membrane/oxidation treatment technologies have been effectively used for the removal of organic, pathogenic, and inorganic pollutants. These technologies have also been adopted on medium-to-large scale for the treatment of severely contaminated organic/pathogenic/inorganic pollutants in practical implementation that is generally hindered by their high energy input and high-cost operations. However, use of advanced oxidation systems has been known to produce secondary pollutants that might consume additional secondary treatment steps and associated overhead costs. On the other hand, traditional flocculation and adsorption methods in their advanced versions that have been modified following nanotechnology protocols offer practical advantages due to their versatility, simplicity, ease of use, and affordability.

On the other hand, use of commercially available synthetic polymers has shown reasonably good performance in wastewater treatment. Most of them have either been extracted from commercially viable synthetic processes or derived from inexpensive monomers, such as AA, acrylamide, and methyl methacrylate. However, a major limitation with these polymers is their nonbiodegradability or they require a long time to achieve biodegradability after their use. This creates limited reuse opportunities and potentially causes serious environmental concerns. In such a situation, biogenic macromolecules showed limited replacement potential. However, commercial aspect of biopolymers is poorer in their robustness, chemical stability, and longevity. Also, challenges related to biogenic materials are their solubility and cost implications in adopting to commercial product development, which requires adjustment of reaction conditions. In most cases, limited biogenic polymer solubility hinders their direct implementation in commercially viable product development. To overcome this demerit, use of several chemical modification techniques has led to the development of useful products. However, use of additional consumables and modification processes has been known to generate secondary pollutants. This not only adds to the cost of the process but also overall product. Even though chemical modification steps have considerably improved the solubility and charge-density characteristics of biogenic materials, however, use of additional synthetic materials has not completely eliminated, which are crucial in inducing surface functionality and pollutant interaction affinity, which helps in remediation of contaminated water.

6.15 Conclusions

Detailed evaluation of the efforts in developing sustainable materials stock for water treatment applications reveals that use of biogenic materials has exponentially increased over the years. Another major reason behind the use of biogenic material is to develop products that are economically viable and environmentally friendly throughout their journey including resource extractions, transportation, operational costs, product fabrications, and disposal strategies. Additionally, most of the efforts in making use of biogenic and catalytic water systems have aimed at removing persistent EPs rather than conventional pollutants like hardness, heavy metal ions, organic matter or their mixtures, and real industrial effluents. The current trends indicate that future materials and technology development in water treatment sector emphasize the use of biogenic and environmentally friendly polymeric materials that are multifunctional, cost-effective, and resistant to severe application conditions and operational interference, and promote their direct use in real-time industrial applications. Nevertheless, several limitations associated with novel materials are also essential to make development activities practical and time-saving. For this, researchers have moved to extensive collaboration to utilize the skilled knowledge that is spread across academia, industry, and innovation sectors. This collaboration further needs to be aimed at addressing lowering technology/innovation cost, strategizing the material regeneration protocols, and reuse feasibilities, which provide a holistic understanding of materials/products/technology performances, and their potential real-world application scenarios.

Therefore, in conclusion, the development of biogenic materials and catalytic materials in various forms and compositions for wastewater treatment presents a great opportunity to address the emerging as well as existing limitations in current materials and technologies, and challenges associated with commercially available water treatment media and technologies. Even though the currently available commercial materials/methods/technologies have demonstrated satisfactory performance, however, issues associated with their nonbiodegradability, limited recyclability, and reusability still pose serious environmental concerns. Nevertheless, increased use of biogenic materials with suitable chemical modifications are often necessary to improve their separation, remediation, and treatment properties targeting commercial applications.

References

1 von Sonntag, C. and von Gunten, U. (2012). *Chemistry of Ozone in Water and Wastewater Treatment from Basic Principles to Applications*. London: IWA Publishing.

2 Courbat, R., Ramseier, S., Walther, J.L. et al. (1999). Utilisation de l'ozone pour le traitement des eaux potables en Suisse. *GWA Gas, Wasser, Abwasser* 78: 843–852.

3 Lim, S., Shi, J.L., von Gunten, U., and Daniel, L. (2022). McCurry, ozonation of organic compounds in water and wastewater: a critical review. *Water Res.* 213: 118053. ISSN 0043-1354.

4 Wolf, C., von Gunten, U., and Kohn, T. (2018). Kinetics of inactivation of waterborne enteric viruses by ozone. *Environ. Sci. Technol.* 52 (4): 2170–2177.

5 von Gunten, U. (2003). Ozonation of drinking water: part I. Oxidation kinetics and product formation. *Water Res.* 37 (7): 1443–1467.

6 Andersson, A., Lavonen, E., Harir, M. et al. (2020). Selective removal of natural organic matter during drinking water production changes the composition of disinfection by-products. *Environ. Sci. Water Res. Technol.* 6: 779–794.

7 Gonsior, M., SchmittKopplin, P., Stavklint, H. et al. (2014). Changes in dissolved organic matter during the treatment processes of a drinking water plant in Sweden and formation of previously unknown disinfection byproducts. *Environ. Sci. Technol.* 48 (21): 12714–12722.

8 Lavonen, E.E., Kothawala, D.N., Tranvik, L.J. et al. (2015). Tracking changes in the optical properties and molecular composition of dissolved organic matter during drinking water production. *Water Res.* 85: 286–294.

9 Li, X.-F. and Mitch, W.A. (2018). Drinking water disinfection byproducts (DBPs) and human health effects: multidisciplinary challenges and opportunities. *Environ. Sci. Technol. 52* (4): 1681–1689.

10 Wang, X., Mao, Y., Tang, S. et al. (2015). Disinfection byproducts in drinking water and regulatory compliance: a critical review. *Front. Environ. Sci. Eng.* 9: 3–15.

11 Meunier, L., Canonica, S., and von Gunten, U. (2006). Implications of sequential use of UV and ozone for drinking water quality. *Water Res.* 40: 1864–1876.

12 Wang, J. and Chen, H. (2020). Catalytic ozonation for water and wastewater treatment: recent advances and perspective. *Sci. Total Environ.* 704: Article 135249.

13 Miklos, D.B., Remy, C., Jekel, M. et al. (2018). Evaluation of advanced oxidation processes for water and wastewater treatment – a critical review. *Water Res.* 139: 118–131.

14 Rose, A.L. and Waite, T.D. (2003). Kinetics of iron complexation by dissolved natural organic matter in coastal waters. *Mar. Chem.* 84: 85–103.

15 Giannakis, S., López, M.I.P., Spuhler, D. et al. (2016). Solar disinfection is an augmentable, in situ-generated photo-Fenton reaction – part 2: a review of the applications for drinking water and wastewater disinfection. *Appl. Catal. B* 198: 431–446.

16 Wang, J. and Bai, Z. (2017). Fe-based catalysts for heterogeneous catalytic ozonation of emerging contaminants in water and wastewater. *Chem. Eng. J.* 312: 79–98.

17 Braslavsky, S.E. and Rubin, M.B. (2011). The history of ozone part VIII. Photochemical formation of ozone. *Photochem. Photobiol. Sci.* 10: 1515–1520.

18 Nagarkatti, M.G. (1991). Ozone in water treatment: application and engineering. *J. Hazard. Mater.* 20: 881–882.

19 Huang, C.P., Dong, C., and Tang, Z. (1993). Advanced chemical oxidation: its present role and potential future in hazardous waste treatment. *Waste Manag.* 13: 361–377.

20 O'Shea, K.E. and Dionysiou, D.D. (2012). Advanced oxidation processes for water treatment. *J. Phys. Chem. Lett. 3* (15): 2112–2113.

21 Hassaan, M.A., El, A., and Madkour, F.F. (2017). Testing the advanced oxidation processes on the degradation of direct blue 86 dye in wastewater. *Egypt. J. Aquat. Res.* 43: 11–19.

22 Wang, Z., Chen, C., Ma, W., and Zhao, J. (2012). Photochemical coupling of iron redox reactions and transformation of dissolved organic matter (DOM). *J. Phys. Chem. Lett.* 3: 2044–2051.

23 Mason, T.J. (2009). Sonochemistry – beyond synthesis educ. *Chem.* 46: 140–144.

24 Contreras, S., Pascual, E., Esplugas, S. et al. (2002). Comparison of different advanced oxidation processes for phenol degradation. *Water Res.* 36: 1034–1042.

25 Venieri, D., Karapa, A., Panagiotopoulou, M., and Gounaki, I. (2020). Application of activated persulfate for the inactivation of fecal bacterial indicators in water. *J. Environ. Manag.* 261: 110223.

26 Glaze, W.H., Kang, J.-W., and Chapin, D.H. (1987). The chemistry of water treatment processes involving ozone, hydrogen peroxide and ultraviolet radiation. *Sci. Eng.* 9: 335–352.

27 Wang, H.-W., Li, X.-Y., Hao, Z.-P. et al. (2017). Transformation of dissolved organic matter in concentrated leachate from nanofiltration during ozone-based oxidation processes (O_3, O_3/H_2O_2 and O_3/UV). *J. Environ. Manag.* 191: 244–251.

28 Rahaman, M.S., Vecitis, C.D., and Elimelech, M. (2012). Electrochemical carbon-nanotube filter performance toward virus removal and inactivation in the presence of natural organic matter. *Environ. Sci. Technol.* 46: 1556–1564.

29 Fu, L., Wu, C., Zhou, Y. et al. (2019). Ozonation reactivity characteristics of dissolved organic matter in secondary petrochemical wastewater by single ozone, ozone/H_2O_2, and ozone/catalyst. *Chemosphere* 233: 34–43.

30 Sedlak, D.L. and Von Gunten, U. (2011). The chlorine dilemma. *Science* 331: 42–43.

31 von Gunten, U. (2003). Ozonation of drinking water: part II. Disinfection and by-product formation in presence of bromide, iodide or chlorine. *Water Res.* 37: 1469–1487.

32 Regmi, P., Stewart, H., Amerlinck, Y. et al. (2019). The future of WRRF modelling – outlook and challenges. *Water Sci. Technol.* 79 (1): 3–14.

33 Liao, J., Lin, S., Zeng, M., and Yang, Y. (2016). A miniature photoelectrochemical sensor based on organic electrochemical transistor for sensitive determination of chemical oxygen demand in wastewaters. *Water Res.* 94 (May): 296–304.

34 Yang, J., Chen, J., Zhou, Y., and Wu, K. (2011). A nano-copper electrochemical sensor for sensitive detection of chemical oxygen demand. *Sensors Actuators B Chem.* 153 (1): 78–82.

35 Liang, L., Yin, J., Bao, J. et al. (2019). Preparation of Au nanoparticles modified TiO2nanotube array sensor and its application as chemical oxygen demand sensor. *Chin. Chem. Lett.* 30 (1): 167–170.

36 Pai, T.Y., Yang, P.Y., Wang, S.C. et al. (2011). Predicting effluent from the wastewater treatment plant of industrial park based on fuzzy network and influent quality. *Appl. Math. Model.* 35 (8): 3674–3684.

37 Fernandez de Canete, J., del Saz-Orozco, P., Baratti, R. et al. (2016). Soft-sensing estimation of plant effluent concentrations in a biological wastewater treatment plant using an optimal neural network. *Expert Syst. Appl.* 63: 8–19.

38 Bagheri, M., Mirbagheri, S.A., Ehteshami, M., and Bagheri, Z. (2015). Modeling of a sequencing batch reactor treating municipal wastewater using multi-layer perceptron and radial basis function artificial neural networks. *Process Saf. Environ. Prot.* 93: 111–123.

39 Qin, X., Gao, F., and Chen, G. (2012). Wastewater quality monitoring system using sensor fusion and machine learning techniques. *Water Res.* 46 (4): 1133–1144.

40 Leitão, C., Fernandes, L., Ribeiro, R. et al. (2016). Development of soft sensors based on analytical and spectral data on a real small size wastewater treatment plant. *CONTROLO*, 04 September 2016, p. 323. Springer, Cham.

41 Hou, S.D., Ling, L., Dionysiou, D.D. et al. (2018). Chlorate formation mechanism in the presence of sulfate radical, chloride, bromide and natural organic matter. *Environ. Sci. Technol.* 52: 6317–6325.

42 Sheng, B., Yang, F., Huang, Y. et al. (2020). Transformation of endogenic and exogenic Cl/Br in peroxymonosulfate-based processes: the importance of position of Cl/Br attached to the phenolic ring. *Chem. Eng. J.* 381: 122634.

43 Hammes, F., Meylan, S., Salhi, E. et al. (2007). Formation of assimilable organic carbon (AOC) and specific natural organic matter (NOM) fractions during ozonation of phytoplankton. *Water Res.* 41 (7): 1447–1454.

44 Van der Kooij, D., Hijnen, W.A.M., and Kruithof, J. (1989). The effects of ozonation, biological filtration and the distribution on the concentration of easily assimilable organic carbon (AOC) in drinking water. *Ozone Sci. Eng.* 11: 297.

45 Ramseier, M.K., Peter, A., Traber, J., and von Gunten, U. (2011). Formation of assimilable organic carbon during oxidation of natural waters with ozone, chlorine dioxide, chlorine, permanganate, and ferrate. *Water Res.* 45 (5): 2002–2010.

46 Leenheer, J.A. and Croue, J.P. (2003). Characterizing aquatic dissolved organic matter. *Environ. Sci. Technol.* 37 (1): 18A–26A.

47 Elovitz, M.S., von Gunten, U., and Kaiser, H.-P. (2000). The influence of dissolved organic matter character on ozone decomposition rates and Rct. In: *Natural Organic Matter and Disinfection by-Products, Characterization and Control in Drinking Water*, ACS Symposium Series (ed. S.E. Barrett, S.W. Krasner, and G. Amy), 761. Washington, DC: American Chemical Society.

48 Korshin, G.V., Li, C.-W., and Benjamin, M.M. (1997). Monitoring the properties of natural organic matter through UV spectroscopy: a consistent theory. *Water Res.* 31 (7): 1787–1795.

49 Aeschbacher, M., Graf, C., Schwarzenbach, R.P., and Sander, M. (2012). Antioxidant properties of humic substances. *Environ. Sci. Technol.* 46 (9): 4916–4925.

50 Bourgin, M., Beck, B., Boehler, M. et al. (2018). Evaluation of a full-scale wastewater treatment plant upgraded with ozonation and biological post-treatments: abatement of micropollutants, formation of transformation products and oxidation by-products. *Water Res.* 129: 486–498.

51 von Gunten, U. (2018). Oxidation processes in water treatment: are we on track? *Environ. Sci. Technol. 52* (9): 5062–5075.

52 von Sonntag, C. and Schuchmann, H.-P. (1997). Peroxyl radicals in aqueous solutions. In: *Peroxyl Radicals* (ed. Z.B. Alfassi), 173–234. John Wiley & Sons Ltd.

53 Hübner, U., von Gunten, U., and Jekel, M. (2015). Evaluation of the persistence of transformation products from ozonation of trace organic compounds – a critical review. *Water Res.* 68: 150–170.

54 Mons, M.N., Heringa, M.B., van Genderen, J. et al. (2013). Use of the threshold of toxicological concern (TTC) approach for deriving target values for drinking water contaminants. *Water Res.* 47 (4): 1666–1678.

55 von Sonntag, C. (2006). *Free-Radical-Induced DNA Damage and Its Repair.* A Chemical Perspective; Springer: Berlin, Heidelberg.

56 Lee, Y. and von Gunten, U. (2016). Advances in predicting organic contaminant abatement during ozonation of municipal wastewater effluent: reaction kinetics, transformation products, and changes of biological effects. *Environ. Sci. Water Res. Technol.* 2 (3): 421–442.

57 Guo, X., Minakata, D., Niu, J., and Crittenden, J. (2014). Computer-based first-principles kinetic modeling of degradation pathways and byproduct fates in aqueous-phase advanced oxidation processes. *Environ. Sci. Technol.* 48 (10): 5718–5725.

58 Lee, M., Blum, L.C., Schmid, E. et al. (2017). A computer-based prediction platform for the reaction of ozone with organic compounds in aqueous solution: kinetics and mechanisms. *Environ Sci Process Impacts* 19 (3): 465–476.

59 Johnson, M.D., Hornstein, B.J., and Wischnewsky, J. (2008). Ferrate(VI) oxidation of nitrogenous compounds. In: *Ferrates*, vol. 985 (ed. V.K. Sharma), 177–188. American Chemical Society.

60 Sharma, V.K. (2010). Oxidation of nitrogen-containing pollutants by novel ferrate(VI) technology: a review. *J. Environ. Sci. Health, Part A: Tox. Hazard. Subst. Environ. Eng.* 45 (6): 645–667.

61 Vikesland, P.J., Fiss, E.M., Wigginton, K.R. et al. (2013). Halogenation of bisphenol-A, triclosan, and phenols in chlorinated waters containing iodide. *Environ. Sci. Technol.* 47 (13): 6764–6772.

62 How, Z.T., Linge, K.L., Busetti, F., and Joll, C.A. (2017). Chlorination of amino acids: reaction pathways and reaction rates. *Environ. Sci. Technol.* 51 (9): 4870–4876.

63 Chen, Y.-T., Chen, W.-R., Liu, Z.-Q., and Lin, T.-F. (2017). Reaction pathways and kinetics of a cyanobacterial neurotoxin β-N-methylamino-L-alanine (BMAA) during chlorination. *Environ. Sci. Technol.* 51 (3): 1303–1311.

64 Criquet, J., Rodriguez, E.M., Allard, S. et al. (2015). Reaction of bromine and chlorine with phenolic compounds and natural organic matter extracts – electrophilic aromatic substitution and oxidation. *Water Res.* 85: 476–486.

65 Sivey, J.D. and Roberts, A.L. (2012). Assessing the reactivity of free chlorine constituents Cl2, Cl2O, and HOCl toward aromatic ethers. *Environ. Sci. Technol.* 46 (4): 2141–2147.

66 Dodd, M.C. and Huang, C.H. (2007). Aqueous chlorination of the antibacterial agent trimethoprim: reaction kinetics and pathways. *Water Res.* 41 (3): 647–655.

67 Duirk, S.E., Desetto, L.M., and Davis, G.M. (2009). Transformation of organophosphorus pesticides in the presence of aqueous chlorine: kinetics, pathways, and structure-activity relationships. *Environ. Sci. Technol.* 43 (7): 2335–2340.

68 Joo, S.H. and Mitch, W.A. (2007). Nitrile, aldehyde, and halonitroalkane formation during chlorination/chloramination of primary amines. *Environ. Sci. Technol.* 41 (4): 1288–1296.

69 Buth, J.M., Arnold, W.A., and McNeill, K. (2007). Unexpected products and reaction mechanisms of the aqueous chlorination of cimetidine. *Environ. Sci. Technol.* 41 (17): 6228–6233.

70 Duirk, S.E. and Collette, T.W. (2006). Degradation of chlorpyrifos in aqueous chlorine solutions: pathways, kinetics, and modeling. *Environ. Sci. Technol.* 40 (2): 546–551.

71 Bedner, M. and MacCrehan, W.A. (2006). Transformation of acetaminophen by chlorination produces the toxicants 1,4-benzoquinone and N-acetyl-p-benzoquinone imine. *Environ. Sci. Technol.* 40 (2): 516–522.

72 Dodd, M.C., Shah, A.D., von Gunten, U., and Huang, C.-H. (2005). Interactions of fluoroquinolone antibacterial agents with aqueous chlorine: reaction kinetics, mechanisms, and transformation pathways. *Environ. Sci. Technol.* 39 (18): 7065–7076.

73 Rebenne, L.M., Gonzalez, A.C., and Olson, T.M. (1996). Aqueous chlorination kinetics and mechanism of substituted dihydroxybenzenes. *Environ. Sci. Technol.* 30 (7): 2235–2242.

74 Rule, K.L., Ebbett, V.R., and Vikesland, P.J. (2005). Formation of chloroform and chlorinated organics by free-chlorine-mediated oxidation of triclosan. *Environ. Sci. Technol.* 39 (9): 3176–3185.

75 Buth, J.M., Steen, P.O., Sueper, C. et al. (2010). Dioxin photoproducts of triclosan and its chlorinated derivatives in sediment cores. *Environ. Sci. Technol.* 44 (12): 4545–4551.

76 Acero, J.L., Piriou, P., and von Gunten, U. (2005). Kinetics and mechanisms of formation of bromophenols during drinking water chlorination: assessment of taste and odor development. *Water Res.* 39 (13): 2979–2993.

77 Heeb, M.B., Criquet, J., Zimmermann-Steffens, S.G., and von Gunten, U. (2014). Oxidative treatment of bromide-containing waters: formation of bromine and its reactions with inorganic and organic compounds – a critical review. *Water Res.* 48: 15–42.

78 Rodriguez, E.M., Acero, J.L., Spoof, L., and Meriluoto, J. (2008). Oxidation of MC-LR and -RR with chlorine and potassium permanganate: toxicity of the reaction products. *Water Res.* 42 (6–7): 1744–1752.

79 Willach, S., Lutze, H.V., Eckey, K. et al. (2017). Degradation of sulfamethoxazole using ozone and chlorine dioxide – compound-specific stable isotope analysis, transformation product analysis and mechanistic aspects. *Water Res.* 122 (Supplement C): 280–289.

80 Miao, H.F., Han, H.H., Ji, X.P. et al. Oxidative transformation of oxcarbazepine by Cl2, ClO2 and O3: characteristics and pathways. *Water Sci. Technol.* Water Supply2017 17 (1): 84.

81 Wang, Y., Liu, H., Xie, Y. et al. (2015). Oxidative removal of diclofenac by chlorine dioxide: reaction kinetics and mechanism. *Chem. Eng. J.* 279 (Supplement C): 409–415.

82 Wang, P., He, Y.-L., and Huang, C.-H. (2012). Oxidation of antibiotic agent trimethoprim by chlorine dioxide: reaction kinetics and pathways. *J. Environ. Eng.* 138 (3): 360–366.

83 Rav-Acha, C. (1998). Transformation of aqueous pollutants by chlorine dioxide: reactions, mechanisms and products. In: *Quality and Treatment of Drinking Water II* (ed. J. Hrubec), 143–175. Berlin, Heidelberg: Springer Berlin Heidelberg.

84 Cwiertny, D.M., Snyder, S.A., Schlenk, D., and Kolodziej, E.P. (2014). Environmental designer drugs: when transformation may not eliminate risk. *Environ. Sci. Technol.* 48 (20): 11737–11745.

85 Glaze, W.H. (1987). Drinking-water treatment with ozone. *Environ. Sci. Technol.* 21 (3): 224–230.

86 Glaze, W.H., Kang, J.-W., and Chapin, D.H. (1987). The chemistry of water treatment processes involving ozone, hydrogen peroxide and ultraviolet radiation.

87 Cho, M., Chung, H., Choi, W., and Yoon, J. (2005). Different inactivation behaviors of MS-2 phage and *Escherichia coli* in TiO2 photocatalytic disinfection. *Appl. Environ. Microbiol.* 71 (1): 270–275.

88 Ikai, H., Nakamura, K., Shirato, M. et al. (2010). Photolysis of hydrogen peroxide, an effective disinfection system via hydroxyl radical formation. *Antimicrob. Agents Chemother.* 54 (12): 5086–5091.

89 Tchobanoglous, G., Burton, F., and Stensel, H. (2003). *Wastewater Engineering*. New York: Metcalf & Eddy Inc.

90 Imoberdorf, G. and Mohseni, M. (2012). Kinetic study and modeling of the vacuum-UV photoinduced degradation of 2,4-D. *Chem. Eng. J.* 187: 114–122.

91 Zoschke, K., Dietrich, N., Bornick, H., and Worch, E. (2012). UV-based advanced oxidation processes for the treatment of odour compounds: efficiency and by-product formation. *Water Res.* 46: 5365–5373.

92 Balcı, S. and Tomul, F. (2023). Catalytic wet peroxide oxidation of phenol through mesoporous silica-pillared clays supported iron and/or titanium incorporated catalysts. *J. Environ. Manag.* 326 (Pt B): 116835.

93 Ibarguen-Mondragon, E., Revelo-Romo, D., Hidalgo, A. et al. (2020). Mathematical modelling of MS2 virus inactivation by Al/Fe-PILC-activated catalytic wet peroxide oxidation (CWPO). *Environ. Sci. Pollut. Res. Int.* 27 (16): 19836–19844.

94 Mesa Medina, S., Rey, A., Durán-Valle, C. et al. (2021). Performance of iron-functionalized activated carbon catalysts (Fe/AC-f) on CWPO wastewater treatment. *Catalysts 11*: 337.

95 Boukhemkhem, A., Rida, K., Pizarro, A.H. et al. (2019). Iron catalyst supported on modified kaolin for catalytic wet peroxide oxidation. *Clay Miner.* 54: 67–73.

96 Rey, A., Hungria, A.B.B., Duran-Valle, C.J.J. et al. (2016). On the optimization of activated carbon-supported iron catalysts in catalytic wet peroxide oxidation process. *Appl. Catal. B Environ.* 181: 249–259.

97 Eksteen, J., Oraby, E., and Tanda, B. (2017). A conceptual process for copper extraction from chalcopyrite in alkaline glycinate solutions. *Miner. Eng.* 108: 53–66.

98 Shin, D., Ahn, J., and Lee, J. (2019). Kinetic study of copper leaching from chalcopyrite concentrate in alkaline glycine solution. *Hydrometallurgy* 183: 71–78.

99 Chen, H., He, J.F., Zhu, L.T. et al. (2022). Eco-friendly oxidation leaching from chalcopyrite powder and kinetics assisted by sodium chloride in organic acid media. *Adv. Powder Technol.* 33: 103547.

100 Chen, X.M., Peng, Y.J., and Bradshaw, D. (2014). The separation of chalcopyrite and chalcocite from pyrite in cleaner flotation after regrinding. *Miner. Eng.* 58: 64–72.

101 Rancaño, L., Rivero, M.J., Mueses, M.Á. et al. (2020). Comprehensive kinetics of the photocatalytic degradation of emerging pollutants in a LED-assisted photoreactor. S-metolachlor as case study. *Catalysts* 11: 48.

102 Kochany, J. and James Maguire, R. (1994). Sunlight photodegradation of metolachlor in water. *J. Agric. Food Chem. 42* (2): 406–412.

103 Li, Y., Yao, S., Wen, W. et al. (2010). Sol-gel combustion synthesis and visible-light-driven photocatalytic property of perovskite LaNiO33. *J. Alloys Compd.* 491: 560–564.

104 Tang, P., Sun, H., Cao, F. et al. (2011). Visible-light driven $LaNiO_3$ nanosized photocatalysts prepared by a sol-gel process. *Adv. Mater. Res.* 279: 83–87.

105 Aman, D., Zaki, T., Mikhail, S., and Selim, S.A. (2011). Synthesis of a perovskite $LaNiO_3$ nanocatalyst at a low temperature using single reverse microemulsion. *Catal. Today* 164: 209–213.

106 Rueda Márquez, J.J., Levchuk, I., and Sillanpää, M. (2018). Application of catalytic wet peroxide oxidation for industrial and urban wastewater treatment: a review. *Catalysts 8*: 673.

107 Cheng, M., Zeng, G., Huang, D. et al. (2018). Efficient degradation of sulfamethazine in simulated and real wastewater at slightly basic pH values using Co-SAM-SCS/H2O2 Fenton-like system. *Water Res.* 138: 7–18.

108 Munoz, M., Mora, F.J., de Pedro, Z.M. et al. (2017). Application of CWPO to the treatment of pharmaceutical emerging pollutants in different water matrices with a ferromagnetic catalyst. *J. Hazard. Mater.* 331: 45–54.

109 Ribeiro, R.S., Silva, A.M.T., Figueiredo, J.L. et al. (2016). Catalytic wet peroxide oxidation: a route towards the application of hybrid magnetic carbon nanocomposites for the degradation of organic pollutants. A review. *Appl. Catal. B Environ.* 187: 428–460.

110 Tchobanoglous, G., Burton, F.L., and Stensel, H.D. (2004). *Wastewater Engineering*. Singapore: Treatment and Reuse; McGraw-Hill Higher Education.

111 Ribeiro, A.R., Nunes, O.C., Pereira, M.F.R., and Silva, A.M.T. (2015). An overview on the advanced oxidation processes applied for the treatment of water pollutants defined in the recently launched Directive 2013/39/EU. *Environ. Int.* 75: 33–51.

112 Mena, I.F., Diaz, E., Moreno-Andrade, I. et al. (2018). Stability of carbon-supported iron catalysts for catalytic wet peroxide oxidation of ionic liquids. *J. Environ. Chem. Eng.* 6: 6444–6450.

113 Bedia, J., Monsalvo, V.M., Rodriguez, J.J., and Mohedano, A.F. (2017). Iron catalysts by chemical activation of sewage sludge with FeCl3 for CWPO. *Chem. Eng. J.* 318: 224–230.

114 Gosu, V., Sikarwar, P., and Subbaramaiah, V. (2018). Mineralisation of pyridine by CWPO process using nFe0/GAC catalyst. *J. Environ. Chem. Eng.* 6: 1000–1007.

115 Mohedano, A.F., Monsalvo, V.M., Bedia, J. et al. (2014). Highly stable iron catalysts from sewage sludge for CWPO. *J. Environ. Chem. Eng.* 2: 2359–2364.

116 Wang, N., Zheng, T., Zhang, G., and Wang, P. (2016). A review on Fenton-like processes for organic wastewater treatment. *J. Environ. Chem. Eng.* 4: 762–787.

117 Linley, S., Leshuk, T., and Gu, F.X. (2013). Magnetically separable water treatment technologies and their role in future advanced water treatment: a patent review. *Clean: Soil, Air, Water* 41: 1152–1156.

118 Paneva, D., Stoilova, O., Manolova, N., and Rashkov, I. (2004). Magnetic hydrogel beads based on chitosan. *e-Polymers* 4: 60.

119 Gómez-Pastora, J., Dominguez, S., Bringas, E. et al. (2017). Review and perspectives on the use of magnetic nanophotocatalysts (MNPCs) in water treatment. *Chem. Eng. J.* 310: 407–427.

120 Maruthupandy, M., Muneeswaran, T., Vennila, T. et al. (2022). Development of chitosan decorated Fe3O4 nanospheres for potential enhancement of photocatalytic degradation of Congo red dye molecules. *Spectrochim. Acta A Mol. Biomol. Spectrosc.* 267 (1): 120511.

121 Xu, J.-K., Zhang, F.-F., Sun, J.-J. et al. (2014). Bio and nanomaterials based on Fe_3O_4. *Molecules* 19: 21506–21528.

122 Maruthupandy, M., Muneeswaran, T., Anand, M., and Quero, F. (2020). Highly efficient multifunctional graphene/chitosan/magnetite nanocomposites for photocatalytic degradation of important dye molecules. *Int. J. Biol. Macromol.* 153: 736–746.

123 Martha, S., Sahooa, P.C., and Parida, K.M. (2015). An overview on visible light responsive metal oxide based photocatalysts for hydrogen energy production. *RSC Adv.* 5: 61535–61553.

124 Li, X., Xiong, J., Tang, Z. et al. (2023). Recent progress in metal oxide-based photocatalysts for CO2 reduction to solar fuels: a review. *Molecules 28*: 1653.

125 Shen, H., Peppel, T., Stunk, J., and Sun, Z. (2020). Photocatalytic reduction of CO2 by metal-free-based materials: recent advances and future perspective. *Sol. RRL* 4: 1900546.

126 Irfan, S., Khan, S.B., Lam, S.S. et al. (2022). Removal of persistent acetophenone from industrial waste-water via bismuth ferrite nanostructures. *Chemosphere* 302: 134750.

127 Domen, K., Naito, S., Soma, M. et al. (1980). Photocatalytic decomposition of water vapour on an NiO–$SrTiO_3$ catalyst. *J. Chem. Soc. Chem. Commun.* 543.

128 Sato, S. and White, J.M. (1980). Photodecomposition of water over Pt/ TiO_2 catalysts. *Chem. Phys. Lett.* 72: 83.

129 Santos, L.M.C.d., Silva, E.S.d., Oliveira, F.O. et al. (2021). Ozonized water in microbial control: analysis of the stability, in vitro biocidal potential, and cytotoxicity. *Biology* 10: 525.

130 Domen, K., Naito, S., Onishi, T., and Tamaru, K. (1982). Photocatalytic decomposition of liquid water on a NiO $SrTiO_3$ catalyst. *Chem. Phys. Lett.* 92: 433.

131 Sayama, K. and Arakawa, H. (1993). Photocatalytic decomposition of water and photocatalytic reduction of carbon dioxide over zirconia catalyst. *J. Phys. Chem.* 97: 531.

132 Kato, H. and Kudo, A. (1998). New tantalate photocatalysts for water decomposition into H_2 and O_2. *Chem. Phys. Lett.* 295: 487.

133 Machida, M., Yabunaka, J., and Kijima, T. (1999). Efficient photocatalytic decomposition of water with the novel layered tantalate $RbNdTa_2O_7$. *Chem. Commun.* 1939.

134 Kato, H. and Kudo, A. (2001). Water Splitting into H_2 and O_2 on Alkali Tantalate Photocatalysts $ATaO_3$ (A = Li, Na, and K). *J. Phys. Chem. B* 105: 4285.

135 Shimizu, K., Tsuji, Y., Kawakami, M. et al. (2002). Photocatalytic water splitting over spontaneously hydrated layered tantalate $A_2SrTa_2O_7{\cdot}nH_2O$ (A=H, K, Rb). *Chem. Lett.* 31: 1158.

136 Galinska, A. and Walendziewski, J. (2005). Photocatalytic water splitting over Pt–TiO_2 in the presence of sacrificial reagents. *Energy Fuel* 19: 1143.

137 Kubacka, A., Fernández-García, M., and Colón, G. (2012). Advanced nanoarchitectures for solar photocatalytic applications. *Chem. Rev.* 112: 1555–1614.

138 Wang, Y., He, Y., Lai, Q., and Fan, M. (2014). Review of the progress in preparing nano TiO2: an important environmental engineering material. *J. Environ. Sci.* 26: 2139–2177.

139 Ugurlu, M. and Karaoglu, M.H. (2009). Removal of AOX, total nitrogen and chlorinated lignin from bleached Kraft mill effluents by UV oxidation in the presence of hydrogen peroxide utilizing TiO2 as photocatalyst. *Environ. Sci. Pollut. Res.* 16: 265–273.

140 Li, H., Zhang, Y., Wang, S. et al. (2009). Study on nanomagnets supported TiO2 photocatalysts prepared by a sol-gel process in reverse microemulsion combining with solvent-thermal technique. *J. Hazard. Mater.* 169: 1045–1053.

141 Lee, D.E., Devthade, V., Moru, S. et al. (2022). Magnetically sensitive TiO2 hollow sphere/Fe3O4 core-shell hybrid catalyst for high-performance sunlight-assisted photocatalytic degradation of aqueous antibiotic pollutants. *J. Alloys Compd.* 902: 163612.

142 Gnanasekaran, L., Hemamalini, R., Rajendran, S. et al. (2019). Nanosized Fe3O4 incorporated on a TiO_2 surface for the enhanced photocatalytic degradation of organic pollutants. *J. Mol. Liq.* 287: 2–8.

143 Madima, N., Kefeni, K.K., Mishra, S.B. et al. (2022). Fabrication of magnetic recoverable Fe3O4/TiO2 heterostructure for photocatalytic degradation of rhodamine B dye. *Inorg. Chem. Commun.* 145: 109966.

144 Suzuki, K., Mizuno, N., and Yamaguchi, K. (2018). Polyoxometalate photocatalysis for liquid-phase selective organic functional group transformations. *ACS Catal.* 8: 10809–10825.

145 Wang, D., Liu, L., Jiang, J. et al. (2020). Polyoxometalate-based composite materials in electrochemistry: state-of-the-art progress and future outlook. *Nanoscale* 12: 5705–5718.

146 Omwoma, S., Chen, W., Tsunashima, R., and Song, Y.-F. (2014). Recent advances on polyoxometalates intercalated layered double hydroxides: from synthetic approaches to functional material applications. *Coord. Chem. Rev.* 258–259: 58–71.

147 Liu, S. and Tang, Z. (2010). Polyoxometalate-based functional nanostructured films: current progress and future prospects. *Nano Today* 5: 267–281.

148 Ma, X., Li, H., Chen, L., and Zhao, J. (2016). The main progress over the past decade and future outlook on high-nuclear transition-metal substituted polyoxotungstates: from synthetic strategies, structural features to functional properties. *Dalton Trans.* 45: 4935–4960.

149 Dolbecq, A., Dumas, E., Mayer, C.R., and Mialane, P. (2010). Hybrid organic-inorganic polyoxometalate compounds: from structural diversity to applications. *Chem. Rev.* 110: 6009–6048.

150 Schmid, P., Hohenschutz, M., Graß, X. et al. (2022). Counterion effect on α-Keggin polyoxometalates in water: the peculiar role of H+ on their salting-in effect and co-assembly with organics. *J. Mol. Liq.* 359: 119214.

151 Ong, W.J., Putri, L.K., and Mohamed, A.R. (2020). Rational design of carbon-based 2D nanostructures for enhanced photocatalytic CO2reduction: a dimensionality perspective. *Chemistry* 26: 9710–9748.

152 Wang, W., Wang, X., Xiang, C. et al. (2019). Orthogonal molecular recognition of chaotropic and hydrophobic guests enables supramolecular architectures. *ChemNanoMat.* 5 (1): 124–129.

153 Jin Lan, Y., Wang, B.H., Xiao, Z., and Pingfan, W. (2021). Application of polyoxometalates in photocatalytic degradation of organic pollutants. *Nanoscale Adv.* 3: 4646–4658.

154 Gao, Y., Choudhari, M., Such, G.K., and Ritchie, C. (2022). Polyoxometalates as chemically and structurally versatile components in self-assembled materials. *Chem. Sci.* 13: 2510–2527.

155 Zhang, Y., Zhao, M., Huang, J. et al. (2023). Controllable synthesis, photocatalytic property, and mechanism of a novel POM-based direct Z-scheme nano-heterojunction α-Fe2O3/P2Mo18. *Molecules 28*: 6671.

156 Yu, J., Wang, T., and Rtimi, S. (2019). Magnetically separable TiO_2/FeOx/POM accelerating the photocatalytic removal of the emerging endocrine disruptor: 2,4-dichlorophenol. *Appl. Catal. B Environ.* 254: 66–75.

157 Bai, L., Pan, X., Guo, R. et al. (2022). Sunlight-driven photocatalytic degradation of organic dyes in wastewater by chemically fabricated ZnO/Cs4Si$W_{12}O_{40}$ nanoheterojunction. *Appl. Surf. Sci.* 599: 153912.

158 Lai, S.Y., Ng, K.H., Cheng, C.K. et al. (2021). Photocatalytic remediation of organic waste over Keggin-based polyoxometalate materials: a review. *Chemosphere* 263: 128244.

159 Zhou, Y., Chen, G., Long, Z., and Wang, J. (2014). Recent advances in polyoxometalate-based heterogeneous catalytic materials for liquid-phase organic transformations. *RSC Adv.* 4: 42092–42113.

160 Granadeiro, C., Cruz, S., Goncalves, G. et al. (2012). Photoluminescent bimetallic-3-hydroxypicolinate/graphene oxide nanocomposite. *RSC Adv.* 2: 9443–9447.

161 Lopes, J.L., Martins, M.J., Nogueira, H.I.S. et al. (2021). Carbon-based heterogeneous photocatalysts for water cleaning technologies: a review. *Environ. Chem. Lett.* 19: 643–668.

162 López-Quintela, M.A. and Rivas, J. (1996). Nanoscale magnetic particles: synthesis, structure and dynamics. *Curr. Opin. Colloid Interface Sci.* 1: 806–819. https://doi.org/10.1016/s1359-0294(96)80085-9.

163 Homocianu, M. and Pascariu, P. (2022). High-performance photocatalytic membranes for water purification in relation to environmental and operational parameters. *J. Environ. Manag.* 311: 114817.

164 Khulbe, K. .C., Feng, C. .Y., and Matsuura, T. (2008). Synthetic membranes for membrane processes. In: *Synthetic Polymeric Membranes*. Berlin, Heidelberg: Springer Laboratory. Springer.

165 Molinari, R., Lavorato, C., Argurio, P. et al. (2019). Overview of photocatalytic membrane reactors in organic synthesis, energy storage and environmental applications. *Catalysts* 9: 239.

166 Koe, W.S., Lee, J.W., Chong, W.C. et al. (2020). An overview of photocatalytic degradation: photocatalysts, mechanisms, and development of photocatalytic membrane. *Environ. Sci. Pollut. Res.* 27: 2522–2565.

167 Zyoud, A.H., Zubi, A., Hejjawi, S. et al. (2020). Removal of acetaminophen from water by simulated solar light photodegradation with ZnO and TiO2 nanoparticles: catalytic efficiency assessment for future prospects. *J. Environ. Chem. Eng.* 8: 104038.

168 Tian, S., He, Y., Zhang, L. et al. (2023). CNTs/TiO2-loaded carbonized nanofibrous membrane with two-type self-cleaning performance for high efficiency oily wastewater remediation. *Colloids Surf. A Physicochem. Eng. Asp.* 656: 130306.

169 Samsudin, M.F.R., Jayabalan, P.J., Ong, W.-J. et al. (2019). Photocatalytic degradation of real industrial poultry wastewater via platinum decorated BiVO4/g-C3N4 photocatalyst under solar light irradiation. *J. Photochem. Photobiol. A Chem.* 378: 46–56.

170 Heidari, S., Haghighi, M., and Shabani, M. (2020). Sunlight-activated BiOCl/BiOBr–Bi24O31Br10 photocatalyst for the removal of pharmaceutical compounds. *J. Clean. Prod.* 259: 120679.

171 Luo, H., Yan, M., Yilin, W. et al. (2021). Facile synthesis of PVDF photocatalytic membrane based on NCQDs/BiOBr/TiO2 heterojunction for effective removal of tetracycline. *Mater. Sci. Eng. B* 265: 114996.

172 Pandey, J. (2020). Biopolymers and their application in wastewater treatment. In: *Emerging Eco-friendly Green Technologies for Wastewater Treatment. Microorganisms for Sustainability*, vol. 18 (ed. R. Bharagava), 245–266. Singapore: Springer.

173 Elgarahy, A.M., Eloffy, M.G., Guibal, E. et al. (2023). Use of biopolymers in wastewater treatment: a brief review of current trends and prospects. *Chin. J. Chem. Eng.* 64: 292–320.

174 Udayakumar, G.P., Muthusamy, S., Selvaganesh, B. et al. (2021). Ecofriendly biopolymers and composites: preparation and their applications in water-treatment. *Biotechnol. Adv.* 52: 107815.

175 Panayotova, M. Water treatment with biopolymer–zeolite nanocomposites. In: *Biorenewable Nanocomposite Materials*, vol. 2:. Desalination and Wastewater Remediation, 75–111. American Chemical Society.

176 Xu, G.-R., Wang, J.-N., and Li, C.-J. (2013). Polyamide nanofilm composite membranes (NCMs) supported by chitosan coated electrospun nanofibrous membranes: preparation and separation performance research. *Desalination* 328: 31–41.

177 Özacar, M. and Şengil, İ.A. (2003). Evaluation of tannin biopolymer as a coagulant aid for coagulation of colloidal particles. *Colloids Surf. A Physicochem. Eng. Asp.* 229 (1–3): 85–96.

178 Maćczak, P., Kaczmarek, H., Ziegler-Borowska, M. et al. (2022). The use of chitosan and starch-based flocculants for filter backwash water treatment. *Materials* 15 (3): 1056.

179 Elwakeel, K.Z. (2010). Removal of Cr(VI) from alkaline aqueous solutions using chemically modified magnetic chitosan resins. *Desalination* 250 (1): 105–112.

180 Yang, F.C. and Yang, P. (2022). Biopolymer-based membrane adsorber for removing contaminants from aqueous solution: progress and prospects. *Macromol. Rapid Commun.* 43 (3): 2100669.

181 Sohouli, E., Irannejad, N., Ziarati, A. et al. (2022). Application of polysaccharide-based biopolymers as supports in photocatalytic treatment of water and wastewater: a review. *Environ. Chem. Lett.* 20 (6): 3789–3809.

182 Yaashikaa, P.R., Senthil Kumar, P., and Karishma, S. (2022). Review on biopolymers and composites – evolving material as adsorbents in removal of environmental pollutants. *Environ. Res.* 212 (Pt A): Article 113114.

183 Elwakeel, K.Z. (2010). Environmental application of chitosan resins for the treatment of water and wastewater: a review. *J. Dispers. Sci. Technol.* 31 (3): 273–288.

184 Elgarahy, A.M., Elwakeel, K.Z., Mohammad, S.H., and Elshoubaky, G.A. (2021). A critical review of biosorption of dyes, heavy metals and metalloids from wastewater as an efficient and green process. *Cleaner Eng. Technol.* 4: 100209.

185 Mruthunjayappa, M.H., Kotrappanavar, N.S., and Mondal, D. (2022). New prospects on solvothermal carbonisation assisted by organic solvents, ionic liquids and eutectic mixtures – a critical review. *Prog. Mater. Sci.* 126: 100932.

186 Nataraj, S.K. (2022). *Emerging Pollutant Treatment in Wastewater*. CRC Press. eBook ISBN 9781003214786.

187 Sharma, V.T., Manohara, H.M., Kamath, S.V. et al. (2020). Fe–Al based nanocomposite reinforced hydrothermal carbon: efficient and robust absorbent for anionic dyes. *Chemosphere* 259: 127421.

188 Yujiao Tan, Xin Wang, Fuquan Xiong, Jiarong Ding, Yan Qing, Yiqiang Wu. Preparation of lignin-based porous carbon as an efficient absorbent for the removal of methylene blue, *Ind. Crop. Prod.*, Volume 171, 2021,113980.

189 Zhu, S., Xu, J., Yu, W. et al. (2022). Simultaneous production of clean water and organic dye from dyeing wastewater by reusable lignin-derived porous carbon. *Ind. Crop. Prod.* 187: 115314.

190 Ahmaruzzaman, M., Roy, P., Bonilla-Petriciolet, A. et al. (2023). Polymeric hydrogels-based materials for wastewater treatment. *Chemosphere* 331: 138743.

191 Zhu, D., Bai, B., and Hou, J. (2017). Polymer gel systems for water management in high-temperature petroleum reservoirs: a chemical review. *Energy Fuel 31* (12): 13063–13087.

192 Abousalman-Rezvani, Z., Roghani-Mamaqani, H., Riazi, H., and Abousalman-Rezvani, O. (2022). Water treatment using stimuli-responsive polymers. *Polym. Chem.* 13: 5940–5964.

193 Kistler, S. (1931). Coherent expanded aerogels and jellies. *Nature* 127: 741.

194 Kistler, S.S. (1932). Coherent expanded-aerogels. *J. Phys. Chem. 36* (1): 52–64.

195 Franco, P., Cardea, S., Tabernero, A., and De Marco, I. (2021). Porous aerogels and adsorption of pollutants from water and air: a review. *Molecules* 26 (15): 4440.

196 Wang, S., Li, X., Liu, Y. et al. (2018). Nitrogen-containing amino compounds functionalized graphene oxide: synthesis, characterization and application for the removal of pollutants from wastewater: a review. *J. Hazard. Mater.* 342: 177–191.

197 Maleki, H. (2016). Recent advances in aerogels for environmental remediation applications: a review. *Chem. Eng. J.* 300: 98–118.

198 Long, L.-Y., Weng, Y.-X., and Wang, Y.-Z. (2018). Cellulose aerogels: synthesis, applications, and prospects. *Polymers* 10: 623.

199 Sheth, Y., Dharaskar, S., Khalid, M., and Sonawane, S. (2021). An environment friendly approach for heavy metal removal from industrial wastewater using chitosan based biosorbent: a review. *Sustainable Energy Technol. Assess.* 43: 100951.

200 Chaudhary, J.P., Nataraj, S.K., Gogda, A., and Meena, R. (2014). Bio-based superhydrophilic foam membranes for sustainable oil–water separation. *Green Chem.* 16: 4552–4558.

201 Chaudhary, J.P., Vadodariya, N., Nataraj, S.K., and Meena, R. (2015). Chitosan-based aerogel membrane for robust oil-in-water emulsion separation. *ACS Appl. Mater. Interfaces* 7 (44): 24957–24962.

202 Kumar, A., Paul, P., and Nataraj, S.K. (2017). Bionanomaterial scaffolds for effective removal of fluoride, chromium, and dye. *ACS Sustain. Chem. Eng.* 5 (1): 895–903.

203 Mruthunjayappa, M.H., Kotrappanavar, N.S., and Mondal, D. (2022). Bioinspired engineering protein nanofibrils-based multilayered self-cleaning membranes for universal water purification. *J. Hazard. Mater.* 424: 127561.

204 Kamath, S.V., Attokkaran, J.R., Maraddi, A.S. et al. (2024). Cello-MOF filters for multifaceted emerging pollutant abatement and their value addition. *Chem. Eng. J.* 479: 147805.

205 Samadi, A., Wang, Z., Shaobin Wang, S.K. et al. (2023). Polyaniline-based adsorbents for water treatment: roles of low-cost materials and 2D materials. *Chem. Eng. J.* 478: 147506.

206 D'Souza, G.B., Kumar, A., Kamath, S.V. et al. (2022). Designing engineered biopolymer mesh filter for robust sequestration of chromium (VI), fluoride and other emerging pollutants: a sustainable approach. *Chem. Eng. J.* 443: 136462.

207 Kovtun, V., Pasovets, V., and Pieczonka, T. (2017). Tribological properties and microstructure of the metal-polymer composite thin layer deposited on a copper plate by electrocontact sintering. *Arch. Metall. Mater.* 62: 51–58.

208 Bloor, D., Donnelly, K., Hands, P.J. et al. (2005). A metalpolymer composite with unusual properties. *J. Phys. D. Appl. Phys.* 38: 2851–2860.

209 Sasmal, D., Maity, J., Kolya, H., and Tripathy, T. (2017). Selective adsorption of Pb (II) ions by amylopectin-g-poly (acrylamide-co-acrylic acid): a biodegradable graft copolymer. *Int. J. Biol. Macromol.* 97: 585–597.

210 Al-Kindi, S., Al-Bahry, S., Al-Wahaibi, Y. et al. (2022). Partially hydrolyzed polyacrylamide: enhanced oil recovery applications, oil-field produced water pollution, and possible solutions. *Environ. Monit. Assess.* 194 (12): 875.

7

Ordered Porous and Nanomaterials-Based Water Treatment Systems

7.1 Introduction to Porous Materials in Water Filtration

Ordered porous and mesoporous materials with well-defined pore architectures have attracted significant interest among material scientists and innovators particularly working in the separation science and technology sector. Ordered porous materials have gained unprecedented importance in separation and purification applications due to their unique tunable structural characteristics such as unique pore pattern, uniform porosity, ordered pore tortuosity, and surface functionality. In addition to this, recent advances in design and development of ordered porous materials reveal their potential in large-scale production applications in environmental remediation, specifically in water treatment applications. Therefore, ordered pore architecture is deemed to provide several advantages over conventional porous materials for designing futuristic and advanced water treatment solutions [1–4].

Environmental applications of ordered nanoporous and mesoporous materials have been broadly adopted in applications like adsorption of organic pollutants, in particular, dyes, heavy metals, and other contaminants from water. The other major feature of ordered materials originates from the presence of high surface area and unique pore size and their architecture, which favorably facilitates the ideal adsorption process as they can effectively attract and trap contaminants via surface wetting, surface interactions, affinity, ion exchange, and other mechanisms. In addition to this, synergic combination of affinity and surface functional moieties can facilitate selectivity and adsorption capacity for specific pollutants. As discussed in the previous Chapter 3 and Chapter 5, photocatalytic applications of several microporous and mesoporous semiconducting materials have been discussed. This chapter, however, provides a larger overview of ordered porosity

Materials and Methods for Industrial Wastewater and Groundwater Treatment,
First Edition. S. K. Nataraj.

aspects of advanced materials covering recent advancements and new trends in use in groundwater and wastewater treatment applications [3, 5]. In the past, most of the highly porous materials were synthesized either following top-down method or bottom-up solid-state material preparation strategies. Amongst hydrothermal synthesis, co-precipitation method, sol-gel method, and physicochemical depositions like chemical vapor deposition methods found commercial attraction. These techniques have been used extensively to prepare a variety of standalone and hybrid materials targeting various applications including energy generation, conversion, storage, and environmental applications like water treatment and wastewater remediation [6–10].

In the past, natural-origin clays that are categorized mainly into bentonite, kaolinite, montmorillonite-smectite, vermiculite, illite chlorite, talc, pyrophyllite, and so on have been widely explored for various applications. Lately, synthetic clays that are of the porous microscale and nanoscale materials having characteristics of metal oxide, hydroxide, oxyhydroxide zeolitic, and so on were generally synthesized by means of hydrothermal techniques. Generally, most of the solid-state material preparation techniques yield small crystal size, typically in the range of nano to few microns, depending on the experimental conditions. Interestingly, these materials have been directly used in the form of powder, deposits of layers, thin-film layers, and packed beds. However, these approaches have been ineffective and inefficient in most of the cases as such strategies may fail due to structural instabilities and lack of substrate compatibility. Further, efforts have also been made to overcome shortcomings associated with loosely packed nano-/microscale porous materials following techniques such as pelletized filters, use of various binders and multilayer architectures, and use of polymers as dispersion media. For example, zeolitic systems have been used for water treatment applications in the past. Zeolitic materials are highly porous systems that are crystalline and nontoxic in nature with three-dimensionally connected hydrated aluminosilicate, which possesses ion exchange properties and acts as an adsorbent for the removal of harmful microbes. Dispersed/colloidal contaminants and soluble toxins from various water sources including groundwater have been remediated using zeolitic porous materials. Later, zeolites were also converted into an engineered module in the form of ceramic membranes used in ultrafiltration (UF) and microfiltration (MF) applications. However, these zeolitic materials required polymeric, a binder that was initially used for creating a stable structure but some of the pores in the zeolitic powder might block a significant porosity [11–16].

On the other hand, advanced synthetic porous materials that include micro-/nanoscale materials like MOFs, COFs, layered double hydroxides (LDHs), transition metal chelonids (TMCs), transition metal dichalcogenides (TMDs), and various types of natural and synthetic clays have been rediscovered for their potentials in water treatment applications. Interestingly, these advanced materials also possess added synergic properties that help in added advantage to overall process of remediating contaminated water. For instance, in addition to high porosity, most of the advanced materials either possess unique surface functionality, affinity, or catalytic properties against target pollutants. From a more fundamental perspective, highly porous materials with a combination of catalytic active sites and functional moieties offer the clear advantage of confinement effect that helps in improved catalytic remediation performance. Even though confinement effects have been noticed in several porous materials in the past, advanced target-specific materials with versatile functionality in porous materials have exhibited improved water remediation properties. Nonetheless, porous materials have been widely used as substrates or support materials for various active sites including catalysts in the past. Interestingly, it is important to note that the pore size optimization in porous materials is a very important parameter in deciding the target-specific application in which pore size of absorbent/substrate needs to be optimized to fit the pollutant. For instance, pore size of zeolite-type materials such as aluminophosphates is nearly 1.23 nm for VPI-5 (pores larger than 10 Å), for instance, advanced family of aluminophosphate-based molecular sieves denoted as VPI-5, which pose disadvantage when targeting pollutant molecules such as pharmaceutical is bulky. Therefore, advanced porous materials are required to possess sufficiently large pore size with a narrow pore size distribution so as to retain the "confinement effect" in advanced porous materials [17–22].

Therefore, advanced ordered porous materials have attracted significant attention from researchers, scientists, and end-users around the world. This interest also led to increased efforts to design and develop novel strategies to prepare large-scale ordered porous materials with mesopores which helps in improved diffusivity within the porous crystals. On the other hand, the "templating" phenomenon, which is a simple idea to use specific constituents to both act as precursor as well as structure directing units such as quaternary ammonium ions, has been the versatile template source used for the zeolite synthesis, in which long chain alkyl moieties linked to the nitrogen that induce a specific impact on the type of zeolite prepared as a tertiary or a composite *Building Units* during hydrothermal preparation step. Interestingly, to alter the pore size and their dimensions, alkyl functions can be substituted with a longer hydrocarbon chain or shorter alkyl giving rise to a porous and crystalline structure that is accessible for utilization. Interestingly, in 1992, Japanese researchers working in Mobil Oil Corp discovered the possibilities of the existence of ordered mesoporous phases. In this study, for the first time, researchers reported the preparation of MCM-41, which possessed pores (3 nm) that were regularly spaced in a hexagonal array. Further, detailed study and optimization procedures lead to the pore size tuned from 1.4 to 10 nm depending upon the experimental conditions and concentration of precursors. This further led to the innovation in finding novel possibilities in obtaining a series of ordered porous materials useful for water treatment applications and fields of catalysis. Therefore, the innovation in porous material and development of a family of large synthetic porous materials was greatly influenced by the discovery of the mesoporous silica MCM-41. Now, a series of synthetic porous materials derived from various origins including inorganic, organic, hybrid, and composites generally referred to as nanostructured materials or ordered mesoporous materials (OMMs) have been explored and interest in understanding these new categories of materials in water treatment applications has led to the extensive research activities in the sector [14, 23, 24].

Immediately after the discovery of MCM-41, a dramatic increase in materials science activities targeting development of clay or clay-type materials has shown wider acceptance in different sectors. In spite of several potentials, initial applications for porous materials emerged from the field of catalysis. However, two important features of mesoporous materials in catalytic application showed limited end results which may be due to (i) lower acidic nature of active sites on these materials than those found in materials such as zeolitic systems that were inefficient in converting hydrogenation reactions, (ii) mesoporous materials (all-silica materials) are, in general, unstable when used in harsh conditions. However, aluminum-based network structures later showed improved performance and stability in various catalytic applications. These two important features round in MCM-41 due to noncrystalline nature of the materials. However, later efforts were dedicated to making skeletal walls of mesoporous materials more crystalline and ordered a structure to cater incorporation of the higher acidity, higher resistance to high-temperature test conditions, steaming, and easier handling [25].

Later studies paid significant attention to not only tuning chemical properties of the ordered porous materials but also physical properties such as pore, pore dimension, pore architecture, and their distribution to extract higher efficiency from the new class of materials. Also, surface chemistry nomenclatures have been widely adopted to describe the pore dimensions and their distribution during the design and development of ordered porous materials. The pores are generally categorized into three main types, namely (i) micropores having pore diameter less than 2 nm, (ii) mesopores having pore diameter ranging from 2 to 50 nm, and (iii) macropores having pore diameter sizes larger than 50 nm as defined by the IUPAC classifications. These nomenclatures were also adopted to distinguish between novel materials based on their three different adsorption phenomena and interaction mechanisms that derive the characteristics of the different pores in novel solid materials. These include volume filling existing porosity by the target (i) molecules having size similar to the pore size of substrate known as micropores, (ii) pore filling through capillary condensation referred to as mesopores, and (iii) unenhanced adsorption in larger pore architecture known as macropores.

Interestingly, despite the enormous activities that triggered development of a large number of nanomaterials with the adoption of nanoscience and nanotechnology approaches, till now nanoporous materials are referred to with pretext as *micro-* to define porosity on which there is still unsolved debate among researchers and direction for future nanomaterials' nomenclature is still unclear [26–28].

7.2 Porous Natural and Synthetic Materials

Advanced water treatment technologies require newer approaches and a new stock of materials that give paramount importance to ensuring the availability of safe and clean drinking water from various sources including groundwater and surface water. Recent advances have shown greater interest in both natural and synthetic clay-based porous materials that have shown significant potential in water treatment applications in the form of adsorbents and filtration barriers. Therefore, this section gives comprehensive information on the use of different types of clays from both natural and synthetic origins having various porosities, pore architecture, and surface properties useful in water treatment. Historically, natural clays including bentonite, kaolin, montmorillonite, and so on have long been in use in water treatment. Each clay has unique pore features that are mainly characterized and derived from its unique mineral composition, constituents, arrangements, surface texture, and structural characteristics. Interestingly, several studies have conducted adsorption mechanisms of various pollutants on clay surfaces that provided an enriching and detailed insight into the intricate interactions between clay substrates and pollutants or contaminants. The detailed studies further revealed the major influencing factors on pollutant adsorption process, which involves mainly concentration, pH, temperature, and contact time. Other minor influences include agitation or stirring of active materials in contaminated water or wastewater and loading of active materials. All these factors greatly influence the adsorption capacity and efficiency of different natural as well as synthetic clay in removing various contaminants from contaminated water. Also, the nature of contaminants influences overall treatment efficiency such as heavy metals; organic pollutants like dyes, pesticides, organic matter, and pharmaceutical ingredients; and emerging contaminants including microplastics, surfactants, hormones, and so on. Nevertheless, the role of surface functionality of the solid substrate also majorly influences the initial interaction leading to the affinity of target pollutants. In such a case, surface modification techniques have been widely used to enhance the adsorption capacity when used in water treatment such as functionalization, introduction of cation exchange moieties, and composite formation. Interestingly, increased efforts have also been made to induce commercially attractive and economically significant properties like material regeneration post-adsorption applications, disposal of spent adsorbents, and cost-effective preparation methodologies [29–31].

On the other hand, nanoporous materials offer fascinating assemblies of useful chemistries, ranging from classical molecular organic to zeolites, inorganic, and hybrid framework solids. Many zeolitic 3D framework structures with porosity ranging from microporous and mesoporous architectures have been initially used for gas separation, catalysis, and water treatment applications. Even though several limitations associated with initial studies on zeolitic systems have been highlighted, continued efforts in producing crystalline material characteristics resulted in fruitful results. Even though efforts were made in the early 1940s to create crystalline inorganic zeolitic frameworks, the usage of aluminosilicate gels started in the 1940s. However, commercial usage of their potential was realized 20 years later with the filing of a patent for the process and product involving silica xerogel. Early efforts were made in creating highly porous zeolites having typically pore sizes less than 8 Å. However, these efforts laid a strong base for the design and development of work on mesoporous silica-based porous solids with pore sizes approaching 100 Å in the 1990s [32–35].

Further, another class of novel ordered porous materials has been undertaken starting with the synthesis of amorphous ordered porous materials in the organic domain. For instance, internal void space is formed with twisted molecular chain assemblies formed within so-called polymers of intrinsic microporosity (PIMs) [36–40]. Ordered porosity in these new classes of molecular architecture originates from the inefficient packing of rigid, contorted polymer chains. Interestingly, order in porosity is induced by linking organic chain units via covalent bond interaction that results in increased rigidity to overall polymer and turns into combination of porous aromatic frameworks (PAFs), which is amorphous in nature and organic framework, which is crystalline due to COFs. Similarly, amorphous organometallic polymers have been lately influencing subsequent creation of ordered three-dimensional framework materials through the self-assembly of organic and inorganic components in larger context; they are referred to as MOFs. Interestingly, till now, over 70,000 MOFs have been prepared with various combinations of organic/inorganic precursors in the pure crystalline state. In all cases, an ordered porous network has been well-designed and explored for specific end applications. Throughout, efforts have also been made to design and synthesize robust and ordered frameworks, which, during application, facilitate foreign material (in case of water treatment, pollutant) to reach the specific pore volumes that are excessive of 5.02 cm^3/g. In all cases, discrete organic moieties have been prepared as well as extended solids that possess cage structures that are favorably used for remediating pollutants from contaminated water. These porous network structures in the form of cages can be assembled into either amorphous or crystalline materials to be favorably used by the pollutants to access the available porosity. Nevertheless, cage-like discrete structures were initially observed in Zeolites and then MOFs further extended in novel hybrid materials leading to the discovery of another class of ordered porous materials like polyoxometalates with the help of coordination bonding between organic and inorganic constituent units. Similarly, metal–organic cages (MOCs) or metal–organic polyhedral (MOP) have also been produced following directed assembly of a range of inorganic/organic "building blocks." Therefore, advances in materials science in association with nanoscience interface have led to the discovery of a series of ordered porous materials that range span from organic to inorganic and hybrid chemistries including MOFs, linked MOP, COFs, aluminosilicate gels, particulate organic carbons (POCs), PIMs, organometallic polymers, extended networks, MOFs and porous coordination polymers (PCPs), functionalized silica spheres, microporous carbons, clays, liquid MOFs, porous liquids, and synthetic zeolites as depicted in Figure 7.1 [41–45].

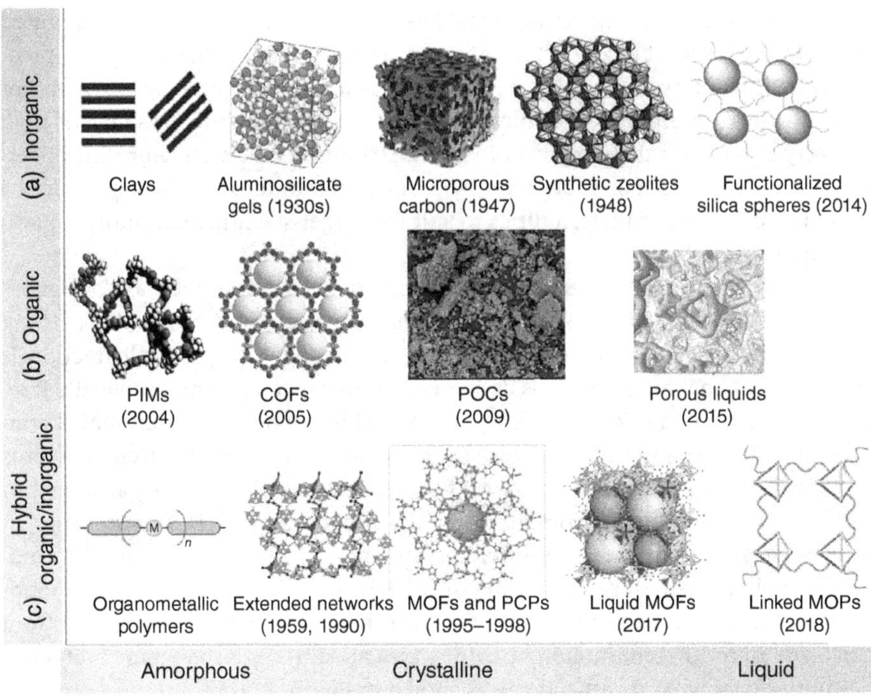

Figure 7.1 The illustration of various clay-based materials having different compositions from various origins and compositions including (a) inorganic origin, (b) organic origin, and (c) hybrid organic/inorganic compositions having amorphous solids to crystalline morphologies. *Source:* Taylor et al., (2016) / with permission of ELSEVIER.

7.3 Block Copolymers as Ordered Nanoporous Precursors

In conventional membrane preparation, phase inversion technique is one of the most commonly used strategies to prepare porous membranes. Phase inversion techniques can be broadly divided into four types, namely (i) nonsolvent-induced phase separation (NIPS), (ii) thermal-induced phase separation (TIPS), (iii) evaporation-induced phase separation (EIPS), and (iv) vapor-induced phase separation (VIPS) preparation method. All these phase inversion techniques can be further described following the same thermodynamic theory, which implies that a phase separation occurs in cast polymer solution when a thermodynamically stable solution system is exposed to an unstable environment. Phase separation results in solidifying of polymer-rich phase from the cast membrane solution as porous membranes. On the other hand, block copolymers can be utilized to form ordered phase-separated patterns in a cast by using incompatible segments present in a copolymer system through self-assembly into well-defined ordered nanoscale structures. These ordered patterns in phase-separated segments can then be converted into porous architecture that has great prospects to enhance both the selectivity and permeance of the membrane. Another major feature emerging out of block copolymer systems is the presence of high porosity along with the narrow pore size distribution and high degree of porosity. Moreover, block polymers with isoporous membranes can also be induced with unique surface functionalities to suit the target-specific membrane separation application [46–51].

Block copolymers are specialty polymers and can be described as “designer” polymers. These are composed of more than one fraction of molecules that are arranged in a linear arrangement. This linear arrangement of each fraction is called a block in which a block is defined as a portion of a polymer molecule fraction that is a monomeric unit characterized by at least one configurational or constitutional feature absent from that of adjacent portions. Therefore, a copolymer is a macromolecule derived from more than one species of monomer arranged in various conformations and compositions. The suitable polymerization step known as copolymerization involving different monomers results in copolymer formation. Copolymers produced following the copolymerization of two varying monomeric species are called dipolymers. On the other hand, polymers obtained from three and four different monomeric units are called tri- or terpolymers and tetra- or quarter-copolymers, respectively as shown in Figure 7.2 [52, 53].

Block copolymers provide a well-ordered pore architecture upon self-assembly phenomenon in their melt due to the confined interaction between different blocks of polymeric fractions within the copolymer system. When block copolymers are used for the application of membranes, block copolymeric systems offer the unique

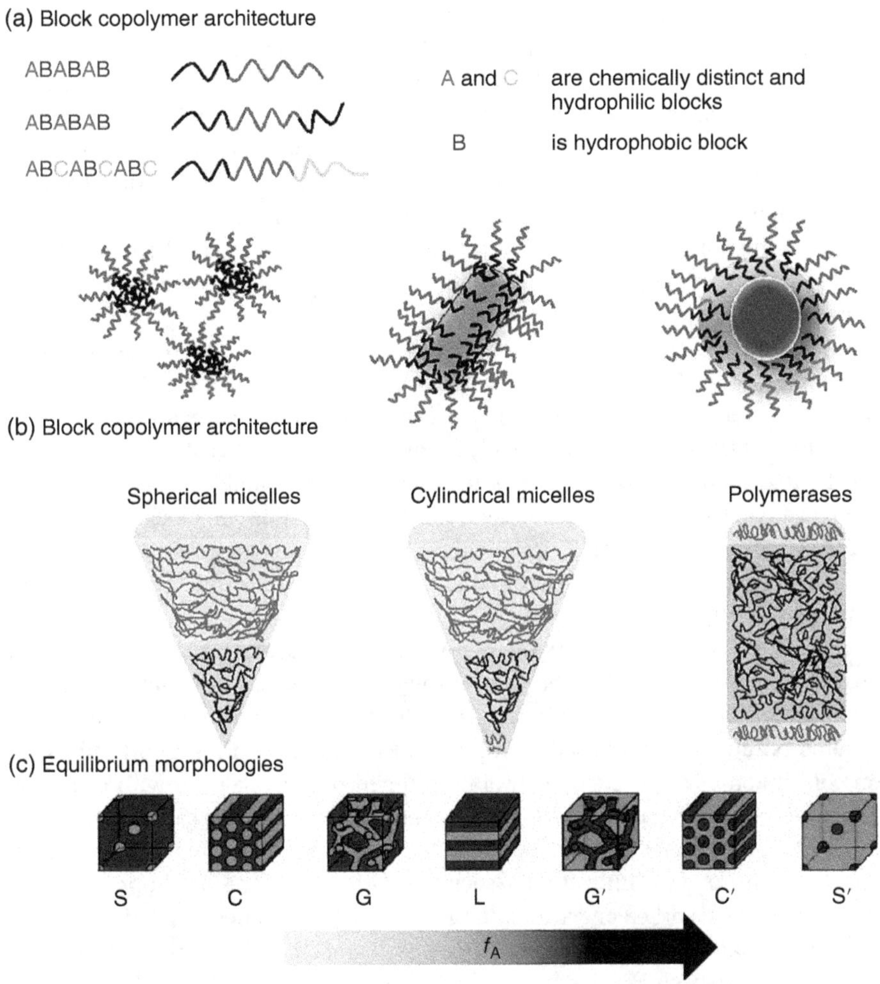

Figure 7.2 (a) Schematic representation of various types of block copolymer architectures prepared with varying monomers following different synthetic procedures, and (b) illustration showing different self-assemblies obtained from amphiphilic block copolymers in a solution based on the varying monomeric fraction and their packing densities, and (c) mean-field pattern prediction in thermodynamic equilibrium phase structures induced due to the symmetric diblock self-assembly in melt for conformationally. Here, phase separations are labelled as: L-lamellar, C-hexagonal cylinders, G-discontinuous cubic, and S-body-centered cubic spheres. f_A represents the denomination of the volume fraction.

feasibility of precisely controlling and tuning porosity in resultant membranes at the nanoscale level and offer several fascinating properties within one membrane. Block copolymers having different polymeric segments provide a pore characteristic based on the length and fraction of constituent polymeric units as repeating units that are covalently linked together in a copolymer chain architecture. For instance, a block copolymer with A and B monomers may result in the structure AA-BB-AA-BB- as a block copolymer, which is typically referred to as a "diblock" copolymer; similarly, AAA-BBB-CCC-AAA-BBB-CCC as triblock copolymer, AA-BB-CC-DD-AA-BB-CC-DD as tetra-block copolymer, and so on [51, 54–57].

The block copolymer possesses a variety of structural characteristics that give them unique properties of attaining morphologies useful for the preparation of isoporous membranes. However, it is a challenging task to tune the pore sizes at nanoscale in block copolymer membrane which mainly depends on the selection of appropriate copolymer and their molar fractions. Further, selecting the appropriate copolymer and conditions for membrane solution preparation and casting play a vital role. Here, the thermodynamic parameters of block copolymers significantly affect their final morphologies. Both conventional and advanced membrane manufacturing processes mostly follow solution-based casting methods and the final morphology often results in nonequilibrium and kinetically trapped structure. In such a situation, order–disorder transition in block copolymers translates into a variety of morphologies resulting in asymmetric to symmetric block forming (i) spherical (S) domains, (ii) cylinders (C), and (iii) lamellae (L) patterns [58–60].

Therefore, block copolymers offer more than one advantageous property and are capable of giving desired porosity through the selection of suitable repeating units and their molar fraction to make them an extraordinary candidate for ordered porous polymer-based membranes. Ever since the first report published on conversion of block copolymers into ordered porous material with a specific focus laid on developing asymmetric superstructure via phase separation in 2007 [61], a new materials stock emerged to make use in the membrane separation sector. Scientists and innovators associated with membrane development since then extensively worked on various combinations of monomeric units in a copolymer extrapolating block copolymer systems to reach new avenues. Alongside, efforts have also been made to understand the polymer assemblies as the combination of more than one monomer in a system has shown versatile self-assemblies. Ever since, different block copolymers having various combinations of monomeric molar fractions and assemblies targeting ordered porous membranes have gained great significance with several promising applications. However, the majority of block copolymer systems have been used in MF, UF, and nanofiltration (NF) categories. As one of the sacrificing constituents is large in size and difficult to prepare the microporous network structures, block copolymer-based membranes have been mainly used for larger constituent separation like suspended solids, colloidal contaminants, cell separation, and gas separation applications [62–65].

The order in block copolymers arises through disorder induced in a melt followed by the self-assembly resulting in desirable patterns. The molecular packing that results in self-assembly of dissimilar polymeric fractions thus thermodynamically gives stable microdomain patterns. These microdomain patterns in block copolymers appear in the bulk state that is governed by the low mixing entropy and positive mixing enthalpy of the constituent segments. Interestingly, these dissimilar polymeric units are covalently bonded between each segment; the overall bulk of the system stabilizes without experiencing any type of macroscopic phase separation. Therefore, this phenomenon minimizes the interfacial energy by adopting well-defined microdomain patterns with constant interfacial curvature and stretched interfacial chain configurations. Typically, in a physicochemical approach, a block copolymer membrane is prepared from a solution that is greatly influenced by the nature of solvent. Solution is cast or spin-coated to obtain polymer film in which the block copolymer is kinetically trapped in a solid state of the material. There exists a favorable selectivity for one polymer fraction over another in a block copolymer system, which leads to the obtaining of a film with dissimilar volume fractions of blocks. This further leads to the formation of varying morphological transitions through self-assembly mechanism into a variety of crystal-like ordered microphase morphologies on a mesoscopic length scale, which typically range between *ca* 10 and a few 100 nm. This self-assembly occurs as a result of segmental incompatibility of dissimilar polymers that are chemically different in nature. Interestingly, these blocks do not undergo macroscopic phase separate due to the existence of covalent bonding between each dissimilar fraction [66].

For instance, a block copolymer with spherical domains of the shorter block or smaller fractions of one polymeric block may transform into a morphology with a cylindrical structure. This is due to the selective swelling and expansion of shorter blocks. Similarly, a cylinder-morphology in a diblock copolymer transforms into a sphere-forming morphological microstructure upon selective swelling of the minor matrix-forming block. Interestingly, as the number of blocks in a copolymer system increases, possibilities of forming novel microstructure separation through complex structural self-assemblies increases such as alternating spheres (core-shell), three-domain lamellae, cylinders, gyroid, perforated structures, and so on (Figure 7.2c). Among the most extensively studied systems are coil-coil-type diblock copolymers. Here, thermostable morphologies of self-assemblies mainly depend on three factors, namely volume fraction of the component, the molecular weight, and the degree of block incompatibility, which is expressed following the Flory-Huggins parameter χ. Therefore, as predicted in the figure, membrane morphological parameters like uniform-sized spheres, cylinders, lamellae, and complex bicontinuous gyroid nanostructures are generally theoretically predicted and experimentally proven by varying these three parameters [67–72].

Conventionally, the NIPS method is the most commonly used technique to prepare polymeric membranes for MF, UF, NF, and RO applications. The phase inversion for polymer cast solution is achieved by immersion precipitation process, which is generally referred to as polymer precipitation. Even though NIPS is a simple and widely adopted technique, achieving control over pore size and distribution is a challenging task. Therefore, recent advances in membrane science have attempted to overcome this significant technological challenge that has limited control over structure and surface properties, which directly influences the membrane performance parameters during their usage in water treatment applications. In this direction, novel approaches like a combination of NIPS with self-assembly phenomenon that occurs with copolymer in solution have been adopted, naming the solution with NIPS technique (SNIPS). This has given greater control over pore size design and manipulation during the membrane preparation [73, 74].

Therefore, ordered microsphere separation achieved through the self-assembly process in copolymer membranes creates patterns and subsequent sacrifice of one segment from the block copolymer assembly following physicochemical step yields well-ordered porous membranes useful for various applications such as wastewater treatment. In one such attempt, researchers have used polystyrene (PS)-block-polyferrocenyldimethylsilane (PFS) block copolymer, which is the epitaxial disorder analogy to create the ordered pattern in the resultant membrane. Similarly, PS-b-poly(4-vinylpyridine) (PS-b-P4VP) has been subjected to self-assembly procedures to yield isoporous membranes. In both cases, important physical parameters like copolymer solution composition, concentration, solvent evaporation time, and size of block copolymer fractions played an important role in determining the ordered morphologies of the resultant membrane. Further, the effect of polymer fraction size on the membrane structure formation was analyzed using poly(isoprene-b-styrene-b-4-vinylpyridine) triblock terpolymers. In this approach, similar block volume fractions were used for the self-assembly as well as for NIPS (SNIPS) method, which showed a favorable ordered pattern from SNIPS technique. SNIPS method has been further extended for a series of block copolymers having different combinations of polymeric segments to achieve appropriate ordered porosity and surface functionality [75–78].

7.4 Block Copolymer-Based Membranes for Wastewater Treatment

In recent times, polymer-based membranes and associated processes have become one of the significant technological achievements in addressing water crisis and achieving environmental remediation due to their high separation efficiency, quick solution, energy conservation, environmentally friendly and technology offering molecular level filtration options. With the advent of new materials and understanding fundamentals of polymer-based material science aspects, researchers have adopted copolymer-based membranes as an effective barrier for separating the number of pollutants from contaminated sources. Copolymer-based membranes initially adopted to develop MF and UF membranes have played an important role in separating organic matter, suspended contaminants, microorganisms, and other biological contaminants from wastewater. Even though flux and rejection trade-offs existing in conventional polymer-based membranes have shown some limitations in achieving optimal separation and permeation performance, uniform pore size, and their unique distribution showed high rejection performance while large numbers of pores facilitate higher flux rates.

For the first time, Peinemann et al. developed block copolymer self-assembly (S) by combining NIPS strategy with SNIPS to prepare asymmetric isoporous membranes. Further, a team led by Nunes et al. studied in detail the pore formation mechanism and revealed that the pore formation proceeds with the generation of micellar assembly followed by phase separation. Further, continued efforts have led to the development of a series of ordered block copolymer membranes with unique surface properties. For instance, amphiphilic PS-*block*-poly (acrylic acid) (*P*S-*b*-PAA) was developed as a classical hydrophilic membrane using dilute solution to achieve a variety of nanostructures like spheres, cylinders, vesicles, lamellae, and so on. These microstructures were initially optimized in the form of micellar structures. This approach has shown greater control over pore size and their distribution. Interestingly, the hydrophilic nature of PAA in a *P*S-*b*-PAA system poses a challenge in preparing isoporous membranes following the SNIPS method due to the difficulty in obtaining optimized precipitation in a nonsolvent. However, subsequent studies have overcome the precipitation issues associated with the hydrophilic PAA segment in the SNIPS process. Considering unique properties associated with PAA such as biocompatibility, ionization, antibacterial property, and hydrophilic nature, which find attraction in water treatment applications, preparation of *P*S-*b*-PAA-based isoporous membranes has been widely explored. Also, to make *P*S-*b*-PAA-based isoporous membranes versatile, alternative strategies have been adopted. For example, researchers have used polyisoprene-*b*-PS-*b*-poly(acylic acid) (PI-*b*-*P*S-*b*-PAA) triblock copolymer by conversing poly(*N*,*N*-dimethylacrylamide) to poly(acrylic acid), which was further turned into isoporous and functional membrane. Such novel conversion strategies involving simple chemistry opened new synthetic procedures routes to improve porosity and improved pore functionalities in resultant ordered porous membranes. Blending and/or compositing is another physicochemical strategy to include PAA in a block copolymer system, which was further converted into isoporous membranes [61, 79].

For instance, a pH-responsive composite isoporous membrane was prepared following phase inversion blending *P*S-*b*-PAA with popularly used polyethersulfone (PES). Further, a team of researchers prepared metal ions-induced isoporous membranes following the self-assembly of *PS-b*-PAA with metal precursors in the SNIPS procedure. Interestingly, the compositing strategy of using metal ion constituents has led to improved mechanical properties, antifouling properties, and tuning of the pore size in the resultant membrane. In addition, blending strategy and conventional and simple additive-free methodologies in asymmetric *PS-b*-PAA were also followed to generate isoporous membranes by selecting a suitable combination of block segments in SNIPS process, which was greatly dependent on PS/PAA block ratio [80, 81].

Even though SNIPS methodology is simple and yields several advantages in preparing isoporous membranes, technique requires a large number of block copolymers, precursors, and water resources for the precipitation process. Similar to conventional membrane process operation, block copolymer-based membranes also require high applied pressure to result in higher permeabilities, which is energy-intensive during wastewater treatment applications. These limitations restrict wider implementation of technology at large-scale operations. Another major technical disadvantage is expected from the solvent evaporation step during which phase separation could lead to unidirection resulting in a block copolymer membrane phase separated in vertical cylinders on the surface. These limitations have passed away for adopting new strategies.

In one such approach, additives were used to control the unidirectional evaporation of solvent resulting in hierarchical porous scaffold, but the pores are only formed after rinsing out the soluble additives. For this, a process named gas-flow self-assembly with NIPS (gSNIPS) was adopted in which solvent evaporation from self-assembly of *PS-b*-P4VP *via* controlled the solvent evaporation conditions using gas flow. This approach was an attempt to optimize pore orientation in a gas-flow-induced self-assembly process. Following this process, PS_{678}-*b*-PAA_{180} was converted to isoporous membrane in which a high molecular weight copolymer was blended with a low molecular weight PS_{144}-*b*-PAA_{22} to attain macroscopic phase separation followed by pore formation through EIPS method. Nevertheless, EIPS process also consumes high polymer concentrations and is nonsolvent for precipitation process. However, the method is promising to attain the desired pore orientation and control [53].

Similar attempts have been made to develop a hybrid process through which isoporous membranes could be prepared. Here, a combination of SNIPS and EIPS was attempted on the composite of $PS_{70,500}$-*b*-$PAA_{13,000}$ and amine-terminated poly(*N*-isopropylacrylamide) (pNIPAM-NH_2) through SNIPS or airflow-accelerated EIPS (acEIPS). While SNIPS for *PS-b*-PAA/pNIPAM-NH_2 membrane consumed a lesser amount of precursor (6–7.8 wt%), whereas acEIPS consumed 5 wt% more than previous technique and precipitation was carried out without water. Isoporous membrane prepared following these two approaches exhibited excellent ordered surface morphologies and remarkable water permeances of 1500 L/bar/m^2/h for SNIPS-membrane and 3000 L/bar/m^2/h for acEIPS-membrane at room temperature [82, 83].

7.5 Block Copolymer-Based Membranes for Desalination

Following novel membrane preparation strategies, block copolymers' various self-assemblies can be achieved. Through the selection of appropriate polymer concentration, molar fractions and phase separation approaches, a large number of pore morphologies can be easily optimized. Also, along with preparation strategies, novel membrane fabrication approaches like stacking the morphologies in successive layers can also yield technological advantages during large-scale membrane preparation. This requires suitable macromolecular design strategies and a combination of parameters can yield several useful nanostructures that could be beneficial for applications like desalination. However, creation of continuous and connected pore architecture in multiple layered block copolymer assemblies is vital to attain the free-low of permeate and achieve higher product flux. Therefore, during the membrane preparation, creation of bicontinuous porous layers plays a significant role that percolates across membrane thickness. Also, multilayered architecture also provides mechanically stable matrix support useful for high-pressure operations particularly used in desalination applications. Conventional membrane technology used for desalination applications has matured over the years. Considering commercial aspects of demand for new membranes, several technologically important operational parameters have been adopted during the development of copolymer-based desalination membranes. Therefore, to fulfil the commercial standards of block copolymers, self-assembly techniques have been modified aiming for bicontinuous or cylindrical morphologies in the resultant isoporous membranes.

Nevertheless, initial attempts to make use of block copolymer systems for desalination applications have shown limited success. Interestingly, earlier approaches used cation exchange and anion exchange membranes, which possess largely different selective affinities for pollutant species but sometimes complementary selectivities for the rejection of salts. Therefore, initial attempts largely used complementary aspects of ion exchange polymers for attracting ionic species from contaminated sources. Thus, following a simple principle, a copolymer containing anionic and cationic blocks potentially exhibits uniform rejection of all salts. In addition to this, the polymer compatibility present in block copolymer systems can be further utilized in the formation of surface-functionalized isoporous membranes carrying randomly distributed charged domains throughout the microstructures. However, efforts to prepare copolymer membranes containing block structures having an acid sequence followed by a base sequence have shown limited tractions for larger purposes [84].

Nevertheless, parallel attempts to develop block copolymer-based membranes as a barrier for molecular and ionic species separation mainly focused on optimizing transport characteristics in nanoporous films. This led to the development of various processes to realize the thin-selective layer formation and subsequent membrane module fabrications. Therefore, these complementary efforts were implemented in processing various approaches, protocols, and routes that resulted in the design and development of porous and permeable copolymer-based materials, not necessarily the formation of membranes but films using scalable processing methods. These thin films were then transferred to a stable substrate and porosity was induced following standard self-assembly protocols. For instance, thin films of block polymers with varying concentrations and combinations can be prepared by spin-coating a solution onto a stable solid substrate (silicon or glass) followed by solvent evaporation and appropriate annealing to achieve melt equilibrium nanostructure (*via* microphase separation). In order to create porosity, thin films are subjected to selective swelling in the melt followed by extraction of the swelling agent (pore-forming agent) or a sacrificial moiety (block copolymer segment) may be physically etched out or leached into solution through solvent treatment to create the percolating network of isoporous on nanoscale range useful for desalination applications. Nanoporous membranes can also be prepared by transferring already processed porous thin film onto a mechanically stable and flexible support for end-use [85–87].

Following this approach, block polymers with sphere and cylindrical orientation can be converted into isoporous membranes. In particular, cylindrical domains possessing long axes aligned normally to the plane of the substrate provide optimal pore-forming capabilities. However, orientation of the cylindrical domains of block segments with their long axes, which are perpendicular to the substrate surface, is required while ensuring that they span the entire membrane thickness is one design challenge. Cylinders that are oriented in parallel planes or pores that are dead-end before distortion in the cast film or membrane result in impermeable domains. Nonetheless, researchers and innovators often practice both parallel and perpendicularly oriented structural features. In all the cases, structural orientations and pore-forming capabilities are carefully controlled through various processing conditions, namely (i) Controlling and regulating the rate of solvent evaporation; (ii) Following surface chemistry principles for tailoring the substrate and block polymer interactions; (iii) Careful selection of annealing technique and rate. These careful considerations help in overcoming the challenges associated with selection of substrate, solvent, film formation technique, and orientation of pore structure in supported thin films. Among all, cylindrical pores created in both selective layer thin film and membrane have demonstrated enhanced performance when used for filtration applications. For instance, PS-*b*-poly(methyl methacrylate) (PS-PMMA) used for the preparation of UF range membranes showed excellent performance for virus filtration. In this case, PS-*b*-PMMA block copolymer was coated on a functionalized flexible substrate, followed by annealing, and self-assembly. Upon phase separation, the pore-forming (PMMA) phase was etched following ultraviolet light exposure which degrades PMMA moieties and crosslinks PS. Subsequently, PMMA was washed or etched from the system by treatment with acetic acid. As a result, stable crosslinked nanoporous PS thin film was then carefully transferred onto a commercially flexible support for further filtration studies. This was the first of its kind study in preparing highly ordered nanoporous block copolymer membranes, which performed better than commercially available samples for separation and permeation parameters demonstrating higher selectivities and fluxes. This study further opened new avenues for researchers to follow and a series of copolymers have since been converted into ordered nanoporous membranes both in thin-film format and standalone membranes. These studies also extended to sphere, gyroid, lamellar, and other orientation phase separations that originated due to presence of various molar fractions [88, 89].

Further, recent advances in synthetic and design strategies have opened possibilities of obtaining polymer chains, monodispersed chains, and lengthy and molecular weight. Also, advanced polymerization techniques have enabled optimization of polymer chain length and molecular weight of each segment in a block copolymer chain with polydispersity index near 1. This remarkable control over polydispersity parameter has led to better control over phase separation domains and subsequent removal from the system to obtain isoporous copolymer membranes with narrow size distribution and nanopores less than 10 nm. These systematic pore size optimization approaches block polymer-based membranes enabling researchers to push the limits associated with pore parameters. Precise control over pore sizes less than 10 nm regime and their uniform distribution in block copolymers gives water treatment technologies a technological advantage. One of the major limitations associated with conventional membranes is the optimization of size-selective filtration process, which can be potentially overcome with the use of isoporous block copolymer membranes. Even though the earlier effort of using block copolymer was only restricted to the UF and MF regime, most of the block polymer membranes till today have been fabricated with pore diameters in the range of 10–50 nm. With the advent of new techniques, a range of block polymer-based membrane devices can easily reach NF and reverse osmosis (RO) regimes, which is essential for desalination applications. In this direction, design and fabrication of block copolymer membranes with smaller pore (<10 nm) features have been now identified which will enable membrane science use of novel material stocks in wider membrane process applications. However, in order to be a perfect desalination membrane, pore size required to be reduced to the 0.7–0.9 nm scale. Nevertheless, RO and NF range pore regimes can still attract enormous industrial interest in block copolymer-based membranes [90–93].

Interestingly, the main feature of pore creation in block copolymers depends on the size of the microphase-separated segment in block polymers. Further, size of the microphase-separated segment or block depends on the overall degree of polymerization of a constituent monomer. Therefore, a simple way to directly reduce the pore size in block copolymer membrane is to reduce molecular weight or use lower molecular weight precursors. Nonetheless, this simple approach alone can lead to limited success and faces technological restrictions to achieve ionic species separations in the desalination range. Particularly, during copolymerization of constituent polymers, if the overall degree of polymerization is reduced to make one block lower molecular weight, the value of χN (total number of polymer segments/monomers) can reduce too low, which will directly affect self-assembly to occur due to lack of thermodynamic driving force. However, controlling polymerization which exhibits higher dissimilarities between individual constituent blocks (high χ block polymers) allows the self-assembly with smaller features. Block polymers containing silicone possess high free energy (χ) values which helps in obtaining pore size below 5 nm due to their hydrophobic nature. For instance, aligned cylinders (5 nm) can be generated using maltoheptaose-*b*-poly(trimethylsilylstyrene) materials. Also, surface functionalities can be induced in isoporous membranes using charged moieties in one block [94].

Nevertheless, a major concern with designing free-standing membrane or thin-film selective layer coated on porous substrate is to impart required mechanical integrity to the isoporous membrane during application. In particular, designing a desalination membrane using the lower-molecular-weight block copolymer is a challenging task in which generally, upon use of low-molecular-weight block segments, membranes become too fragile for practical applications. Particularly, in desalination membranes, presence of a subnanoscale porous structure requires higher applied pressure (>20 bar) to overcome the osmotic pressures of saline water or seawater during the desalination process.

It is important to note that most of initial studies on copolymer-based membranes primarily used diblock materials. Subsequently, efforts have been made to include a third block in the form of rubbery moiety in the membrane design to improve mechanical properties. For instance, the mechanical property of nanoporous membranes was attained with the use of triblock copolymer, polyisoprene-*b*-PS-*b*-poly(4-vinylpyridine), which showed ~3-times higher than that of membranes prepared from the diblock counterpart, Also, precursors can be carefully designed to incorporate a reactive or functional matrix that can be further crosslinked attain high mechanical stability to preventing the collapse of nanostructured pores during the high-pressure operations. Another approach to improve the mechanically stable membrane with a pore diameter below 2 nm is to composite the precursor with composite materials such as assembly of liquid crystals, which is further stabilized with a magnetic field followed by crosslinking of the block copolymer system [95].

Another post-chemical modification approach is also used to prepare mechanically robust and functionally active block polymer membranes. In this approach, additive materials are used that can be post-synthetically modified by introduction of chemical functionality that not only narrows the pore size in useful for desalination purposes but also imparts desired surface/pore functionality. For instance, stitching of polyelectrolyte brushes through the covalent interaction on the surface and within the nanopores of a block polymer membrane. These polyelectrolytic moieties then undergo swelling upon hydration during desalination applications leading to pores packing (less than 5 nm in diameter) from mechanically robust block copolymer segments. In one such attempt, a pore diameter of 2 nm was achieved during desalination filtration experiments when polyelectrolytes used for pore coating using high-density poly(2-acrylamido-ethane-1,1-disulfonic acid) (PADSA) brushes. These brushes were covalently attached to a PI-PS-PDMA material, which served as a precursor for isoporous membranes [96, 97].

Lately, computational studies such as molecular dynamics (MD) simulations have been utilized to obtain critical information, which would help in designing nanoporous membranes targeted for desalination application that provides molecular-level insights into the selection of an ideal water transportation and nanostructures for ion rejection. Detailed simulation studies not only track detailed dynamic trajectories of water molecules that permeate through isoporous membranes but also help in understanding ionic species around and within model membranes. These interactive simulations are useful for accounting for rejection patterns of solute from the desalination process. For this, solute rejection pattern is determined by using effective pore size of the block copolymer membrane. During the MD simulating membranes targeted for desalination application, an optimized pore size is identified and selected depending upon the size of the hydrated solute in the feed solution. In the case of saline feed or salt ions, a hydration shell or cage of water molecules surrounds the solute ion. Hence, the permeation or diffusion of ionic species into small pores is difficult as the solvation shell or water molecule cage surrounding ionic species first has to be removed to facilitate the ion to enter the nanopore. For example, the diameters of hydrated sodium (Na^+) cations and chloride (Cl^-) anions are 6.4 Å and 7.7 Å, respectively, while the thickness of the water shell or water cage surrounding an ion is between 2.5 and 3.3 Å. Therefore, it is important to start with the designing pore size of target ionic species in desalination applications to achieve optimal salt rejection. This implies that pore size should be optimized to be smaller than the hydrated ion. However, the sizes of the water molecules and ionic species from monovalent salts are relatively close. It is expected to have similar interactions from each one with the membrane surface and experience different interactions at pore walls. Therefore, effective salt rejection in nanoporous block polymer membranes is not solely predicted by the sizes of the hydration cages or shells of the ionic species. In such situations, a more precise estimation for the salt rejection patterns may be established following the potential of mean force (PMF) for ions technique along the pore estimation from an MD simulation. In general, the potential of the mean force of a solute gives the free energy differences that occur upon changes in location or diffusion pathway of the solute. This understanding reveals the information on molecular-level rearrangements of solute species that must follow, for instance, changes in solvent structure and degree of hydration during translation of the solute. Through this procedure, the PMF gives a measure of the free energy barrier that exists for the entry or diffusion of ions into membrane pores, important insights into the various molecular or ionic species, and their structural rearrangements. These estimations include both information on the water within the hydration cage surrounding an ionic species and the polymers within the membrane system that occur as the ion enters a nanoporous channel [98–100].

Interestingly, the PMF evaluation earlier carried out on carbon nanotubes using MD simulation techniques revealed a strong correlation between diameters of pores (range from 0.66 to 1.09 nm) and the free energetic barrier, which exists for ion entry. Accordingly, for CNT with pore diameter of 0.66 nm, the free energy change occurs upon entry of Na^+ to 23 kcal/mol, whereas the free energy barrier is noted to be 3 kcal/mol for CNT with a pore diameter of 0.93 nm. Therefore, membranes with smaller pores exhibit higher resistance to the entrance of the water shells that surround the ions, thereby yielding expectedly high resistance to the ion transportation as well. Contrarily, membrane pore resistance drastically decreases with an increase in the pore diameter.

Therefore, pushing limits of block polymer microphase-separated assemblies to create nanoporous membranes from the MF or UR or NF regime to the RO regime is a challenging task. However, the creation of subnanoporous membranes to achieve higher salt rejection potentially attracted enormous attention for desalination purposes. Understanding pore formation mechanisms and parameters associated with the selection of block copolymers to create RO membranes that selectively allow for the passage of water molecules while completely rejecting the dissolved solutes is still the main task. However, the several parameters discussed in this section will likely not be sufficient in obtaining efficient membrane parameter criteria in fabricating technologically advanced members to meet the existing polyamide-based RO membrane's commercial standards in both economic and technological terms. But, realization of highly ordered nanoporous membranes with a higher number of pores per unit area can well address the world's thirst for desalinated water in a sustainable manner. The larger number of uniquely distributed subnanoporous membranes can easily address the issues associated with the high energy requirements in desalination processes. Upon realization, the energy requirement drastically reduces in producing desalinated water useful for both potable and nonpotable applications and in the larger industrial interest, the resource recovery from various manufacturing processes. For this, a well-defined nanostructure and array of emerging chemistries associated with block polymer preparation, and self-assembly parameters will allow the development of target-specific designer separation devices with desired surface properties that can regulate solute transport based on membrane and solute interactions.

7.6 Advanced Porous Materials

According to IUPAC, mesoporous and microporous materials are porous materials with pore size ranges of 2–50 nm and less than 2 nm, respectively. Interest in mesoporous materials arose in the 1980s as the commercially popular Zeolites with micropores (1.2 nm) could not meet the increasing demands in practical water treatment applications involving adsorption or mass transportation of large molecules. This triggered the exploration of natural clays and an innovative series of synthetic mesoporous clays. Even though innovation in the mid-1990s to make use of silica gel and pillared clays was explored, commercially successful ordered mesoporous clays later opened a new avenue for the new class of materials research. In this direction, first highly ordered mesoporous M41S-silicate molecular sieves were discovered with the use of quaternary ammonium cationic surfactants. This innovation also led to the use of the popular templating concept, which was widely adopted to prepare a series of mesoporous materials. Later in 1998, preparation of a large mesoporous silica SBA-15 using triblock copolymers as a template fundamentally changed the materials synthesis concepts. More significantly, the higher thermal and chemical stability of SBA-15 and their quick preparation and reproducibility features gained wider popularity among material researchers leading to the adaptation of mesostructured materials in various sectors [101].

The unique pore architecture in the mesoporous materials with high surface area, tunable pore volume, large pore size, and mesoporous channels in OMMs offer several advantages useful for a wide range of applications. Particularly, in water treatment applications, large pore sizes enable pollutants to reach an effective surface via various interactive access and diffusion mechanisms. For example, the ordered large mesopore frameworks enable favorable conditions for internal mass transportation. On the other hand, large surface areas provide enhanced adsorption and loading capabilities to target molecules and ions. Lately, mesoporous materials have significantly contributed to the performance of active material applications that include advanced technologies like energy storage and conversion, catalysis, adsorption, separation, and purification. Even though several parameters associated with mesoporous have been well accounted, however, two critical issues, namely: (i) precise control in atomic, nano-, and microscale, and (ii) "unfulfilled" elements and components are yet to be critically evaluated (Figure 7.3) [102–104].

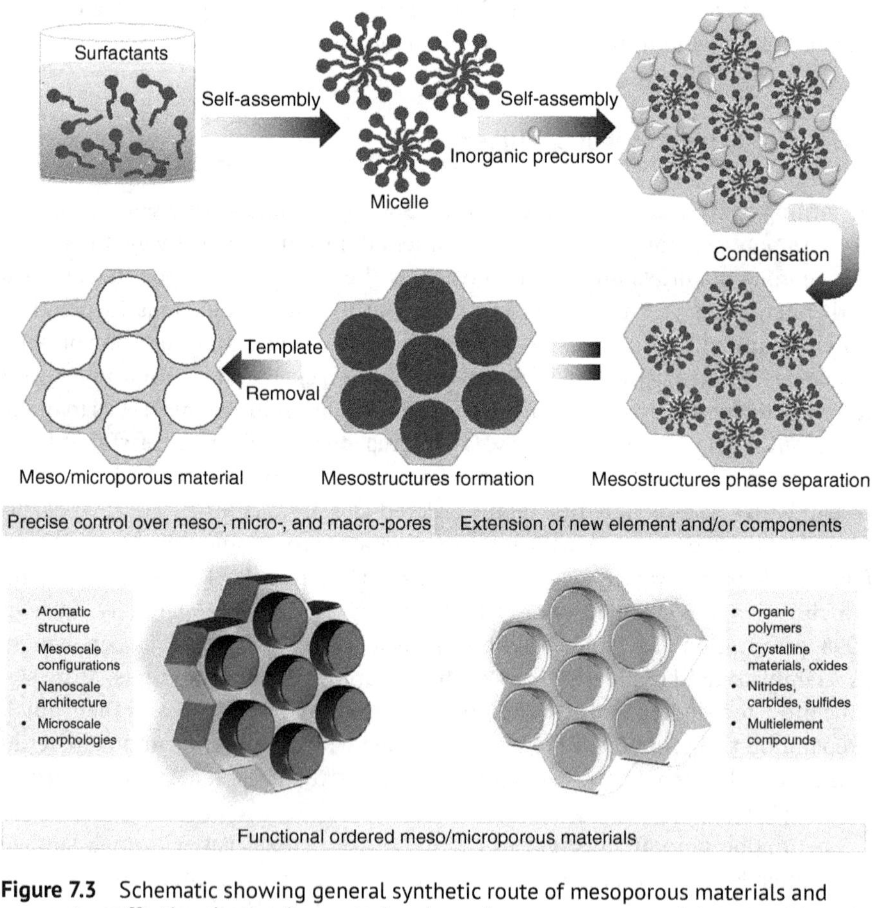

Figure 7.3 Schematic showing general synthetic route of mesoporous materials and parameters affecting the precise control and novel mesoporous material systems that can be derived from the type of porous materials.

Procedures to prepare the precise mesoporous materials to accommodate a wide range of properties useful for water treatment applications is the greatest challenge. However, it is essential to design mesoporous materials that are functionally active toward attracting a variety of pollutants. Even though advanced synthetic practices currently followed give great control over the parameter optimization in mesostructures, however, it remains quite challenging to manipulate each morphological and meso-/microstructural parameter in porous materials such as pore size, mesophase, wall thickness, pore characteristics, surface functionality, and nano-/microscale materials' dimension with highly reproducible and tuneable parameters.

In this direction, the last two decades have witnessed tremendous advances in the synthesis of ordered and well-organized mesostructures with tunable pore size, surface area, and wall/surface. For instance, the pore size in resultant mesoporous materials can be increased by altering surfactants and their concentrations at a high molecular weight or by the use of a swelling agent. Also, surface and wall properties can be optimized (thickened) through elongating hydrophilic block of amphiphilic block copolymers in the precursor. Following these strategies, many useful materials have been designed in the last decade such as core–shell architecture structures and asymmetric mesoporous materials are among many. For instance, a surfactant-assisted coating method was followed to fabricate multifunctional core–shell $Fe_3O_4@SiO_2$ nanoparticles in which mesopore channels have been *in situ* aligned. Thus, making this approach a simple, reproducible, and commercially viable interfacial assembly technique, which is a landmark template to adopt for the preparation of versatile mesoporous composites. Therefore, the following surfactant-assisted fabrication technique series of mesoporous nanocomposites having anisotropic configurations have been developed with morphologies like arc, Janus, multipod shape, and so on [105–108].

Apart from inorganic-based mesoporous materials, carbonaceous nanomaterials with versatile pore architecture have emerged in recent times prepared following organic–organic assembly approach. Most of these approaches follow soft-templating preparation techniques (monomicelle-based assembly methods) to obtain mesoporosity in organic polymers such as monolayers, mesoporous nanospheres, and spirals. Therefore, monomicelle assembly technique has been adopted as a powerful tool for precise control over pores in the building blocks resulting in composite spherical micelles instead of conventional surfactant/block copolymer molecules. On the other hand, porous carbonaceous materials with specific pore characteristics have been designed as target-specific porous materials useful for water treatment applications. Carbonaceous materials also offer several advantages over inorganic systems in terms of microporosity, which results in large surface area. Pore geometric structures allow pollutants to diffuse in large numbers exploiting rich active sites. Thermal conductivity also helps in adopting porous carbon materials in advanced desalination technologies like capacitive deionization (CDI) for desalination applications. On the other hand, microporous carbon has been in use for decades through the adsorption process. Lately, microporous carbon has now been used as substrate coating various functional nanomaterials to induce a variety of functionalities. These functional micro-/mesoporous carbons have been used for the adsorptive removal of many toxic pollutants, such as Zn(II), Cu(II), Cr(VI), CN, and so on. However, functional sites on the carbonaceous materials surface block effective porosity, thereby limiting their performance. Nevertheless, carbonaceous materials were hybridized with mesosilica in order to improve the adsorptive removal performance of micro-/mesoporous carbonaceous materials [109–111].

7.7 MOFs, COFs, and POPs in Wastewater Treatment

Water is essential to all living beings on the earth. However, by every passing day, water pollution is increasing due to increased anthropological activities, which is causing stress on water resources. Use of conventional water treatment technologies including adsorption, boiling, catalytic degradation, sedimentation, membrane process, and distillation have been in practice to address the increasing concerns. Even though conventional technologies have shown promising results, modern pollutants are posing serious challenges and resistance to conventional treatment techniques. However, recent advanced chemical and biological treatment processes have shown significant removal efficiencies for modern and emerging pollutants. Nevertheless, even advanced techniques have also proven inadequate to redeem the increased pollution-related concerns such as emerging and persistent contaminants such as microplastics, pharmaceutical waste, surfactants, hormones, and trace organic contaminants (TrOCs). In this situation, advanced adsorption and adsorptive filtration techniques are believed to provide effective sequestration to both persistent and emerging contaminants from various polluted waters [112–115].

For this purpose, meso-/microporous materials contain both active functional and affinity sites on their surface and permanent cavities. These permanent active sites on meso-/microporous substrates historically exist on materials like zeolites, porous carbons, and mesoporous silica, and are considered the ideal platform for capturing complex pollutants. In this direction, recent advances have explored novel porous materials such as MOFs, covalent organic polymers (COPs) that include amorphous porous organic polymers (POPs), polymeric carbon nitride (CN), and their crystalline derivative COFs as affordable, effective, and task-specific adsorbents having greater affinities compared to the conventional porous materials [30, 112, 116–119].

MOFs are considered highly porous systems whose frameworks are built up via well-designed and controlled coordination bonds in a reproducible preparation technique forming a crystalline network. These well-defined porous architectures allow the versatile manipulation of functional moieties and precise pore control, which helps in water purification applications. On the other hand, crystalline COFs are different from MOFs in that they are prepared using pure organic/organic monomers unlike in MOFS organic/inorganic precursors via robust covalent bonding. These systems also offer several benefits as functionalized building blocks and pore surfaces. On the other hand, porous materials prepared with only organic backbones however lacking long-range order are generally categorized as amorphous POPs. Therefore, high porosity, functional groups, and affinity characteristics give MOFs, COPs, and POPs unique capabilities for robust contaminant capture.

Therefore, MOFs, COPs, and POPs have been widely used for treatment of industrial wastewater of different nature including textile industry, pharmaceutical waste, and industrial wastewater containing trace organic matter. In particular, use of large quantities of organic dyes for coloring cloths has a long history. As the demand for colorful clothing and textiles at low prices is increasing, the use of artificial dyes as low-cost raw materials become prevalent in modern industry. At present, it is estimated that more than 10,000 varieties of dyes are in use in the market and estimated over 700,000 tons of synthetic dyes are being consumed every year [120]. Even though textile dyeing process consumes less amounts of dye/pigments, however, unfortunately, considerable amount of unused coloring agents as end up effluent. These untreated dye effluents are discharged into water reservoirs like lakes, rivers, and surface water quantity of which is estimated to account for more than 15% of dyes produced annually. Another major concern with dye effluents is their chemical stability due to their chemical structure. Therefore, dyes are rigid and difficult to degrade easily in water bodies. Before these complex and sturdy organic molecules mix in larger aquatic resources, it is important to remove them from the contaminated sources for a healthy ecosystem. For this purpose, high surface area mesoporous MOFs, COFs, and POPs have been considered promising candidates for selective dye separation from industrial wastewaters. Even though filtration technologies (ceramic and polymer membranes) have opted to selectively remove dye from wastewater with limited success. However, over a period of time, filters get clogged, which leads to fouling due to the organic nature of pollutants. Also, small molecular weights of dye easily penetrate membranes (UF and NF), which mainly relies on size exclusion. In this situation, use of highly selective porous materials through adsorptive removal has been highly recommended.

7.8 Ordered Porous Materials in Dye Wastewater Treatment

MOFs with pore ranges between 0.3 and >10 nm are most suitable for selectively reject dyes from wastewater in NF process. The inherent crystalline properties of MOFs with appropriate pore morphologies act as selective molecular sieving when used as composite constituents in mixed matrix membranes (MMMs). With this strategy, researchers have successfully prepared MOF-/polymer-mixed matrix NF membrane for higher rejection of organic dyes. However, polymers herein clog the pores in the long term that are present in MOFs which drastically reduces the contaminant pore filling. However, to overcome limitations associated with polymer interference, further pore-accessible strategies known as coordination-driven *in situ* self-assembly have been followed for the preparation of continuous and defect-free MOF/polymer hybrid membranes. Following this strategy, ZIF-8/PSS membranes were prepared using 2-methylimidazole solution and poly(sodium 4-styrenesulfonate) (PSS) into a Zn^{+}-loaded polyacrylonitrile (PAN) dope solution, which was poured on a dead-end filtration cell [121–124]. These membranes (ZIF-8/PSS) were further used for organic dye contaminants such as methyl blue (MB) with ~98% rejection and a flux more than 260 $L/m^2/h/MPa$. In a similar approach, UiO-MOF-MMMs were prepared in a facile synthesis procedure which was used for removal of various organic dyes wastewater sources [125].

On the other hand, using flexible monomers and utilizing potentials of diverse covalent bond interactions, a series of POPs have been synthesized. Following these strategies, flexible monomers help in fine-tuning pore apertures in resulting POPs that are useful for removal of target pollutants from wastewater sources. This strategy was adopted in the preparation of COFs, which offer ordered pore architecture and uniform pore channels. In this direction, highly flexible, crystalline and defect-free, COF membranes (COMs) were prepared using *p*-toluene sulfonic acid (PTSA) as precursor as well as the catalyst which intern serve as the binder of amine and aldehyde monomers, and when used, acted as a water reservoir for the Schiff base reaction. Similarly, following this strategy, a series of COMs were prepared for filtration applications with detailed organic matter rejection and permeation assessment NF membrane set-up. These membranes have been optimized to possess pore dimensions below 1 nm to achieve all the organic molecules such as dyes, organic matter, and other emerging pollutants such as pharmaceuticals and surfactants. The results reveal complete rejection of various pollutants as the pore size is narrow (<1 nm) and uniform micropores exist in resulting COM. These membranes have further shown greater potential in the treatment of textile industry or dye industries [126, 127].

In addition to these strategies, simple molecular sieving abilities of COF-based membranes have been utilized which is facilitated by altering the charge properties of the covalent framework. Most of the dye molecules are relatively large and charged. Utilizing these characteristics, thin cationic-CF membrane such as EB-COF:Br prepared through the interfacial polymerization of (EB), cationic monomer, and 1,3,5-triformylphloroglucinol (TFP). The resulting cationic-COF membranes showed better rejection against positively charged dye molecules from wastewater. Interestingly, the thin-films produced at the liquid/liquid interface then transferred on a substrate in a layer-by-layer stacking process to obtain stable, large-scale, membranes in dense interfacial EB-COF:Br membranes. Using these membranes several dye contaminated wastewaters with high efficiencies in their rejection due to the combined effect of size exclusion of interfacial layer and electrostatic interactions between charged membrane and charged pore walls. In all cases remarkable rejection of the anionic dyes contaminants such as methyl orange (MO), fluorescein sodium salt (FSs), and potassium permanganate (PP) have been removed up to ~99%, ~99%, and ~98%, respectively. However, cationic and neutral dyes showed lower rejection rates due to similar charges under identical operational conditions which revealed that anionic molecular rejection was achieved through the combination of size exclusion and selective molecular sieving through EB-COF:Br membrane [128, 129].

Even though membrane processes like NF and RO have shown efficient removal capabilities for organic dye molecules, organic fouling- and pore-clogging affects their large-scale and long-term operations. In this situation, highly stable COPs due to their organic backbones and hydrolytic properties were used as effective filters for complex water treatment as COFs show greater stability over MOFs in water. Therefore, novel porous COFs with remarkably large surface areas and adjustable porosity have been used for wastewater treatment via sorption mechanism. Further, a dense iron-terephthalate having (MOF-235), a cationic framework has been developed as an efficient barrier for the removal of both anionic dye (MO), and cationic dye (MB) from wastewater sources. Similarly, mesoporous MOF Zirconium-Metalloporphyrin PCN-222 (PCN-222), a zirconium-metalloporphyrin with dual functions and large pore architecture and suitable surface charge/zeta potentials, which offer favorable interaction have been used for the adsorptive removal of various pollutants (MO and MB) from wastewater. Also, PCN-222, and MOF-545 a high surface area of (2336 m^2/g) with a large open channel having a pore diameter of 3.7 nm, have been favorably for the retention of organic dye molecules in the nano-pockets. These pore architectures provide favorable adsorptive sites than a MOF which possesses smaller micropore (<2 nm). The zeta potentials of PCN-222 ranged from ~23 mV at pH 3 to ~−13 mV at pH 10 which is relatively large positive and large negative potentials. This range of charged surface characteristics make PCN-222 suitable absorbent for both anionic and cationic MB (1239 mg/g) and MO (1022 mg/g) organic pollutants in wastewater [130–134].

On the other hand, two dimensional COFs (COFs (TPT-azine-COF and TPT-TAPB-COF) with abundant accessible 1D channels have been extensively utilized for toxic organic pollutants like MO, MB, and rhodamine B (RhB). Here, both TPT-azine-COF and TPT-TAPB-COF facilitated with various amine monomers and a heteroatom-rich linker TPT-CHO successfully used for RhB removal. In case of RhB, an adsorption capacity of TPT-azine-COF (surface area 957 m^2/g) was 725 mg/g and TPT-TAPB-COF (surface area 1020 m^2/g) showed 970 mg/g. In another study, a compressed polycrystalline TAPB-PDA COF film in the form of a disc-shaped pellet was used as a filter in a continuous flow for the removal of RhB in various contaminated sources. Similarly, amorphous POPs with target specific functionalization were used for the retention of water-soluble organic pollutants through their unique pore architecture. However, pore dependent retention activities were restricted due to their disordered rigid polymeric backbone that deemed to obstruct the accessibility of organic molecule entering the pore space leading to entrapment [135–138].

In recent times, due to increased economic activities, oil consumption and transportation has extensively increased which led to the spill over and leakage of oil and related substances. These oil contamination due to spill over and leakage across the world threatening the ecosystem and environment. Due to the environmental severity that arise due to oil spill over, cleaning processes consume tons of cleaning chemical which lead to the generation of secondary pollutant costing over 10 billion dollars annually [139–141]. Most of the conventional (centrifugation and *in situ* burning) have proven limited success. Therefore, to overcome the limited removal efficiencies of conventional methods or to replace the inefficiently oil wastewater treatment techniques, novel nanoporous POP adsorbents targeting selective capture of organic liquids/oil from water have been designed and successfully utilized. In one such attempt a porphyrin-based POP (PCPF-1) prepared via homocoupling of 5,10,15,20-tetrakis(4-bromophenyl)porphyrin (tbpp) has been successfully used for the removal of oil contamination. Here, the hybrid 2D layered structure which was conjugated with porphyrin, and phenyl rings that resulted in the formation of high surface area POP and superhydrophobicity have been used for selective extraction water from petroleum oil (gasoline and saturated C5–C8 hydrocarbons) contaminated wastewater leaving being concentrated oil source. Here, PCPF-1 recorded higher oil take-up efficiencies compared to conventional materials used for the oil-spill clean-up. However, preparation of PCPF-1 is a cumbersome process and results in a powder product, which is difficult to recycle in a real field application making it commercially unattractive [142, 143].

Therefore, innovative routes of materials preparation and applications have been followed to overcome the limitations of the previously known porous materials used for oil clean-up. In one such approach, a simple COF-V having superhydrophobicity has been engineered to obtain appropriate porosity suitable for selective oil separation from oil-contaminated sources. In this case, vinyl functionalized COF via a thiol–ene click-reaction to obtain COF-VF. To make the overall system practically viable, COF-VF was integrated with commercial melamine foams resulting in a practically viable foam superhydrophobic-COF-VF@foam. Further, to make overall product robust, various monomeric precursors also have been altered by submerging monolithic melamine foam in 1,3,5-tris(4-aminophenyl)-benzene and 2,5-divinylterephthalaldehyde, which was then catalyzed with acetic acid to obtain end-hydrophobic foam. These COF-VF@foams were rigorously tested for their selective oil-uptake or removal efficiencies from multiple organic contaminated sources. The uptake performance of the resulting COF-VF@foam was examined with multiple organic liquids afterward. The performance of COF-VF@foams has remarkable in achieving sorption capacities for oil/organic liquids from 67 to 142 times owing to their superoleophobicity and superhydrophobicity nature [143].

In another approach, a new class of lipophilic networks of functional porous polymers, namely conjugated microporous polymers (CMPs) have been designed and developed for the treatment of complex oil/water separation applications. In a first-of-its-kind approach, the surface superhydrophobicity of CMPs generally known as hypercrosslinked CMP (HCMP-1) was prepared using 1,3,5-triethynylbenzene in a homocoupling polymerization, whereas analogues of HCMP-1 were prepared following binary monomers 1,4-diethynylbenzene and 1,3,5-triethynylbenzene. Both HCMPs showed superoleophilic characteristics having a water contact angle (CA) of 167° for HCMP-1 and 157° for HCMP-2 due to their large pore architecture and superhydrophobic nature of precursors. These characteristics in HCMPs facilitate the selective capturing of oily contaminants and nonpolar organic solvents present in water sources with enhanced adsorption kinetics for oil exceeding the capacities ~10-fold than their weight. For HCMP-1, a maximum absorbency for oils that reached up to 10 times their weight. On the other hand, adsorption for different organic solvents from water ranged from ~700 to 1500 wt%. Further, to make CMPs commercially viable for real-field applications, CMPs have been converted into practically feasible aerogels having hierarchical porosity, which facilitate faster and selective adsorption of oil while at the same time allowing passage of water at higher flux rates. In this case, adsorption of oil and various organic solvents has been trapped in the wider porosity of CMP aerogels network resulting in higher removal efficiencies for oil contaminants ranging from ~20 to 53 times their weight. This study signifies the importance of larger porous filters compared to their nonporous and dense counterpart in which the oil adsorption was only restricted to 16 times [144, 145].

The wettability in MOFs can be manipulated through various functionalization techniques in which surface modifications are suitable for inducing affinity for oil. Even though initial applications were developed for wettable MOFs, fluorous MOFs (FMOFs) were for adsorption of aliphatic and aromatic oil components. However, later researchers extended the use of new MOF derivatives to water treatment applications. In one such study, FMOF-1 which was prepared using silver(I) 3,5-bis(trifluoromethyl)-1,2,4 triazolate (AgTz) showed great removal capabilities for oil components in the oil-spill contaminated water. In this clean-up approach, FMOFs showed improved oil removal efficiencies due to their hydrophobic pore surface. Later, popular MOFs, HKUST-1 were also tested for their oil removal capabilities from emulsion (300 mg/L)-type milky aqueous solution in which dispersion of HKUST-1 with optimal stirring and shaking resulted in complete removal of the oil portion from contaminated water. In addition to this, a detailed study on HKUST-1 showed its remarkable removal capacity of ~4000 mg/g [143, 146–150].

In addition to direct use of porous materials in water treatment applications, nano-/microporous systems have also been used as surface modifying agents for various substrates like fabrics, carbon cloth, foams, sponges, metal mesh, polymer mesh, nanofibers, and so on to impart superhydrophilic or hydrophobic properties. For instance, hydrophobic MOF-based FG-HKUST-1 sponges or foams were developed to clean up oil-contaminated wastewater with exceptional oil uptake capacity of ~8000 mg/g, which is more than two times higher than pristine HKUST-1. Similarly, melamine foam surface modified with 2D MOF kgd-Zn resulting in kgd-Zn@MF showed superior adsorption capacities for oil and various organic solvents between 5077—13,786 wt%. This exceptional performance makes nano-/microporous MOF-modified porous supports of commercially viable and technologically attractive filtration media in realizing complex wastewater treatment in real-world practical applications [149–153].

7.9 Ordered Porous Materials for TrOC Treatment

Groundwater, surface water reservoirs like lakes and rivers, and urban streams are greatly contaminated with high concentrations of organic contaminants such as trace organic compounds (TrOCs). The TrOCs occur in various forms and molecular weights in groundwater, surface water, and widespread occurrence of organic matter in industrial wastewater threatens the quality of drinking water and other water resources. This situation in future is expected to worsen as partially treated and untreated wastewaters continue to be discharged into the environment. TrOCs occur in low concentrations ranging from the ppt (one part per trillion) to ppb (one part per billion) level in groundwater and surface water. However, it is estimated that prolonged ingestion of TrOCs even in trace amounts will threaten human health. Therefore, historically, there have been several technologies that have been developed over the past decades such as physicochemical methods and biological treatment techniques. However, most of the TrOCs are rigid and complex and can be removed by these conventional water treatment techniques. In such a situation, advanced and ordered porous materials have emerged as effective alternatives through their adsorption/desorption properties for the efficient treatment of hazardous TrOCs. More so, several unique tuneable properties of ordered porous materials such as structures and functionalities enable them to turn into effective media for the treatment and removal of TrOCs such as advanced oxidation reactions, the Fenton reaction, photocatalytic degradation, photolysis, and ozonation [154–159].

The TrOCs make up a specific class of hazardous substances that include polyfluorinated alkyl, perfluorinated substances (PFAS), perfluorooctanesulfonic acid (PFOS), and perfluorooctanoic acid (PFOA). Today these substances are widely utilized in household products, fire-fighting foams, food packaging, and several other industrial productions due to their durability and rigidity. Nonetheless, rigidity and long-term stability are a pitfall as they tend to stay in the environment for a very long time making them "forever chemicals." These characteristics make them nonbiodegradable in nature resulting in continued accumulation either in living organisms, soil, and aquatic reservoirs. Nonetheless, prolonged ingestion and exposure to PFASs can result in serious health issues in humans and living beings such as kidney disease, thyroid conditions, cancer, autoimmune disease, and so on. Therefore, it is important to develop rapid, effective, and efficient treatment and removal techniques for trace PFAS from various water sources. Given the nature and complexities involved, conventional techniques have shown limited success against these persistent pollutants [160].

On the other hand, macrocyclic compounds such as β-cyclodextrin (β-CD) possess well-defined intrinsic porosities and a unique host–guest interaction that can be utilized to implant various foreign moieties including organic substances. The functional units in porous organic macrocyclic compounds can be tuned to attract or intercalate PFAS for subsequent capture and removal from the contaminated water sources. In this direction, design of ordered porous β-CD-based POP material has been utilized for the rapid capture and removal of PFAS from contaminated water. To make this realistic, rigid aromatic groups were utilized to crosslink the β-cyclodextrin to form ordered mesoporous polymers, which impart a specific affinity to porous networks of β-CD's (DFB-CDP) toward TrOCs. Even though surface area of DFB-CDP $<10\,m^2/g$, but affinity and functional pore characteristics allowed ~95% capturing of PFOA from the water solution. Therefore, use of DFD-CDP drastically reduced PFOA concentration in the treated water from 1 μg/L to 20 ng/L within 24 hours. In contrast, porous β-CD-based (P-CDP) polymer with a surface area of $\sim 260\,m^2/g$ showed $\approx 0\%$ PFOA removal capabilities for organic TrOCs like bisphenol A (BPA). This is due to the absence of suitable functional affinity toward the contaminant moiety. Nevertheless, researchers are dedicating their efforts to a detailed understanding of host-guest and affinity interactions [154, 161].

On the other hand, β-CD COFs have been prepared using the linkage of NH_2-β-CD, (heptakis(6-amino-6-deoxy)-β-CD (Am7CD)) and aldehyde (terephthalaldehyde (TPA)) monomers. As produced, COF exhibited high selectivity toward a variety of TrOCs in water including naproxen, ibuprofen, BPA, and 4-nonyl phenol. Further detailed studies also revealed efficiencies of COFs in removing 13 different PFAS molecules using amine-functionalized 2D-COFs in which 1,3,5-tris(4-aminophenyl)benzene with the mixture of TPA/dialdehyde with azide-functionalized ethylene glycol (EG) side chains were used as precursors to prepare azide-functionalized COFs. Further, the azide groups were reduced using PPh_3 to produce X%[NH_3]-COFs wherein X ranges from 0 to 100%. Subsequently, to evaluate the adsorption performance for Ammonium perfluoro-2-propoxypropionate (GenX), [NH_3]-COFs having different densities of amine groups were selected for application purposes. Here, a COF with 20% amine content exhibited the highest uptake of GenX amongst all functional COFs, whereas 28% [NH_3]-COF showed wider application potential for PFAS selective remediation with adsorption of >90% for 12 out of 13 PFASs [115, 162].

Further, a new class of PAF prepared with task-specific functionalities has been developed aiming to apply for TrOCs' removal. In particular, PAFs showed great potential in removing TrOCs such as PFOA from contaminated water. Following a similar approach, three hydrophobic chains with chains of different lengths were functionalized using quaternary ammonium moieties to create a series of frameworks including PAF-1, -NDMB, PAF-1-NDMP, and PAF-1-NDMH. Upon thorough characterization and optimization, functionalized PAF-1s were assessed for their effectiveness in extracting PFOA from contaminated water. On the other hand, PAF-1-NDMB showed an optimal PFOA sorption performance of 99.99%, which is better than that of the other two PAF-1 derivatives. Interestingly, PAF-1-NDMB showed 99.99% removal efficiencies for PFOA within two minutes of contact time, which is a remarkable decline in the concentration of PFOA from 1000 ppb to 54 ppt. To assess the maximum uptake capacities, PAF-1-NDMB was subjected to accelerated adsorption assessment test through which it was estimated that functionalized PAF (PAF-1-NDMB) showed 2000 mg/g uptake at the equilibrium concentration of 600 ppm, which is recorded high among similar class materials. Based on these estimations, real-world applications of PAF-1-NDMBs were explored for PFOA (500 ppb) contaminated water treatment applications, which showed using 300 mg adsorbent over 3.5 L of PFOA contaminated water can be purified bringing the contamination below permissible level of <53 ppt. Studies were also conducted for DFB-CDP that produced only ~200 mL of drinking quality water making PAF-1-NDMB a superior choice for treatment of PFAS-contaminated water [161, 163].

On the other hand, crystalline MOFs like MIL-101(Cr) have shown promising adsorption capabilities for TrOCs like PFASs. In one such application, a quaternary and tertiary amine-functionalized MIL-101(Cr) were used for the selective adsorption of PFOA. In this approach, different amine groups following PSM method were grafted onto MIL-101(Cr) to obtain optimal adsorption capacity for PFOA (1.89 mmol/g). Similarly, two zirconium-based MOFs, UiO-66 and UiO-67 were prepared with optimal cavity, which was built to accommodate pollutant moieties using large ligand biphenyl-dicarboxylic acid (bpdc) for the preparation of UiO-67, while benzene-dicarboxylic acid (bdc) a shorter ligand used for the preparation of UiO-66. This larger cavity in UiO-67 facilitated a higher adsorption capacity of PFOA with 743 mg/g, which is higher than that of UiO-66 and which showed 388 mg/g uptake efficiencies. Furthermore, photocatalytic activities of MOFs were also utilized to degrade PFAS present in contaminated water sources. In such an approach, titanium-based MOF (MIL-125-NH_2) showed a PFOA degradation efficiency of ~98.9% under Hg-lamp irradiation in 24 hours, through light-induced hydroxyl ($^{\bullet}OH$) radical mechanism from MIL-125-NH_2 [164, 165].

Another set of TrOCs has also been categorized as persistent chemicals or emerging pollutant pharmaceuticals and personal care products (PPCPs) and endocrine-disrupting substances in water. Both conventional and advanced water treatment techniques like membrane-based separation processes have shown limited success against these persistent chemicals. However, selective adsorption techniques have shown greater potential for their remediation from contaminated wastewater. However, due to the complex nature and cost involved in treatment, it is important to design a treatment technique, which is affordable, recyclable, and easy to prepare. These requirements have been easily met with the use of functionalized porous UiO and MIL-series MOFs. Thus, UiO and MIL series MOFs have been shown the removal of PPCPs including ciprofloxacin (CPX), ketoprofen (KTP), sodium diclofenac (DCF), amoxicillin (AMX), acetaminophen (ACE), carbamazepine (CBZ), naproxen (NPX), and so on. Among these, UiO-66 showed the highest adsorption capacity against IBU was ~638.24 mg/g, and for KTP was ~606.49 mg/g [166, 167].

Similarly, COFs such as TPB-DMTP-COF have also been also been explored for the remediation of PPCPs (sulfamerazine [SMT]) from the contaminated water with an adsorption capacity of 209 mg/g. These COFs have shown a fast adsorption equilibrium and reliable recyclability, making them commercially viable and attractive in the future. On the other hand, a spherical COF (TPB-DMTP-COF) showed favorable affinity toward sulfonamides (SAs) generally found in water sources and food. Another class of persistent chemicals such as antibacterial drugs are widely used and discharged in excess to water resources. However, antibacterial drugs in water persist for long term causing hazardous effects on exposed living beings. Therefore, to remove such stable drug molecules from contaminated water, a spherical COF mounted in the SPE (solid-phase extraction) column has been used as both detection and removal media for TrOCs like SAs in trace levels. The removal of various SAs such as sulfadiazine (SD), sulfamerazine (SM_1), sulfathiazole (ST), sulfamethazine (SM_2), sulfapyridine (SP), and sulfamethoxazole (SMX) have also been effectively removed from contaminated sources using magnetic COFs [168].

The trace amounts of pharmaceutical pollutants from contaminated water have been successfully removed using magnetic COFs. Here, magnetic COFs were prepared using *p*-phenylenediamine (PDA), TFP, and $-NH_2$-functionalized Fe_3O_4@ SiO_2 nanoparticles (MNP-NH_2) as precursors to prepare high surface area and crystalline magnetic COFs (MCOFs). Further, a modified version of MCOF-2 with excellent thermal and chemical stability were used for the removal of a nonsteroidal anti-inflammatory drug DCF found in various water supplies with uptake capacity (565 mg/g). On the other hand, UiO-66 showed 189 mg/g and ZIF-8 PCDM-1000 was capable of adsorbing 360 mg/g. Similarly, porous 3D COFs have also been used for the robust removal of ofloxacin, a commonly used antibiotic. The self-assembled 3D framework of porous oligomer produced through collaborative intermolecular hydrogen bonding (HOF-TAM-BDA) impart 3D COFs (COF-320) flexible pore structures and open channels, which were then used for the entrapment of larger TrOCs such as ofloxacin. Here, the HOF-TAM-BDA showed adsorption rate of ~140.92 mg/g/min, while for COF-320, adsorption rate was restricted to 0.95 mg/g/min. Nevertheless, among these two, the HOF-TAM-BDA showed maximum adsorption capacity of 151.5 mg/g with remarkable recyclability [169].

Interestingly, most of the porous materials are not only used as adsorbent but also as photocatalyst/catalyst for task-specific organic pollutant degradation purposes making them a versatile class of treatment materials. One such porous material is MOF-5, which not only acts as an adsorbent but also as a photocatalyst for the degradation of methyl viologen dichloride in water. Similarly, Ti–Zr MOFs showed photocatalytic degradation through ACE oxidation under solar-simulated radiation. The optimized porous system which is MOF, TiZr15, demonstrated maximum photocatalytic degradation activity through complete ACE conversion only after 90 minutes during which catalytic process was initiated with the generation of $O_2^{\bullet-}$ radicals. Simultaneously, traces of photogenerated $^{\bullet}OH$ radicals were also involved in the degradation of TrOCs. In another system, sulfate radicals ($^{\bullet}SO_4^-$) produced from MOFs were also explored for the target-specific degradation of TrOCs that include pesticides like malathion, atrazine, and 2,4-dichlorophenoxyacetic acid and antibiotics such as sulfamethoxazole, tetracycline, and sulfamethoxazole. Similarly, popularly known as Fe-MOF, MIL-100(Fe) with peroxydisulfate activation was used for the treatment of water samples contaminated with antibiotics traces under irradiation which generated Fe^{III} surface on the structure leading to the reduction to Fe^{II}, thereby resulting the formation of $\bullet SO_4^-$radicals. These $\bullet SO_4^-$ then promoted the generation of $O_2\bullet^-$ and •OH radicals leading to complete progradation of antibiotics [170, 171].

In addition to this, COPs with extended π-conjugated structures such as CMPs and COFs have been widely used for the treatment of TrOCs in contaminated water. In one such attempt, strategies like incorporating single Cu sites in the COF (COF-909) alter the photocatalytic properties of the modified porous systems. On the other hand, presence of abundant terpyridyl constituents in COF-909 provides the modified system an open triple coordinate bonding interaction to transitional metals like Cu. Incorporation of Cu through a simple soaking the COF in copper chloride/methanol solution for 12 hours at RT which turns it into COF-909(Cu) imparting unique chemical characteristics. The COF-909(Cu) was then used for the photocatalytic degradation of SMX in presence/absence visible-light irradiation which showed over 98% elimination within 30 minutes under light and no visible change in absence. Similarly, photocatalytically active UPC-CMP-1 prepared through a Sonogashira–Hagihara coupling interaction between iron(III) 5,10,15,20-tetrakis-(4-bromophenyl)porphyrin and 1,4-diethynylbenzene showed excellent degradation activity against Congo red. Similarly, advanced ordered porous materials have been extensively designed and tested against the various contaminants for their remediation efficiencies. Detailed studies reveal that appropriate surface functionalization, and doping incorporated task-specific properties, which were then directly used for the effective treatment of innated wastewaters [172].

7.10 Ordered Porous Materials Heavy Metal Adsorption

Heavy metal contamination is another major concern for aquatic ecosystems and has been considered one of the most prevalent environmental issues. There are several toxic heavy metal ions such as Cr^{VI}, As^{III}, As^{V}, Hg^{II}, Pb^{II}, and Cu^{II} present in groundwater due to natural causes and in surface water generally originating from the discharge of industrial wastewater. It is well documented that the higher concentration of heavy metals in drinking water or food chain causes life-threatening diseases in both humans and living beings. Therefore, it is important to eliminate various heavy metals found in water sources using suitable treatment techniques. However, conventional water treatment techniques such as ion exchange method, chemical precipitation, membrane filtration, and adsorption have proven limited success in recovery and removal of heavy metals. In such a situation, advanced porous materials have shown promising removal efficiencies against a series of heavy metals [173–175].

Among many advanced materials and techniques, ordered porous materials can be tuned to attain specific functionality through various processes. In one such attempt, a series of PAF-based nanoporous traps that were functionalized with diverse chelators have been effectively utilized for the sorption of dangerous heavy metal ions. Here, mercury has been selectively entrapped using thiol groups in PAF-1, which showed a strong affinity toward heavy metal ions and PAF-1 nanotrap showed exceptional mercury binding capacity. To make a functionalized nanotrap, first PAF-1 was treated HCl for the chloromethylation resulting in PAF-1-CH2Cl. Subsequently, the product was treated with NaHS to attach the thiol functionality (PAF-1-SH). Finally, PAF-1-SH when used for mercury adsorption showed a remarkable uptake capacity for mercury (~1000 mg/g) with an extraordinary distribution coefficient (K_d) value of 5.76×10^7 mL/g. In addition to this, PAF-1-SH also showed exceptional adsorption capacity for other co-pollutants like NaI, Ca^{II} Mg^{II}, and Zn^{II} in a wide range of pH (3–11), making it one of the versatile water treatment materials. However, complex material preparation techniques hinder their immediate real-world application [176].

To overcome the limitations associated with synthesis of nanoporous materials through conventional methods, researchers have devised advanced and feasible techniques for the preparation of mercury nanotraps. Therefore, in a facile preparation method, a monomer combined with a chloromethyl group subsequent polymerization to yield POPs comprising higher grafting of chelators. Here, thiol-functionalized POP and 3,5-divinylbenzyl chloride first interact through free-radical polymerization along with azobisisobutyronitrile (AIBN). Further, as produced POP is treated with chloride moiety and NaHs to replace the thiol group resulting in thiol-functionalized porous material (POP-SH) having a high surface area of 1061 m^2/g and readily accessible chelating sites. As a result, POP-SH showed a maximum mercury uptake capacity of over 1200 mg/g at an equilibrium concentration of ≈200 ppm, making it a feasible alternative. Nevertheless, detailed X-ray absorption fine structure (XAFS) on mercury adsorbed POP (Hg-loaded POP-SH) revealed a multimodal distribution of Hg–S distances, which indicates that distinct Hg–S scattering paths that mainly occur due to inefficient coordination of binding sites exist in amorphous material with a rigid polymeric backbone [177].

Further, it addresses the limitations of effective coordination interaction in POPs that hinder the pollutant uptake efficiency. Therefore, to realize the efficient POPs, a 2D COF-V was then designed by grafting different thiol and thioether chelating branches via thiol–ene "click" reactions. As produced, COF-V showed exceptionally high adsorption capacities for Hg^{II} from contaminated water (~863 mg/g), which proved the usefulness of anchored chelators in

POPs and their orientation. Similarly, a thioether-functionalized COF was prepared using a thioether monomer (2,5-bis(3-(ethylthio)propoxy)terephthalohydrazide) as the main precursor, which resulted in COF-LZU8. Further, COF-LZU8 showed simultaneous real-time detection of Hg^{II} ions and adsorption (concentration reduced from 10 to 0.2 ppm). Following this encouraging result, a series of MOFs functionalized with thiol and thioether groups have been successfully designed and used for mercury elimination from the contaminated water. Similarly, Zr-MSA prepared using mercaptosuccinic acid (MSA) as the ligand showed an excellent adsorption capacity for Hg^{II} (~730 mg/g) from contaminated water, which was higher than their counterpart (sulfurized MOF) materials. Here also, along with mercury remediation, the thioether-functionalized ordered porous materials have shown an effective affinity for other metal ions such as copper (PAF-1-SMe showed ~600 mg/g capacity for Cu) in water [178, 179].

Further, chromium exists in the environment in two states, namely Cr^{III} and Cr^{VI} that are considered most stable and exist more predominantly. Among these, Cr^{III} ion state is considered nontoxic and Cr^{VI} is toxic, noxious, and carcinogenic in nature. There have been several efforts made to develop effective techniques to selectively remove Cr^{VI} from contaminated sources with limited success. Among many, adsorptive removal has been relatively effective compared to other advanced methods including membrane-based separation processes. Nevertheless, advanced porous materials have been a choice of materials in recent times due to their versatility. In this direction, dual-pore COFs having hydroxyl groups COF-BTA-DHBZ have been successfully designed for the rapid removal of Cr^{VI} with adsorption capacity of ~272 mg/g within 10 mintues. Cationic POPs COF-BTA-DHBZ also showed a unique ion exchange characteristic in uptaking Cr^{VI}. Further, cationic POP was prepared using an imidazolium-based 3D COF for the effective adsorption of Cr^{VI} from wastewater sources. Interestingly, minimizing toxicity of Cr^{VI} was considerably reduced by converting it to Cr^{III} ions through novel strategies in which use of 2D Zn-MOF prepared using pamoic acid, 4,4′-bipyridine, and $Zn(ClO_4)_2{\cdot}6H_2O$. As prepared, Zn-MOF converted toxic Cr to nonhazardous Cr through a photocatalytic reduction of Cr^{VI} to Cr^{III} [180–182].

7.11 Porous Materials in Nuclear Waste Sequestration

Tsunami-triggered accident in 2011 damaged nuclear power plant in Fukushima, which led to the potential leakage of radionuclides such as natural carbon-14 (^{14}C), radioactive cesium-137 (^{137}Cs), anthropogenic ^{137}Cs], and tritium (^{3}H) in the form of tritiated water (HTO), which have feared to enter Pacific Ocean, which further feared to emit high-energy γ rays and β particles upon decay. These contaminants are estimated to cause serious biological effects in living beings, particularly severe damage to DNA making them some of the most dangerous contaminants. Even though nuclear energy is deemed to be an alternative to fossil fuels as a greener energy source, however, the inadvertent leakage or release of traces of radioactive materials into the environment and aquatic reservoirs poses serious health risks to living beings. Among many, traces of uranium have been detected in nuclear wastewater as one of the significant nuclear residues. Similarly, radionuclides like ^{99}Tc and ^{137}Cs are found in discharged sources of nuclear fission of ^{235}U, ^{239}Pu, Sr^{II}, and Th^{IV}, and radiological iodine account for a major proportion of nuclear wastewater. There have been several attempts made in the past for the safe disposal of radioactive waste following several protocols and separation technologies. However, present technologies have been ineffective against the majority of the radionuclides. However, advanced porous materials have shown promising adsorption trends in recent times [183, 184].

In one such attempt, amidoxime ($-C(NOH)NH_2$, AO), a popular chelating agent for uranium binding, has now been used in combination with porous materials for radionuclide sequestration. The advanced porous materials (COP with AO) have been designed to suit the selective extraction of uranium from wastewater. PAF-1 was developed as an ideal system for the sequestration of uranium through adsorption process. In this approach, PAF-1 was surface-modified with AO moieties following a three-step post-preparation modification. Here, post-surface modification with chloromethylation on PAF-1, NaCN was incorporated to replace −Cl with −CN and hydroxylamine to transform the (CN) group into amidoxime, resulting in PAF-1-CH_2AO. Further, PAF-1-CH_2AO was used for saturated uranium uptake up to 304 mg/g and PAF-1-CH_2AO also showed rapid removal efficiencies of uranium from water, which drastically reduced 4100–0.81 ppb within 90 minutes. As amidoxime characteristically binds to uranyl ions through an η^2 coordination, grafting of amidoxime chains with different lengths on PAF-1 was carried out to prepare affordable absorbents, namely PAF-1-CH_2NHAO and PAF-1-$NH(CH_2)_2$AO. Further, these sorbents were used for the uptake of uranium in which PAF-1-$NH(CH_2)_2$AO showed fourfold higher capabilities than PAF-1-CH_2NHAO. This is due to improved binding abilities of PAF-1-$NH(CH_2)_2$AO existing as a result of flexible AO moieties. Further, advanced POP-based sorbents POP-*o*NH_2-AO and POP2-PO_3H_2 were prepared among which POP_1-AO showed extremely high affinity toward uranium. This POP showed excellent uptake capabilities for uranium found in natural seawater with 4.36 for POP-*o*NH_2-AO, 5.01 for POP2-PO_3H_2, and 8.40 mg/g for POP_1-AO [185–187].

On the other hand, MOFs having catalytic characteristics capable of reducing and immobilizing have been used for sequestration of uranium species and selective extraction from contaminated sources. For this, a novel type of MOF was prepared using polyoxometalates (POMs) resulting in POM-based MOF, SCU-19 useful for selective uptake of uranium. Further, SCI-19 having rich redox sites showed a strong interaction with U^{VI}, and further exhibited photocatalytic activity for in uranium uptake (~728.34 mg/g) and upon visible light irradiation showed over 99% removal (from ≈10 ppm within two hours) of uranium from contaminated water. On the other hand, technetium-99 generated in a typical reactor is generally found in a large portion of nuclear fission products (~840 g/ton). TcO_4^- is one of the extremely hazardous radiation-derived substances due to its long half-life ($t_{1/2}$=2.13 × 10^5 years) and is one of the major species estimated in waste streams. Due to the super acidity and highly radioactive nature of TcO_4^-, the process of removing radionuclide traces following adsorbent materials is a difficult task. However, following suitable ion exchange material can effectively remove the dangerous radionuclide from wastewater. Therefore, it seemed that ion exchange technology could be the most promising technique for segregating and subsequent selective removal of TcO_4^- from nuclear waste. This ion exchange method was also effectively used for selective radionuclides in presence of a high concentration of various counter acid anions such as NO_3, Cl^-, and SO_4^{2-}. This is due to the larger molecule size of TcO_4^- and its lower hydration energy that allows the preferential exchange of radionuclide over counter anions [188, 189].

Porous materials with task-specific functionality are also designed to remove radionuclides like technetium (Tc) through the ion exchange technique. For this purpose, trimethylammonium hydroxide groups grafted porous PAF-1 was prepared (PAF-1-$CH_2N^+(CH_3)_3OH^-$) as technologically feasible material that helps in ion exchange adsorption with outstanding uptake capabilities. This task-specific ion exchange adsorbent showed remarkable uptake capacity for TcO_4^- surpassing currently existing standard commercial products such as PVBTAH-ZIF-8, SLUG-21 and an LDHs Amberlyst-A26 under the identical operating conditions. Similarly, in another attempt, ion exchange sites in porous materials were manipulated to preferentially attract TcO_4^- through an affinity interaction with POPs for affordable remediation. For this purpose, a series of secondary-sphere modifying agents in POPs have been utilized through distributions of charge density around aromatic ring vicinity (*para* position of pyridinium-POPs) to alter the local environment of cation site, which further enhances the ion exchange capabilities of resultant material. Thus, chemically manipulated PQA-*p*N$(Me)_2$Py-Cl showed excellent radionuclide uptakes of efficiencies. In this case, ReO_4^- a common nonradioactive surrogate for TcO_4^- was removed with ~1127 mg/g uptake capacity. Further, theoretical calculations and estimations were carried out to understand the uptake phenomenon at molecular levels in porous materials following DFT calculation. The theoretical estimation revealed that the cationic polymer backbone in PQA-*p*N$(Me)_2$Py-Cl has shown greater electrostatic ion-pairing attraction energies toward ReO_4^- ions which, in turn, facilitates the faster and higher uptake capacity in porous materials, which encouraged researchers to design TcO_4^- removal studies from real radioactive TcO_4^--contaminated water samples. Even though an optimized POP, PQA--*p*N(Me)$_2$Py-Cl was used for selective removal of TcO_4^- using stimulated Hanford low activity waste (LAW) melter recycle stream, however, in this solution, concomitant acidic anions such as Cl^-, NO_3^-, and NO_2^- in this source was extremely high amounting to 300 times that of target TcO_4^-. Nevertheless, selective adsorption studies on PQA-*p*N(Me)$_2$Py-Cl showed ~95% uptake of TcO_4^- from the nuclear wastewater at a phase ratio of 200 mL/g [190–192].

In all cases, performance of advanced porous materials mainly depends on the presence of various counter-ions or concomitant anions and their concentration and nature in nuclear waste while remediation is either through adsorption, ion exchange, or direct uptake. Therefore, designing suitable porous materials needs to consider these major challenges for affordable and effective radionuclide sequestration.

7.12 Future Perspective and Conclusions

Even though research on developing various ordered nanoporous materials and membranes discussed in this chapter is a work in progress at different stages of maturation, however, a substantial amount of proof-of-concept studies have shown a series of potential materials stock useful for effective water treatment. However, at this, it is important to discuss some of the challenges and possibilities in overcoming the shortcomings associated with new materials stock. In the case of ordered porous films/membranes obtained from block polymer systems through physicochemical methods, several process parameters affect the pore size control and their distribution. Understanding film/membrane formation at a larger scale to make copolymer-based systems commercially viable in competition with conventional membranes produced from NIPS process is still the major challenge as the cost of producing block copolymers with specific molar fraction distribution is a costly process. Nevertheless, efforts will continue to optimize the copolymer preparation and their process control will encourage the production of modified versions of precursor that are suitable for target-specific applications in emerging areas including desalination and wastewater treatment.

Nevertheless, this chapter has given details of some of the physical and chemical modification methods developed in obtaining ordered porous membranes with precise control over physical properties of a film or membranes prepared using block polymers in which excellent pore size control through which selective separation abilities were achieved. Nonetheless, the block polymer system as membrane precursors can also be modified following simple and straightforward membrane fabrication protocols for inducing controlled properties that go beyond just pore size and distribution. Through this approach, incorporation of reactive entities, surface modifying agents, or composite precursors to induce pore/surface functionality along the pore walls and surface to convert an ordered porous membrane into an active reactor with an optimal density of active sites on the surface as well as within a small volume, which greatly help in selective separation of contaminants. In addition to this, use of coordination compounds with suitable metal and organic ligands has been known to impact the selective retention and removal of contaminants such as heavy metals and charged species from the contaminated water sources.

There have been several strategies followed to improve the overall performance in ordered porous membranes compared to currently practiced techniques as well as conventional membranes. One simple and earliest method of obtaining separation performances was by creating a transverse direction of the pores, which simulates the overall process as the solution/feed in membrane case interfaces or transports through the membranes. In this design, the nanoscale dimensions of the pore result in a short diffusion length from the center of the pore to the pore wall, which drastically reduces the mass transfer limitations making it a low-flux end product. Even though the narrow pore size distribution in block polymer membranes gives them a unique advantage over conventional membranes, the absence of a large number of pores in a per unit area of the membrane results in narrow distribution to the residence times to each pore. Nevertheless, in some cases, it is estimated that due to the presence of a large number of functional groups with high packing density that are confined within the pores, the uniform distribution of mass flow allows the block polymer membranes to modulate the mass transfer. This also helps in achieving high permeate flux as the high packing density of functional pores prevents channelling and instead utilizes the functionality for mass transfer process more effectively. For instance, functional membranes having well-defined contact/residence time allowed mass transfer to be controlled through precisely aligned pore architecture regulating the higher water flux. Studies also reveal that the membranes in the near future can also be designed to enhance the permeation of selective solutes from a mixture of feed across the membrane by utilizing favorable enthalpic interactions between pore walls and solute moieties. This type of control over permeation of both solvent/water and solutes in a specific application lays the foundation for the design and development of futuristic membranes useful for addressing future challenges in water treatment applications including desalination, selective mixture separation, removal of toxic contaminants, and the selective concentration of value-added products dissolved or present in wastewater or contaminated sources.

Therefore, functional self-assembled and highly ordered porous membranes can be fabricated following several strategies, namely: (i) Use of functional nanomaterials or incorporating the functional moieties possessing either organic, inorganic or hybrid organic/inorganic chemistries into the membrane precursor; (ii) Direct use of self-assembled porous membrane as a template on which post-chemical modification in following grafting active molecules to inside the pore walls or into the bulk of the pores with functional materials; and (iii) Designing an optimal reactive moiety with specific dose and reactivity into the membrane precursor, which later facilitate the post-assembly/fabrication functionalization of precursors. However, each of these strategies carries several advantages and disadvantages in their approach. Therefore, designing high-performance ordered porous membranes using block copolymers following self-assembly technique is still a challenging task. Nevertheless, the success of strategy mainly depends on the optimal utilization of strategy through which membrane is fabricated by introducing a high-density active functional groups and pore geometry and their distribution that helps in producing high-quality water in a quick time that is consistent or better than commercial membranes both in their manufacturing processes as well as real-world performances.

On the other hand, the combination of well-defined and ordered porous architecture with tailored functionality in porous materials makes them materials of choice for potential applications in desalination and decontamination of wastewater sources. As the need for high-performance materials is increasing every passing day due to of lack of resources materials and ever-declining materials stocks, designing highly efficient materials for total elimination of ionic species in salt/seawater and hazardous pollutants in wastewater is directly connected to well-being of human health, society, global ecosystem, and environment at large. In this direction, families of porous materials that include COFs, MOFs, and POPs are expected to play a vital role in water treatment applications due to their unique and unparalleled merits in filtration, catalytic, and sorption applications such as rapid degradation, adsorption kinetics, high uptake capacity, providing selective passage to water molecules, and exceptional selectivity. In this direction, current chapter has given a detailed analysis of various types of porous materials designed, developed, and utilized for remediating various types of contaminants.

Despite several proof-of-concept studies that have proven greater potentials of the porous materials already exist in the vast literature. However, one major challenge associated with use of functional porous is the difficulty in the large-scale preparation and processing of material. This aspect directly affects their commercial as well as real-world application prospects due to which innovations in designing more efficient, cost-effective synthesis methods are essential to substantially improve their bulk preparation yield. Detailed literature studies reveal that currently followed porous material preparing methods are typically energy intensive and generate a large amount of greenhouse gases, which makes overall followed strategies unsustainable. In addition to this, preparation of task-specific porous materials involves cumbersome synthetic procedures that further complicate their use in real-world scale-up activities and applications. To overcome this important demerit, researchers and technologies should invest efforts to design and develop affordable and innovative, sustainable, greener precursors and inexpensive, energy-less preparation techniques. Following a combination of greener materials stock and effacement, synthetic technique would enable large-scale sustainable and cost-effective manufacturing of porous materials, which is crucial to meet industrial standards and requirements. These two basic requirements only will ensure the large-scale and real-world success of porous materials in water treatment applications.

In terms of large-scale real-world applications, major hurdles arise in implementing advanced porous materials' nanotoxicity and nano-related health risks in direct use for desalination and drinking water production. Direct use of novel porous nano-/microscale materials for drinking water application may have to follow US EPA and EU REACH registration norms. As discussed in the chapter in detail, development of novel porous materials involves use of relatively toxic precursors and nonfriendly preparation techniques. Therefore, large-scale adoption has to consider hazardousness involved in a potential product, chemical composition, usage, risks, public benefits, and so on. Currently, several studies have also implemented these evaluation criteria carefully during the materials design and synthesis to ensure the product composition should not contain or bear unreasonable negative effects on human health and the environment upon usage and discharge. Without the safety approval and confidence in the novel materials effects, it is difficult to attract attention and enough confidence from the end-user or public at large in using novel water treatment materials, particularly in drinking water purification applications. However, as the technology is in early stages of realization, efforts are being made to adopt all safety measures and protocols to produce safe materials stocks [193].

Nevertheless, the world has witnessed significant progress in design, development, fabrication, and technology development aspects of advanced and ordered porous materials in the last few decades due to their versatility, unparalleled potential, and, most importantly, demand for newer technologies that are effective and efficient from the water treatment sector. However, the challenges in finding greener, sustainable, and efficient water treatment and environmental remediation techniques for real-world application will remain. However, despite a substantial number of porous materials having been tried and tested for remediating different contaminants from the water sources, there still remains an untapped potential for porous materials, which still exists to explore.

References

1 Kickelbick, G. (2003). Concepts for the incorporation of inorganic building blocks into organic polymers on a nanoscale. *Prog. Polym. Sci.* 28: 83–114.

2 Zheng, J., Chen, C., and Goto, A. (2020). Temperature-directed micellar morphological transformation using CABC-block copolymers and its applications in encapsulation and hidden segment. *Angew. Chem.* 132: 1957–1965.

3 Li, C., Li, Q., Kaneti, Y.V. et al. (2020). Self-assembly of block copolymers towards mesoporous materials for energy storage and conversion systems. *Chem. Soc. Rev.* 49: 4681–4736.

4 Feng, H., Lu, X., Wang, W. et al. (2017). Block copolymers: synthesis, self-assembly, and applications. *Polymers (Basel)* 9: 494.

5 Rzayev, J. and Hillmyer, M.A. (2005). Nanoporous polystyrene containing hydrophilic pores from an ABC triblock copolymer precursor. *Macromolecules* 38: 3–5.

6 Huang, Y.B., Liang, J., Wang, X.S., and Cao, R. (2017). Multifunctional metal-organic framework catalysts: Synergistic catalysis and tandem reactions. *Chem. Soc. Rev.* 46: 126–157.

7 Jiao, J.J., Gong, W., Wu, X.W. et al. (2019). Multivariate crystalline porous materials: synthesis, property and potential application. *Coord. Chem. Rev.* 385: 174–190.

8 Zeng, H., Xie, M., Wang, T. et al. (2021). Orthogonal-array dynamic molecular sieving of propylene/propane mixtures. *Nature* 595: 542–548.

9 Jin, F.Z., Liu, J.J., Chen, Y., and Zhang, Z.J. (2021). Tethering flexible polymers to crystalline porous materials: a win-win hybridization approach. *Angew. Chem. Int. Ed.* 60: 14222–14235.

10 Zhou, J.W. and Wang, B. (2017). Emerging crystalline porous materials as a multifunctional platform for electrochemical energy storage. *Chem. Soc. Rev.* 46: 6927–6945.

11 Gil, A., Korili, S.A., and Vicente, M.A. (2008). Recent advances in the synthesis and catalytic application of pillared clay catalysts. *Catal. Rev.* 50: 153–221.

12 Gil, A., Gandia, L.M., and Vicente, M.A. (2000). Recent advances in the synthesis and catalytic applications of pillared clays. *Catal. Rev. Sci. Eng.* 42: 145.

13 Lan, Y., Liu, Y., Li, J. et al. (2021). Natural clay-based materials for energy storage and conversion applications. *Adv. Sci.* 8: 2004036.

14 Beck, J.S., Vartuli, J.C., Roth, W.J. et al. (1992). A new family of mesoporous molecular sieves prepared with liquid crystal templates. *J. Am. Chem. Soc.* 114: 10834.

15 Dariani, R.S. and Nazari, M. (2016). Comparison of stress, strain, and elastic properties for porous silicon layers supported by substrate and corresponding membranes. *J. Mol. Struct.* 1119: 308.

16 Romanowska, M., Maciag, A., Smith, A.L. et al. (2007). DNA damage, superoxide, and mutant K-ras in human lung adenocarcinoma cells. *Free Radic. Biol. Med.* 43: 1145.

17 Côté, A.P., Benin, A.I., Ockwig, N.W. et al. (2005). *Porous, Crystalline, Covalent Organic Frameworks*. Science, American Association for the Advancement of Science.

18 Bisbey, R.P. and Dichtel, W.R. (2017). Covalent organic frameworks as a platform for multidimensional polymerization. *ACS Cent. Sci.* 3 (6): 533–543.

19 Stock, N. and Biswas, S. (2012). Synthesis of metal-organic frameworks (MOFs): routes to various MOF topologies, morphologies, and composites. *ChemInform* 43 (16): 933–969.

20 Wang, Y., Zhang, W., Wu, X. et al. (2017). Conducting polymer coated metal-organic framework nanoparticles: facile synthesis and enhanced electromagnetic absorption properties. *Synth. Met.* 228: 18–24.

21 Davis, M.E., Saldarriaga, C., Montes, C. et al. (1988). VPI-5: the first molecular sieve with pores larger than 10 Ångstroms. *Zeolites* 8 (5): 362–366.

22 van Bekkum, H., Flanigen, E.M., Jacobs, P.A., and Jansen, J.C. (2001). *Introduction to Zeolite Science and Practice*. Amsterdam, The Netherland: Elsevier.

23 Kresge, C.T., Leonowicz, M.E., Roth, W.J. et al. (1992). Ordered mesoporous molecular sieves synthesized by a liquid-crystal template mechanism. *Nature* 359 (6397): 710–712.

24 Kruk, M. and Jaroniec, M. (2001). Gas adsorption characterization of ordered organic-inorganic nanocomposite materials. *Chem. Mater.* 13 (10): 3169–3183.

25 Firouzi, A., Kumar, D., Bull, L.M. et al. (1995). Cooperative organization of inorganic-surfactant and biomimetic assemblies. *Science* 267 (5201): 1138–1143.

26 Rouquerol, J., Avnir, D., Fairbridge, C.W. et al. (1994). Recommendations for the characterization of porous solids. *Pure Appl. Chem.* 66 (8): 1739–1758.

27 Everett, D.H. (1972). Definitions, terminology, and symbols in colloid and surface chemistry. *Pure Appl. Chem.* 31 (4): 579–638.

28 Mays, T.J. (2007). A new classification of pore sizes. *Stud. Surf. Sci. Catal.* 160: 57–62.

29 Song, Y., Phipps, J., Zhu, C., and Ma, S. (2023). Porous materials for water purification. *Angew. Chem. Int. Ed. Eng.* 62 (11): e202216724.

30 Sun, Q., Aguila, B., Song, Y., and Ma, S. (2020). Tailored porous organic polymers for task-specific water purification. *Acc. Chem. Res.* 53 (4): 812–821.

31 Dong, J., Wee, V., Peh, S.B., and Zhao, D. (2021). Molecular-rotor-driven advanced porous materials. *Angew. Chem. Int. Ed. Eng.* 60 (30): 16279–16292.

32 Cronstedt, A.F. (1756). Natural zeolite and minerals. *Svenska Vetenskaps Akademiens Handlingar Stockholm* 17: 120.

33 Milton, R.M. (1989). Molecular sieve science and technology. In: *Zeolite Synthesis*, vol. 398. Acs Symposium Series (ed. M.L. Occelli and H.E. Robson), 1–10. American Chemical Society.

34 Puppe, L. and Weisser, J. (1984). (Bayer Aktiengesellschaft) crystalline aluminosilicate PSH-3 and its process of preparation. US Patent 4,439,409, 27 March 1984.

35 Roth, W.J., Nachtigall, P., Morris, R.E., and Cejka, J. (2014). Two-dimensional zeolites: current status and perspectives. *Chem. Rev. 114* (9): 4807–4837.

36 McKeown, N.B., Budd, P.M., Ghanem, B., and Msayib, K. (2005). Microporous polymer material. International Patent PCT WO 2005/012397.

37 McKeown, N.B. and Budd, P.M. (2006). Polymers of intrinsic microporosity (PIMs): organic materials for membrane separations, heterogeneous catalysis and hydrogen storage. *Chem. Soc. Rev.* 35: 675–683.

38 McKeown, N.B. (2017). The synthesis of polymers of intrinsic microporosity (PIMs). *Sci. China Chem.* 60: 1023–1032.

39 Budd, P.M., Szekely, G., and Livingston, A. (ed.) (2020). *Sustainable Nanoscale Engineering*, 231–264. Elsevier.

40 Kim, S. and Lee, Y.M. (2015). Rigid and microporous polymers for gas separation membranes. *Prog. Polym. Sci.* 43: 1–32.

41 Ehrling, S., Reynolds, E.M., Bon, V. et al. (2021). Adaptive response of a metal-organic framework through reversible disorder-disorder transitions. *Nat. Chem.* 13: 568–574.

42 Katsoulidis, A.P., Antypov, D., Whitehead, G.F. et al. (2019). Chemical control of structure and guest uptake by a conformationally mobile porous material. *Nature* 565: 213–217.

43 Yang, Y., Li, L., Lin, R.B. et al. (2021). Ethylene/ethane separation in a stable hydrogen-bonded organic framework through a gating mechanism. *Nat. Chem.* 13: 933–939.

44 Balestri, D., Mazzeo, P.P., Carraro, C. et al. (2019). Stepwise evolution of molecular nanoaggregates inside the pores of a highly flexible metal-organic framework. *Angew. Chem. Int. Ed.* 131: 17503–17511.

45 Gu, H., Yu, Z., Li, N. et al. (2023). Structural flexibility in cationic metal–organic framework for boosting ReO4– capture. *Chem. Eng. J.* 466 (143139): 1385–8947.

46 Dong, X., Lu, D., Harris, T.A.L., and Escobar, I.C. (2021). Polymers and solvents used in membrane fabrication: a review focusing on sustainable membrane development. *Membranes (Basel)* 11 (5): 309.

47 Guillen, G.R., Pan, Y., Li, M., and Hoek, E.M.V. (2011). Preparation and characterization of membranes formed by nonsolvent induced phase separation: a review. *Ind. Eng. Chem. Res.* 50: 3798–3817.

48 Loeb, S. and Sourirajan, S. (1963). *Sea Water Demineralization by Means of an Osmotic Membrane*, 117–132. American Chemical Society.

49 Mas, A., Jaaba, H., Sledz, J., and Schue, F. (1996). Membranes en PHB, P(HB-co-9% HV), P(HB-co-22% HV) pour la microfiltration ou la pervaporation propriétés filtrantes et état de surface. *Eur. Polym. J.* 32 (4): 435–450.

50 Wang, H.H., Jung, J.T., Kim, J.F. et al. (2019). cA novel green solvent alternative for polymeric membrane preparation via nonsolvent-induced phase separation (NIPS). *J. Membr. Sci.* 574: 44–54.

51 Suzana Pereira Nunes (2016). Block copolymer membranes for aqueous solution applications. *Macromolecules* 49: 2905–2916.

52 Clodt, J.I. (2022). Block copolymer membranes-progress and challenges. *Membranes (Basel)* 12 (4): 354.

53 Sankhala, K., Koll, J., and Abetz, V. (2020). Facilitated structure formation in isoporous block copolymer membranes upon controlled evaporation by gas flow. *Membranes* 10: 83.

54 Jackson, E.A. and Hillmyer, M.A. (2010). Nanoporous membranes derived from block copolymers: from drug delivery to water filtration. *ACS Nano* 4: 3548–3553.

55 Wang, Y. (2016). Nondestructive creation of ordered nanopores by selective swelling of block copolymers: toward homoporous membranes. *Acc. Chem. Res.* 49: 1401–1408.

56 Radjabian, M. and Abetz, V. (2020). Advanced porous polymer membranes from self-assembling block copolymers. *Prog. Polym. Sci.* 102: 101219–101256.

57 Zhou, J. and Wang, Y. (2020). Selective swelling of block copolymers: an upscalable greener process to ultrafiltration membranes? *Macromolecules* 53: 5–17.

58 Mehta, A. and Zydney, A.L. (2005). Permeability and selectivity analysis for ultrafiltration membranes. *J. Membr. Sci.* 249: 245–249.

59 Sperschneider, A., Scacher, F., Gawenda, M. et al. (2007). Towards nanoporous membranes based on ABC triblock terpolymers. *Small* 3: 1056–1063.

60 Peinemann, K.-V., Abetz, V., and Simon, P.F.W. (2007). Asymmetric superstructure formed in a block polymer via phase separation. *Nature* 6: 992–996.

61 Peinemann, K.V., Abetz, V., and Simon, P. (2007). Asymmetric superstructure formed in a block copolymer via phase separation. *Nat. Mater.* 6: 992–996.

62 Gu, Y. and Wiesner, U. (2015). Tailoring pore size of graded mesoporous block copolymer membranes: moving from ultrafiltration toward nanofiltration. *Macromolecules* 48: 6153–6159.

63 Duong, P.H.H., Chung, T.-S., Jeyaseelan, K. et al. (2012). Planar biomimetic aquaporin-incorporated triblock copolymer membranes on porous alumina supports for nanofiltration. *J. Membr. Sci.* 409: 34–43.

64 Ma, D., Zhou, J., Wang, Z., and Wang, Y. (2020). Block copolymer ultrafiltration membranes by spray coating coupled with selective swelling. *J. Membr. Sci.* 598: 117656.

65 Lee, D.K., Park, J.T., Choi, J.K. et al. (2008). Proton conducting crosslinked membranes by polymer blending of triblock copolymer and poly (vinyl alcohol). *Macromol. Res.* 16: 549–554.

66 Wang, Z., Yao, X., and Wang, Y. (2012). Swelling-induced mesoporous block copolymer membranes with intrinsically active surfaces for size-selective separation. *J. Mater. Chem.* 22: 20542–20548.

67 Ruokolainen, J., Saariaho, M., Ikkala, O. et al. (1999). Supramolecular routes to hierarchical structures: comb-coil diblock copolymers organized with two length scales. *Macromolecules* 32 (4): 1152–1158.

68 Breiner, U., Krappe, U., Thomas, E.L., and Stadler, R. (1998). Structural characterization of the "Knitting Pattern" in polystyrene-block-poly(ethylene-co-butylene)-block-poly(methyl methacrylate) triblock copolymers. *Macromolecules* 31 (1): 135–141.

69 Shibayama, M., Hasegawa, H., Hashimoto, T., and Kawai, H. (1982). Microdomain structure of an ABC-type triblock polymer of polystyrene-poly[(4-vinylbenzyl)dimethylamine]-polyisoprene cast from solutions. *Macromolecules* 15 (2): 274–280.

70 Matsushita, Y., Yamada, K., Hattori, T. et al. (1983). Morphologies of ABC-type triblock copolymers with different compositions. *Macromolecules* 16 (1): 10–13.

71 Mogi, Y., Mori, K., Matsushita, Y., and Noda, I. (1992). Tricontinuous morphology of triblock copolymers of the ABC type. *Macromolecules* 25 (20): 5412–5415.

72 Mai, Y. and Eisenberg, A. (2012). Self-assembly of block copolymers. *Chem. Soc. Rev.* 41: 5969–5985.

73 Zhao, Y.-F., Zhu, L.-P., Yi, Z. et al. (2013). Improving the hydrophilicity and fouling-resistance of polysulfone ultrafiltration membranes via surface zwitterionicalization mediated by polysulfone-based triblock copolymer additive. *J. Membr. Sci.* 440: 40–47.

74 Rajasekhar, T., Trinadh, M., Babu, P.V. et al. (2015). Oil–water emulsion separation using ultrafiltration membranes based on novel blends of poly (vinylidene fluoride) and amphiphilic tri-block copolymer containing carboxylic acid functional group. *J. Membr. Sci.* 481: 82–93.

75 Nunes, S.P., Behzad, A.R., Hooghan, B. et al. (2011). Switchable pH-responsive polymeric membranes prepared via block copolymer micelle assembly. *ACS Nano* 5: 3516–3522.

76 Abetz, V. (2015). Isoporous block copolymer membranes. *Macromol. Rapid Commun.* 36: 10–22.

77 Wang, X., Chen, J., Hong, K., and Mays, J.W. (2012). Well-defined polyisoprene-b-poly(acrylic acid)/polystyrene-b-polyisoprene-b-poly(acrylic acid) block copolymers: synthesis and their self-assembled hierarchical structures in aqueous media. *ACS Macro Lett.* 1 (6): 743–747.

78 Saleem, S., Rangou, S., Abetz, C. et al. (2019). Isoporous membranes from novel polystyrene-b-poly(4-vinylpyridine)-b-poly(solketal methacrylate) (PS-b-P4VP-b-PSMA) triblock terpolymers and their post-modification. *Polymers (Basel)* 12 (1): 41.

79 Nunes, S.P., Sougrat, R., Hooghan, B. et al. (2010). Ultraporous films with uniform nanochannels by block copolymer micelles assembly. *Macromolecules* 43: 8079–8085.

80 Foroutani, K. (2021). Seyed Morteza Ghasemi, Behzad Pourabbas, Molecular tailoring of polystyrene-block-poly (acrylic acid) block copolymer toward additive-free asymmetric isoporous membranes via SNIPS. *J. Membr. Sci.* 623: 119099.

81 Frommer, M.A. and Messalem, R.M. (1973). Mechanism of membrane formation. VI. Convective flows and large void formation during membrane precipitation. *Ind. Eng. Chem. Prod. Res. Dev.* 12: 328–333.

82 Cao, X., Chen, Y., Chai, W. et al. (2015). Thermoresponsive self-assembled nanovesicles based on amphiphilic triblock copolymers and their potential applications as smart drug release carriers. *J. Appl. Polym. Sci.* 132: 41361.

83 Liang, K., Li, X., Wang, L. et al. (2023). Double stimuli-responsive isoporous block copolymer membranes upon phase separation strategies. *J. Membr. Sci.* 669: 121277.

84 Zhang, Y., Almodovar-Arbelo, N.E., Weidman, J.L. et al. (2018). Fit-for-purpose block polymer membranes molecularly engineered for water treatment. *npj Clean Water* 1: 2.

85 Yin, J., Yao, X., Liou, J.Y. et al. (2013). Membranes with highly ordered straight nanopores by selective swelling of fast perpendicularly aligned block copolymers. *ACS Nano* 7: 9961–9974.

86 Wei, M., Sun, W., Shi, X. et al. (2015). Homoporous membranes with tailored pores by soaking block copolymer/homopolymer blends in selective solvents: dissolution versus swelling. *Macromolecules* 49: 215–223.

87 Albert, J.N. and Epps, T.H. (2010). Self-assembly of block copolymer thin films. *Mater. Today* 13: 24–33.

88 Ferrarese Lupi, F., Giammaria, T.J., Seguini, G. et al. (2017). GISAXS analysis of the in-depth morphology of thick PS-b-PMMA films. *ACS Appl. Mater. Interfaces* 9: 11054–11063.

89 Bürger, J., Kunnathully, V.S., Kool, D. et al. (2020). Characterisation of the PS-PMMA interfaces in microphase separated block copolymer thin films by analytical (S)TEM. *Nanomaterials* 10 (1): 141.

90 Kim, S.H., Kwak, S.-Y., and Suzuki, T. (2005). Positron annihilation spectroscopic evidence to demonstrate the flux-enhancement mechanism in morphology-controlled thin-film-composite (TFC) membrane. *Environ. Sci. Technol.* 39: 1764–1770.

91 Shintani, T., Shimazu, A., Yahagi, S., and Matsuyama, H. (2009). Characterization of methyl-substituted polyamides used for RO membranes by positron annihilation lifetime spectroscopy and MD simulation. *J. Appl. Polym. Sci.* 113: 1757–1762.

92 Wei, T., Zhang, L., Zhao, H. et al. (2016). Aromatic polyamide reverse-osmosis membrane: an atomistic molecular dynamics simulation. *J. Phys. Chem. B* 120: 10311–10318.

93 Harder, E., Walters, D.E., Bodnar, Y.D. et al. (2009). Molecular dynamics study of a polymeric reverse osmosis membrane. *J. Phys. Chem. B* 113: 10177–10182.

94 Cushen, J.D., Otsuka, I., Bates, C.M. et al. (2012). Oligosaccharide/silicon-containing block copolymers with 5 nm features for lithographic applications. *ACS Nano* 6: 3424–3433.

95 Feng, X., Kawabata, K., Kaufman, G. et al. (2017). Highly selective vertically aligned nanopores in sustainably derived polymer membranes by molecular templating. *ACS Nano* 11: 3911–3921.

96 Zhang, Y., Mulvenna, R.A., Boudouris, B.W., and Phillip, W.A. (2017). Nanomanufacturing of high-performance hollow fiber nanofiltration membranes by coating uniform block polymer films from solution. *J. Mater. Chem. A* 5: 3358–3370.

97 Mulvenna, R.A., Weidman, J.L., Jing, B. et al. (2014). Tunable nanoporous membranes with chemically-tailored pore walls from triblock polymer templates. *J. Membr. Sci.* 470: 246–256.

98 Wang, Y., He, Z., Gupta, K.M. et al. (2017). Molecular dynamics study on water desalination through functionalized nanoporous graphene. *Carbon* 116: 120–127.

99 Corry, B. (2008). Designing carbon nanotube membranes for efficient water desalination. *J. Phys. Chem. B* 112: 1427–1434.

100 Thomas, M. and Corry, B. (2016). A computational assessment of the permeability and salt rejection of carbon nanotube membranes and their application to water desalination. *Philos. Trans. R. Soc. Lond. Ser. A* 374: 1–20.

101 Kresge, C.T., Vartuli, J.C., Roth, W.J., and Leonowicz, M.E. (2004). The discovery of ExxonMobil's M41S family of mesoporous molecular sieves. In: *Studies in Surface Science and Catalysis*, vol. 148 (ed. O. Terasaki), 53–72. Elsevier.

102 Samanta, P., Desai, A.V., Let, S., and Ghosh, S.K. (2019). Advanced porous materials for sensing, capture and detoxification of organic, pollutants toward water remediation. *ACS Sustain. Chem. Eng.* 7 (8): 7456–7478.

103 Sakamaki, Y., Tsuji, M., Heidrick, Z. et al. (2020). Preparation and applications of metal-organic frameworks (MOFs): a laboratory activity and demonstration for high school and/or undergraduate students. *J. Chem. Educ.* 97 (4): 1109–1116.

104 Ahmed, I. and Jhung, S.H. (2014). Composites of metal–organic frameworks: preparation and application in adsorption. *Mater. Today* 17 (3): 136–146.

105 Zhao, X., Zhao, H., Yuan, H., and Lan, M. (2014). Multifunctional superparamagnetic fe3O4@SiO2 core/shell nanoparticles: design and application for cell imaging. *J. Biomed. Nanotechnol.* 10 (2): 262–270.

106 Li, X., Zhou, L., Wei, Y. et al. (2014). Anisotropic growth-induced synthesis of dual-compartment Janus mesoporous silica nanoparticles for bimodal triggered drugs delivery. *J. Am. Chem. Soc.* 136: 15086–15092.

107 Li, X., Zhao, T., Lu, Y. et al. (2017). Degradation-restructuring induced anisotropic epitaxial growth for fabrication of asymmetric diblock and triblock mesoporous nanocomposites. *Adv. Mater.* 29: 1701652.

108 Zhao, T., Chen, L., Wang, P. et al. (2019). Surface-kinetics mediated mesoporous multipods for enhanced bacterial adhesion and inhibition. *Nat. Commun.* 10: 4387.

109 Meng, Y., Gu, D., Zhang, F. et al. (2005). Ordered mesoporous polymers and homologous carbon frameworks: amphiphilic surfactant templating and direct transformation. *Angew. Chem. Int. Ed.* 44: 7053–7059.

110 Fang, Y., Gu, D., Zou, Y. et al. (2010). A low-concentration hydrothermal synthesis of biocompatible ordered mesoporous carbon nanospheres with tunable and uniform size. *Angew. Chem. Int. Ed.* 49: 7987–7991.

111 Fang, Y., Lv, Y., Che, R. et al. (2013). Two-dimensional mesoporous carbon nanosheets and their derived graphene nanosheets: synthesis and efficient lithium ion storage. *J. Am. Chem. Soc.* 135: 1524–1530.

112 Bhadra, B.N., Ahmed, I., Kim, S., and Jhung, S.H. (2017). Adsorptive removal of ibuprofen and diclofenac from water using metal-organic framework-derived porous carbon. *Chem. Eng. J.* 314: 50–58.

113 Mellah, A., Fernandes, S.P.S., Rodríguez, R. et al. (2018). Adsorption of pharmaceutical pollutants from water using covalent organic frameworks. *Chem. Eur. J.* 24 (42): 10601–10605.

114 Guan, X., Li, H., Ma, Y. et al. (2019). Chemically stable polyarylether-based covalent organic frameworks. *Nat. Chem.* 11 (6): 587–594.

115 Ji, W., Xiao, L., Ling, Y. et al. (2018). Removal of GenX and perfluorinated alkyl substances from water by amine-functionalized covalent organic frameworks. *J. Am. Chem. Soc.* 140 (40): 12677–12681.

116 Huang, Y., Zeng, X., Guo, L. et al. (2018). Heavy metal ion removal of wastewater by zeolite-imidazolate frameworks. *Sep. Purif. Technol.* 194: 462–469.

117 He, C., Xia, W., Zhou, C. et al. (2022). Rational design to manganese and oxygen Co-doped polymeric carbon nitride for efficient nonradical activation of peroxymonosulfate and the mechanism insight. *Chem. Eng. J.* 430: 132751.

118 Wang, Z., Zhang, S., Chen, Y. et al. (2020). Covalent organic frameworks for separation applications. *Chem. Soc. Rev.* 49 (3): 708–735.

119 Hao, S., Jiang, L., Li, Y. et al. (2020). Facile preparation of COF composite membranes for nanofiltration by stoichiometric spraying layer-by-layer self-assembly. *Chem. Commun.* 56 (3): 419–422.

120 Song, Y., Phipps, J., Zhu, C., and Ma, S. (2023). Porous materials for water purification. *Angew. Chem. Int. Ed.* 2023 (135): e202216724.

121 Berijani, K., Morsali, A., and Garcia, H. (2023). Synthetic strategies to obtain MOFs and related solids with multimodal pores. *Microporous Mesoporous Mater.* 349: 112410.

122 Chen, L., Wang, J., Shen, X. et al. (2019). ZIF-67@ Co-LDH yolk–shell spheres with micro-/meso-porous structures as vehicles for drug delivery. *Inorg. Chem. Front.* 6: 3140–3145.

123 Zhao, Y., Zhang, J., Han, B. et al. (2011). Wang, metal–organic framework nanospheres with well-ordered mesopores synthesized in an ionic liquid/CO2/surfactant system. *Angew. Chem. Int. Ed.* 50: 636–639.

124 Bai, Y., Dou, Y., Xie, L.H. et al. (2016). Zr-based metal–organic frameworks: design, synthesis, structure, and applications. *Chem. Soc. Rev.* 45 (8): 2327–2367.

125 Denny, M.S. and Cohen, S.M. (2015). In situ modification of metal–organic frameworks in mixed-matrix membranes. *Angew. Chem. Int. Ed.* 54: 9029–9032.

126 Kong, X., Wu, Z., Strømme, M., and Xu, C. (2024). Ambient aqueous synthesis of imine-linked covalent organic frameworks (COFs) and fabrication of freestanding cellulose nanofiber@COF nanopapers. *J. Am. Chem. Soc.* 146 (1): 742–751.

127 Kandambeth, S., Biswal, B.P., Chaudhari, H.D. et al. (2017). Selective molecular sieving in self-standing porous covalent-organic-framework membranes. *Adv. Mater.* 29: 1603945.

128 Zhang, W., Zhang, L., Zhao, H. et al. (2018). A two-dimensional cationic covalent organic framework membrane for selective molecular sieving. *J. Mater. Chem. A* 6: 13331–13339.

129 Zhang, C., Bai-Heng, W., Ma, M.-Q. et al. (2019). Ultrathin metal/covalent–organic framework membranes towards ultimate separation. *Chem. Soc. Rev.* 48: 3811–3841.

130 Düren, T., Bae, Y.-S., and Snurr, R.Q. (2009). Using molecular simulation to characterise metal–organic frameworks for adsorption applications. *Chem. Soc. Rev.* 38: 1237–1247.

131 Canivet, J., Fateeva, A., Guo, Y. et al. (2014). Water adsorption in MOFs: fundamentals and applications. *Chem. Soc. Rev.* 43: 5594–5617.

132 Ghanbari, T., Abnisa, F., and Daud, W.M.A.W. (2020). A review on production of metal organic frameworks (MOF) for CO2adsorption. *Sci. Total Environ.* 707: 135090.

133 Haque, E., Jun, J.W., and Jhung, S.H. (2011). Adsorptive removal of methyl orange and methylene blue from aqueous solution with a metal-organic framework material, iron terephthalate (MOF-235). *J. Hazard. Mater.* 185: 507–511.

134 Li, H., Cao, X., Zhang, C. et al. (2017). Enhanced adsorptive removal of anionic and cationic dyes from single or mixed dye solutions using MOF PCN-222. *RSC Adv.* 7: 16273–16281.

135 Al-Gheethi, A., Azhar, Q.M., Kumar, P.S. et al. (2022). Sustainable approaches for removing Rhodamine B dye using agricultural waste adsorbents: a review. *Chemosphere* 287: 132080.

136 Al-Buriahi, A.K., Al-Gheethi, A.A., Kumar, P.S. et al. (2022). Elimination of rhodamine B from textile wastewater using nanoparticle photocatalysts: a review for sustainable approaches. *Chemosphere* 287: 132162.

137 Fenton, J.L., Burke, D.W., Qian, D. et al. (2021). Polycrystalline covalent organic framework films act as adsorbents, not membranes. *J. Am. Chem. Soc.* 143: 1466–1473.

138 Li, Y., Chen, W., Hao, W. et al. (2018). Covalent organic frameworks constructed from flexible building blocks with high adsorption capacity for pollutants. *ACS Appl. Nano Mater.* 1: 4756–2761.

139 Chaudhary, J.P., Vadodariya, N., Nataraj, S.K., and Meena, R. (2015). Chitosan-based aerogel membrane for robust oil-in-water emulsion separation. *ACS Appl. Mater. Interfaces* 7 (44): 24957–24962.

140 Chaudhary, J.P., Nataraj, S.K., Gogda, A., and Meena, R. (2014). Bio-based superhydrophilic foam membranes for sustainable oil–water separation. *Green Chem.* 16: 4552–4558.

141 Slater, G. and Cooper, A.I. (2015). Function-led design of newporous materials. *Science* 348: aaa8075.

142 Sun, Q., Aguila, B., Perman, J.A. et al. (2018). Integrating superwettability within covalent organic frameworks for functional coating. *Chem* 4: 1726–1739.

143 Yang, C., Kaipa, U., Mather, Q.Z. et al. (2011). Fluorous metal-organic frameworks with superior adsorption and hydrophobic properties toward oil spill cleanup and hydrocarbon storage. *J. Am. Chem. Soc.* 133: 18094–18097.

144 Nguyen, D.D., Tai, N.-H., Lee, S.-B., and Kuo, W.-S. (2012). Superhydrophobic and superoleophilic properties of graphene-based sponges fabricated using a facile dip coating method. *Energy Environ. Sci.* 5: 7908–7912.

145 Li, H.-X., Sun, D.-Z., Tan, W.-J. et al. (2011). Deng, superhydrophobic conjugated microporous polymers for separation and adsorption. *Energy Environ. Sci.* 4: 2062–2065.

146 Dawson, R., Laybourn, A., Clowes, R. et al. (2009). Functionalized conjugated microporous polymers. *Macromolecules* 42: 8809.

147 Kuhn, P., Krüger, K., Thomas, A., and Antonietti, M. (2008). "Everything is surface": tunable polymer organic frameworks with ultrahigh dye sorption capacity. *Chem. Commun.* 44: 5815.

148 Lin, K.-Y.A., Yang, H., Petit, C., and Hsu, F.-K. (2014). Removing oil droplets from water using a copper-based metal organic frameworks. *Chem. Eng. J.* 249: 293–301.

149 Yogapriya, R. and Kasibhatta, K.R.D. (2020). Hydrophobic-superoleophilic fluorinated graphene nanosheet composites with metal–organic framework HKUST-1 for oil–water separation. *ACS Appl. Nano Mater.* 3: 5816–5825.

150 Gai, S., Fan, R., Zhang, J. et al. (2021). Fabrication of highly stable metal–organic frameworks and corresponding hydrophobic foam through a reticular chemistry strategy for simultaneous organic micropollutant and insoluble oil removal from wastewater. *J. Mater. Chem. A* 9: 3369–3378.

151 Shin, K. and Chung, H. (2015). A Au nanoparticle-incorporated sponge as a versatile transmission surface-enhanced Raman scattering substrate. *Analyst* 140: 5074–5081.

152 Gai, S., Zhang, J., Fan, R.Q. et al. (2020). Highly stable zinc-based metal–organic frameworks and corresponding flexible composites for removal and detection of antibiotics in water. *ACS Appl. Mater. Interfaces* 12: 8650–8662.

153 Chen, Y., Zhang, S., Cao, S. et al. (2017). Roll-to-roll production of metal-organic framework coatings for particulate matter removal. *Adv. Mater.* 29: 1606221–1606227.

154 Alsbaiee, A., Smith, B.J., Xiao, L. et al. (2016). Rapid removal of organic micropollutants from water by a porous ß-cyclodextrin polymer. *Nature* 529: 190–194.

155 Aus der Beek, T., Weber, F., Bergmann, A. et al. (2016). Pharmaceuticals in the environment-global occurrences and perspectives. *Environ. Toxicol. Chem.* 35: 823–835.

156 Ben, W., Zhu, B., Yuan, X. et al. (2018). Occurrence, removal and risk of organic micropollutants in wastewater treatment plants across China: comparison of wastewater treatment processes. *Water Res.* 130: 38–46.

157 Schwarzenbach, R.P., Escher, B.I., Fenner, K. et al. (2006). The challenge of micropollutants in aquatic systems. *Science* 313: 1072–1077.

158 Ali, I., Asim, M., and Khan, T.A. (2012). Low cost adsorbents for the removal of organic pollutants from wastewater. *J. Environ. Manag.* 113: 170–183.

159 Bottoni, P., Caroli, S., and Caracciolo, A.B. (2010). Pharmaceuticals as priority water contaminants. *Toxicol. Environ. Chem.* 92: 549–565.

160 Glüge, J., Scheringer, M., Cousins, I.T. et al. (2020). An overview of the uses of per- and polyfluoroalkyl substances (PFAS). *Environ. Sci. Processes Impacts* 22: 2345–2373.

161 Xiao, L., Ling, Y., Alsbaiee, A. et al. (2017). β-cyclodextrin polymer network sequesters perfluorooctanoic acid at environmentally relevant concentrations. *J. Am. Chem. Soc.* 139: 7689–7692.

162 Wang, R.-Q., Wei, X.-B., and Feng, Y.-Q. (2018). β-cyclodextrin covalent organic framework for selective molecular adsorption. *Chem. Eur. J.* 24: 10979–10983.

163 Liu, X., Zhu, C., Yin, J. et al. (2022). Installation of synergistic binding sites onto porous organic polymers for efficient removal of perfluorooctanoic acid. *Nat. Commun.* 13: 2132.

164 Li, R., Alomari, S., Islamoglu, T. et al. (2021). Systematic study on the removal of per- and polyfluoroalkyl substances from contaminated groundwater using metal–organic frameworks. *Environ. Sci. Technol.* 55: 15162–15171.

165 Sini, K., Bourgeois, D., Idouhar, M. et al. (2019). Metal-organic frameworks cavity size effect on the extraction of organic pollutants. *Mater. Lett.* 250: 92–95.

166 Ahmadijokani, F., Molavi, H., Rezakzemi, M. et al. (2022). UiO-66 metal–organic frameworks in water treatment: a critical review. *Prog. Mater. Sci.* 125: 100904.

167 Lin, S., Zhao, Y., and Yun, Y.-S. (2018). Highly effective removal of nonsteroidal anti-inflammatory pharmaceuticals from water by Zr(IV)-based metal–organic framework: adsorption performance and mechanisms. *ACS Appl. Mater. Interfaces* 10: 28076–28085.

168 Huang, L., Mao, N., Yan, Q. et al. (2020). Magnetic covalent organic frameworks for the removal of diclofenac sodium from water. *ACS Appl. Nano Mater.* 3: 319–326.

169 Hou, L., Shan, C., Song, Y. et al. (2021). Highly stable single crystals of three-dimensional porous oligomer frameworks synthesized under kinetic conditions. *Angew. Chem. Int. Ed.* 60: 14664–14670.

170 Wen, Y., Feng, M., Zhang, P. et al. (2021). Metal organic frameworks (MOFs) as photocatalysts for the degradation of agricultural pollutants in water. *ACS ES&T Eng.* 1: 804–826.

171 Yin, R., Chen, Y., He, S. et al. (2020). *In situ* photoreduction of structural Fe(III) in a metal–organic framework for peroxydisulfate activation and efficient removal of antibiotics in real wastewater. *J. Hazard. Mater.* 388: 121996.

172 Xiao, Z., Zhou, Y., Xin, X. et al. (2016). Iron(III) porphyrin-based porous material as photocatalyst for highly efficient and selective degradation of congo red. *Macromol. Chem. Phys.* 217: 599–604.

173 Srivastava, N. and Majumder, C. (2008). Novel biofiltration methods for the treatment of heavy metals from industrial wastewater. *J. Hazard. Mater.* 151 (1): 1–8.

174 Sud, D., Mahajan, G., and Kaur, M. (2008). Agricultural waste material as potential adsorbent for sequestering heavy metal ions from aqueous solutions–a review. *J. Biomol. Tech.* 99 (14): 6017–6027.

175 Fu, F. and Wang, Q. (2011). Removal of heavy metal ions from wastewaters: a review. *J. Environ. Manag.* 92 (3): 407–418.

176 Li, B., Zhang, Y., Ma, D. et al. (2014). Mercury nano-trap for effective and efficient removal of mercury(II) from aqueous solution. *Nat. Commun.* 5: 5537.

177 Aguila, B., Sun, Q., Perman, J.A. et al. (2017). Efficient mercury capture using functionalized porous organic polymer. *Adv. Mater.* 29: 1700665.

178 Ding, S.-Y., Dong, M., Wang, Y.-W. et al. (2016). Thioether-based fluorescent covalent organic framework for selective detection and facile removal of mercury(II). *J. Am. Chem. Soc.* 138: 3031–3037.

179 Yang, P., Shu, Y., Zhuang, Q. et al. (2019). A robust MOF-based trap with high-density active alkyl thiol for the super-efficient capture of mercury. *Chem. Commun.* 55: 12972–12975.

180 Shen, X., Ma, S., Xia, H. et al. (2018). Cationic porous organic polymers as an excellent platform for highly efficient removal of pollutants from water. *J. Mater. Chem. A* 6: 20653–20658.

181 Li, X., Jin, L., Huang, L. et al. (2021). Imidazolium-based cationic polymeric nanotraps for efficient removal of Cr2O72. *J. Environ. Chem. Eng.* 9: 106357.

182 Kaur, H., Sinha, S., Krishnan, V., and Koner, R.R. (2020). Photocatalytic reduction and recognition of Cr(VI): new Zn(II)-based metal–organic framework as catalytic surface. *Ind. Eng. Chem. Res.* 59: 8538–8550.

183 Nogrady, B. (2023). Is Fukushima wastewater release safe? What the science says. *Nature* 618 (7967): 894–895.

184 Ma, H., Shen, M., Tong, Y., and Wang, X. (2023). Radioactive wastewater treatment technologies: a review. *Molecules* 28 (4): 1935.

185 Sun, Q., Aguila, B., Perman, J. et al. (2018). Bio-inspired nano-traps for uranium extraction from seawater and recovery from nuclear waste. *Nat. Commun.* 9: 1644.

186 Sun, Q., Song, Y., Aguila, B. et al. (2021). Spatial engineering direct cooperativity between binding sites for uranium sequestration. *Adv. Sci.* 8: 2001573.

187 Song, Y., Zhu, C., Sun, Q. et al. (2021). Nanospace decoration with uranyl-specific "hooks" for selective uranium extraction from seawater with ultrahigh enrichment index. *ACS Cent. Sci.* 7 (10): 1650–1656.

188 Zhang, H., Liu, W., Li, A. et al. (2019). Three mechanisms in one material: uranium capture by a polyoxometalate–organic framework through combined complexation, chemical reduction, and photocatalytic reduction. *Angew. Chem. Int. Ed.* 58: 16110–16114.

189 Li, H., Zhai, F., Gui, D. et al. (2019). Powerful uranium extraction strategy with combined ligand complexation and photocatalytic reduction by postsynthetically modified photoactive metal-organic frameworks. *Appl. Catal. B* 254: 47–54.

190 Li, B., Zhang, Y., Ma, D. et al. (2016). Creation of a new type of ion exchange material for rapid, high-capacity, reversible and selective ion exchange without swelling and entrainment. *Chem. Sci.* 7 (3): 2138–2144.

191 Banerjee, D., Elsaidi, S.K., Aguila, B. et al. (2016). Removal of pertechnetate-related oxyanions from solution using functionalized hierarchical porous frameworks. *Chemistry* 22 (49): 17581–17584.

192 Sun, Q., Zhu, L., Aguila, B. et al. (2019). Optimizing radionuclide sequestration in anion nanotraps with record pertechnetate sorption. *Nat. Commun.* 10 (1): 1646.

193 Ferreira, F.V., Souza, A.G., Ajdary, R. et al. (2023). Nanocellulose-based porous materials: regulation and pathway to commercialization in regenerative medicine. *Bioact. Mater.* 29: 151–176.

8

Future of Desalination and Wastewater Treatment Technologies

8.1 Demand for Future Water Treatment Technologies

Emerging drinking water crisis may force water management agencies, policy makers, and a larger population to adopt novel seawater desalination and wastewater treatment technologies to fulfil daily needs. On the other hand, dwindling freshwater and ever-increasing population are expected to greatly affect the accessibility of drinking water. Seawater is abundant. However, extremely high salinity poses a series hurdle for many of the existing technologies including reverse osmosis and thermal water purification processes. Providing desalinated water to increasingly water-scarce urban population, particularly those living near coastlines is deemed to be an alternative solution. On the other hand, wastewater generation and unregulated disposal are increasing exponentially due to various demand-driven industrial, manufacturing, and economical activities. In such a situation, segregation, monitoring, and management of large quantities of wastewater is a major global concern. In addition to this, modern wastewater has been characterized by complex and less-known hazardous substances. This creates several challenges such as analysis of nature of substance, qualitative measurements, monitoring, suitability of the treatment technique as well as handling of the complex wastewater during treatment [1–7].

Specifically, modern and emerging water pollutants are critical since they are less known and their ineffective handling and treatment may cause negative impacts on ecosystems and human health. Considering gravity of the situation, the organizations such as United Nations (UN), EPA, and several national agencies have issued guidelines concerning emerging and persistent pollutants as a global challenge for years to come. Accordingly, the world is now recognizing problems in the water resource, supply, and management sector in close association with water quality, consumption, and contaminations as a global problem affecting

Materials and Methods for Industrial Wastewater and Groundwater Treatment, First Edition. S. K. Nataraj.

global populations and environment at large. Therefore, addressing water pollution due to major new and emerging contaminants, less-known substances, and related aspects will be prioritized soon. Another major issue is the reclamation of resources from the wastewater. Over a century of industrial activities have used millions of litters of water and disposal of partially treated or untreated water has carried enormous useful contaminants in them during their disposal activities. So far, limited efforts have been made to recycle or reclaim the useful resources lying in wastewater. Therefore, future water treatment technologies will focus on recovering value-added products from wastewater sources making it a sustainable process. The design and development of an integrated water treatment system are both challenging and time-consuming tasks. However, complexity of this global problem challenges can be addressed following various strategic approaches, such as regulating actions at local as well as national level, setting up decision-making processes at very diverse levels, engaging different stakeholders from ordinary citizens to policy decision makers and, with obvious social responsibility, the scientific community [8–11].

In this direction, modern and advanced water treatment technologies are being developed incorporating diverse innovative features and approaches. For instance, several nanotechnology-based solutions have been developed having potential for both detection, identification, and implementation of integrated systems to contribute for water remediation and quality improvement. Some of these approaches are simple and rely on the use of nanoengineered materials that are surface-modified to achieve task-specific results in the envisaged application by taking care of target contaminants to be eliminated. These nanotechnology-based approaches can be either complementary to currently available water treatment technologies or standalone based on future developments, which have shown several advantages and are quite effective against a series of emerging pollutants from certain contaminated aqueous environments. Also, water quality monitoring tools with advanced and smart features will play an important role in communicating seriousness of problem and implementing appropriate treatment platform [12–16].

In addition to this, conventional membrane processes having polymeric selective barriers are expected to undergo transformational changes both in terms of materials and their preparation techniques. For instance, polymer-based membranes have been traditionally prepared following solvent-induced phase inversion technique (SIPS), which is considered a solvent-intensive and environmentally not-so-friendly technique. As a replacement for chemical-intensive technique SIPS, block polymer-based membrane systems, which follow a physical etching route will be preferred to obtain precisely controlled pore size and their distributions in newer approaches. Therefore, futuristic approach would lead to innovation in both materials selection and process development. Therefore, alterations in the chemically intensive and complex processes may be replaced with sustainable and greener approaches in the membrane technology sector [17–22].

In addition, the block polymer-based membranes can be effectively prepared using straightforward schemes with controlled pore and surface properties that go beyond selective barriers. Further, block copolymer-based membranes can be easily functionalized following the incorporation of various reactive entities on the surface as well as along the pore wall side to convert a membrane into an interactive site. Alternatively, use of coordination compounds in the membrane matrix allows the effective separation of contaminants such as heavy metal ions through adsorption process. Further, effective control over pore sizes in the block polymer membranes offers a unique advantage over conventional membranes in specialized applications such as biomacromolecule separation (DNA, protein) as the narrow pore size distribution enables the residence times for target molecule with each pore. This allows the uniform distribution of flow in which block polymer membranes regulate mass transfer in an effective manner as it prevents channelling, however utilizes the high packing density of active functional groups decorated within the pores to yield optimal separation and permeation performance. For instance, in a functional block copolymer membrane, a well-defined residence time allowed for solute to navigate precisely through regulating the water flux. Likewise, well-defined block copolymer membranes can be used to significantly improve the water flux along with transportation of selective solutes across using favorable enthalpic interactions between solutes and functional pore walls. This target-specific control over permeation of solutes allows the opening of new avenues for the development of futuristic membranes for water treatment applications including the remediation of emerging pollutants as well as conventional toxic contaminants in addition to the concentration of value-added materials dissolved in wastewater sources.

8.2 Smart Tools in Future Water Treatment Technologies

The terminology "smart technology" can be traced to the acronym word "Smart" which implies "self-managing, monitoring, analysis, and automated reporting technology". As larger population access their smartphones, computers, and other gadgets, overdependency on modern tools demands smarter technology to be able to be managed, monitored, and controlled remotely. Conventional manufacturing and service tools, equipment, and instruments such as pumps, motors, lighting, security systems, and so on need smoother handling to enhance their capabilities and efficiencies. Last decade has witnessed the development of revolutionary smart devices and technologies that have made life of the common man and household easier. Also, consumer products have changed the household through smart technologies. Therefore, smart technologies have changed the way of living, which encouraged every sector to innovate and continuously adopt smarter tools, devices, equipment, and machinery.

Thus, the recent revolution in the development of smarter technologies and digital transformation, and robust adaptation of new technologies have long been revolutionized in several sectors including industrial manufacturing and consumer products and now the water treatment sector is no exception. The complexities in wastewater sources are increasingly threatening the environment, which requires immediate attention as managing wastewater treatment plants (WWTPs) becomes untenable in such situations. Nevertheless, water treatment technologies have evolved over the years. However, the overall management of treatment technologies still faces challenges associated with efficient and sustainable monitoring, handling, and management. Therefore, like every other technology, water treatment sector is also at a critical juncture. It is imperative that use of smart technologies in water purification and wastewater treatment is inevitable. Also, decisions on adaptation of smarter tools and actions in this direction taken today in the design and development of water treatment materials and technologies will shape the overall sector in years to come. Therefore, smarter water treatment technologies are destined to completely transform water purification and wastewater treatment and are spearheaded by smart and digital transformation to significantly improve the performance efficiencies and quality of product-treated water.

Nevertheless, today there is a greater demand than ever for efficient and high-performance water treatment technologies. As digitization and Internet of Things (IoT) technologies are gearing up, intelligent water treatment processes are being optimized to reduce waste, operational costs, and process efficiencies. Real-time monitoring of water quality is essential for effective water treatment and waste management. Real-time contaminated water quality, their level monitoring, and data acquisition can significantly reduce the water leakage by a tank overflow and engineering malfunction. Therefore, water management tools and devices can help fix water leakages in a smart home through intelligent monitoring throughout the day [23–29].

In this direction, smart IoT-based water purification and treatment with cutting-edge features are being employed to build intelligent and self-sustained water treatment processes and systems. The smarter technologies therefore deemed to increase the treatment efficiencies and efficacies through integrated devices like sensors, data analytical tools, and machine learning (ML) algorithms, which help in continuous monitoring and enable the controlling of several aspects of the water treatment system. Therefore, use of advanced technologies has unique characteristics and advantages and future treatment technologies with smart, IoT, and ML-based tools might revolutionize the water treatment sector forever.

A low-cost interconnected network of systems known as "IoT" that is capable of collecting, transmitting, and receiving data has already been well adopted with highly positive outcomes in water management systems. For example, IoT devices have been used for monitoring water levels in real-time as they enable automated data exchange without the assistance or interference of a human. In addition to this, innovators have already designed task-specific devices suitable for monitoring water resources not only for their levels and quality but also including pH, conductivity, salinity, temperature, biochemical oxygen demand (BOD), chemical oxygen demand (COD), and total dissolved solids (TDS). In several smart cities, real-time monitoring and remediation of water quality have become more prevalent. A typical IoT-based water management system is shown schematically in Figure 8.1 [30–34].

A typical smart water management system might include a set of advanced sensors, actuators, controllers, and data visualization software. The controller can remotely monitor and assess variables like conductivity, pH, and turbidity with the help of various sensors. The various data emanating from different sensors are then gathered by the organizer/controller for even analysis. This type of operation is typically useful for managing complex water resources, and supply and distribution networks like community, urban, and/or metropolitan water management public entities and private companies. Further, an appropriate IoT tool is then connected to water management technology through which a stakeholder can access the information and data for the interpretation and

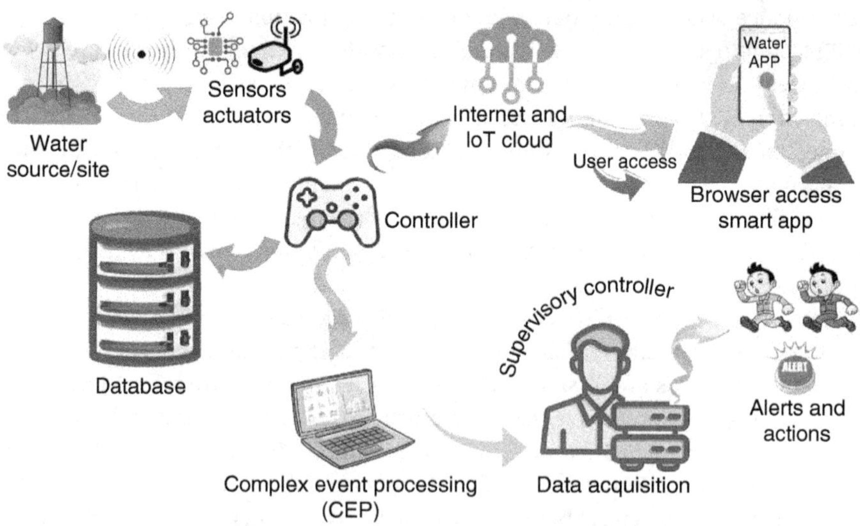

Figure 8.1 A typical remote-accessed water management system using smart and IoT-based technologies.

measurement of relevant real-time database that helps controllers to identify water quality parameters and at the supply side, managers can analyze the water consumption trends sector wise and extent. Similarly, smart water management systems can enable the collection of large sums of information, which empowers the policymakers, managers, and decisionmakers to sort the information and take best possible action through remote access and control over every type of water resource including water tanks, storage systems, wells, dams, valves, wastewater disposal sites, and drainage network both in rural and urban localities [35–43].

On the other hand, water treatment technology implementation mainly relies on the level of contamination and type of contaminants. Traditionally, water quality measurement, analysis, and treatment have been tedious tasks and time-consuming. In such a situation, smart analytical tools help in quicker data collection, monitoring, and decision-making process. In this process, three major tasks are involved, namely (i) water quality testing, (ii) wastewater treatment, and (iii) waste disposal. Conventional manual water quality testing procedures are time-consuming and often vulnerable to human error leading to unreliable data quality. Following standardized smart water quality analysis in real-time without the interference of an operator or even in absence of operator on site, collecting data and sending feed data back to remotely monitored center is envisaged future. In such systems, smart sensors have already been in use for water quality across a period of time consistently and effectively for parameters like conductivity, temperature, pH (acidity), dissolved oxygen (DO), and salinity measurables that majorly impact water management. These parameters are vital even in low concentrations as well as in high levels that may greatly affect their domestic or industrial use. On the other hand, wastewater treatment and management are a crucial step in overall modern process development in which IoT-based management systems can be used for a range of applications such as monitoring of water levels, water quality, water flow, reduction of contaminant levels, leaks in piping systems, and so on. Here reliable and consistent information communication between operators and wastewater management systems can help resolve emergency issues and make overall treatment process more efficient, cost-effective, and time-saving. These systems also help in addressing major leakage-related issues, which generally cause loss of large amounts of water resources. Therefore, these end-to-end solutions help both operators and engineers resolve any issues and/or emergencies if any occurring during the operation of a treatment plant before they become too serious. Finally, water dispersal is another major issue which consumes large resources due to the presence of stringent laws and regulations. Smart technologies ensure the appropriate amount of water is being dispersed throughout the treatment tank, which is a key in water treatment management. To overcome any discrepancies in the proper mixing and selective separation of water components in the treatment tank, pressure sensors are being used to ensure the proper water distribution lines, monitoring of water flow in real-time remotely that helps in meeting supply demand and sustainable operation of treatment and supply plants [44–48].

IoT devices in wastewater treatment plants (WWTPs) such as smart meters that are strategically placed at key points of the plant enable them to collect accurate and relevant data in real-time consistently. Thus, information collected and processed by the sensors and interfaces is then transmitted to a central processing platform, where it is processed and transpired into high-value data, which gives in detailed and real-time picture of the state of the process and operating system. IoT-based tools enable wireless access and monitoring of operations providing greater control, optimization, and supervision beyond the plant site. However, the most valuable advantage of the smart technologies has been real-time monitoring of the waste disposal and sewer network in the WWTP's catchment basin. This helps in pinpointing early detection of discharges into the catchment basins or natural water bodies, thereby enabling taking preventive actions to mitigate the adverse impact upon discharge. In some cases, extended network of tools also helps in effective investigations of sources and detection of pollutants in the network. Most importantly, IoT-based technologies have driven highly efficient resource management in plants and operation networks in addition to significantly upgrading the quality of the water treatment process, such as reagents, consumables, energy, and miscellaneous involved in the operation through the optimized consumption and minimizing waste leading to lower operating costs [49–52].

Therefore, smart water treatment processes designed for producing drinking water, wastewater treatment, automation of irrigation systems, and industrial water management both during consumption as well as waste management post-usage will significantly impact the effectiveness of the technology. On the other hand, automation and smart technologies play a vital role in achieving sustainable industrial activities with paramount importance to safety and optimal operation. This smart revolution over a decade of innovation has been driven by the convergence of information technology (IT) and traditional engineering operational technology (OT) creating a synergic combination between technological innovation and efficiency in the water purification and wastewater treatment sector. At the same time, the innovation in the variety of standards of sensors and their affordability has led to the rapid implementation of new technologies facilitating creation of large volumes of data that measure water quality and the efficiencies of the tools involved in the processes during treatment. These consistent data include detailed analysis of types and concentrations of contaminants including their chemical, physical, biological, and lately, information on emerging pollutant parameters. Further, convergence of different tools and information generated helps in monitoring number of variables and processing capacity either through edge-computing or in the cloud designed to address complex problems with robust and cutting-edge algorithms [53–56].

8.3 Advanced Treatment and Innovative Technologies

Modern water treatment technologies equipped with smart devices help in generating vast amounts of information through the operations of WWTPs that can be leveraged to optimize the economic operation of plants through advanced data analysis techniques. Vast amounts of data enable near-precise reading of trends, patterns, and directions that the process is indicating, which can be correlated to identify the future course of action. Similarly, vast amounts of data can also be exploited to assess the economics involved in wastewater treatment, reclamation of reusable quality water, and extraction of value-added products. In this direction, several smart tools-assisted technologies are making inroads in the water purification and municipal wastewater treatment sector including [57–59]:

1) *Reverse Osmosis (RO), Nanofiltration (NF), and Ultrafiltration (UF)* have been historically used for desalination and wastewater treatment, respectively. In these processes, reusable quality water is forced through membranes under applied pressure leaving behind contaminants, which can be further concentrated to extract as value-added products. In all cases, over the years, smart devices in the separation processes have improved both performance as well as economics of the operations [60, 61].
2) *Forward Osmosis (FO) Process*, on the other hand, has been popularly used as pretreatment technique for RO and other pressure-driven membrane processes. However, due to the innovations in use of efficient draw solute/solutions and effective semipermeable membranes, the process is now being used as standalone technique for both water treatment and resource recovery [62, 63].
3) *Advanced oxidation* process uses efficient catalytic nanomaterials in which reagents produce free radicals that are used for the breakdown of emerging as well as persistent organic contaminants in wastewater. Innovations in nanomaterials have led to higher efficacies and economical adaptation of the process in modern wastewater treatment. Here, also deeper understanding of nanomaterials and their properties has led to faster implementation of the novel materials in the process as effective treatment tools [64–66].
4) *Photocatalytic Oxidation technique* uses an efficient catalytic material such as titanium dioxide, which is activated by ultraviolet (UV) or visible light to break down hazardous pollutants like organic compounds and microorganisms in wastewater. Even though oxidation technique has been in use for decades, however, the effectiveness and efficiencies of the process have been greatly improved with advent of nanoscience and nanotechnology concepts. These active materials under exposure to UV light trigger the generation of free radicals leading to oxidation and followed by degradation of the pollutants [67, 68].

5) *Ultrasonic Reactors* are considered one of the futuristic water treatment techniques, which make use of high-frequency ultrasonic waves to treat wastewater. The high-frequency ultrasonic waves create microbubbles that are highly vibrant and later collapse violently producing high temperatures and pressures, which cause disintegration of contaminants and microorganisms in wastewater tanks leading to decontamination of the source water [69].
6) *Genetically or Naturally Enhanced Microorganisms* are a new class of microorganisms developed for the treatment of wastewater. In this approach, naturally enhanced microorganisms are being developed to treat wastewater with refractory total organic carbon (TOC)/ COD or specific contaminants. The process starts with the selection of microorganisms, then producing improved variants followed by feeding them into the system where contaminants are being treated [70, 71].
7) *Electrocoagulation and Electrooxidation* are other advanced methods being used for water treatment applications in which an electric current is applied to eliminate pollutants through coagulation and oxidation processes [72].

8.4 Renewable Energies in Water Treatment

Historically, for centuries, drinking water purification practices in remote locations have depended on biomass and fossil fuel-based sources. On the other hand, small-to-medium water treatment and supply systems both in remote and rural sites mainly depended on power sources like gasoline, diesel, kerosene pumps, and so on. Later, innovation in hydrology-related technologies enabled identification of water distribution networks beneath the earth's surface leading to use of hand pumps, animal-drawn pumps, wind pumps, solar pumps, biofuel pumps, and so on. However, as economic activity expanded, electrification/ connectivity enabled grid-connected electric pumps. Nonetheless, even today, several remote and rural locations use hand pumps and animal-drawn pumps as the cheapest and simplest means of extracting water from groundwater sources as well as shallow wells. However, electric and diesel-run pumps have been widely adopted for community supply networks as well as small irrigation activities. However, fossil fuel-operated (diesel) pumps, solar pumps, wind pumps, and grid-connected electric pumps have been widely considered convenient and economic options in low-to-medium-income communities. Interestingly, initial policy decisions related to issues of distribution of power sources for pumping systems were complicated. There were several issues related to subsidies and restrictions on using certain water pumping technologies, fuel, spare parts, equipment standardization, taxation on imported equipment, and installation at remote locations delayed the implementation of modern technologies in water treatment sector. However, indigenous design and technologies have sprung up across the globe in the mid-19th century. This enables the development of faster-standardized equipment contributed to both technological innovations as well as easier access to water sources to larger populations. Lately, grid-connected electric pumps, wind pumps, mechanical wind pumps (windmills), electrical wind pumps, solar (photovoltaic) pumps, biofuel pumps, and so on have been widely developed and implemented for water pumping and water purification applications [73–75].

On the other hand, very significant innovations are taking place in design and operations of WWTPs in which wastewater treatment designers, researchers, and professionals contemplate the energy generation potentials. Therefore, future of WWTPs expands not only beyond water treatment plants but also energy generation sources with features having onsite digester gas combustion or cogeneration. This would lead to extensive innovation activities in the design of advanced WWTPs, which might transform wastewater as an important source of sustainable energy. Some of the transformative hybrid plant designs might include use of co-generators, modified digestors for sewage sludge, use of sludge for incineration, enzymatic biofuel cells, and biofuel cell-assisted electricity generation that can generate a significant amount of energy. In addition to this, zero liquid and zero solid targets can also be implemented aiming at recovery of value-added products in addition to onsite energy generation at WWTPs. This could allow not only recycling and recovery of highly precious nutrients but also identification of disposal solids from wastewater and reuse of treated water for non-potable applications like irrigation and industrial purposes. Thus, future WWTP research focuses on multipurpose operating units with integrated innovative devices that control components in water recovery, energy generation, and distribution systems.

Even though initial attempts to generate electricity from WWTPs have shown great potentials, however, improved attempts and technological adaptation are yet to be made to achieve threshold contributions in energy generation from wastewater plants. Now, research efforts have been redirected to achieve optimal and self-sufficient electricity generation goals from WWTPs. For instance, thermal energy generated through incineration method has shown limited output. On the other hand, use of anerobic digestion technique has shown a high level of electricity generation efficiency through which self-sufficiency can be achieved to operate WWTP and cogeneration attempts can further yield surplus production. Nevertheless, a series of attempts to optimize various hybrid WWTPs have revealed three primary feasible approaches that can lead to optimal electricity generation or heat recovery, namely (i) Direct use of digester gas for combustion via cogeneration that can be used in in-sewer heat exchangers, (ii) Deployment of external heat pumps that can recover heat from wastewater, and (iii) Following solar thermal generation technique. Another minor contribution toward electricity generation could come from the direct incineration of sewage sludge [76, 77].

Nevertheless, multidimensional approach in wastewater treatment could help in significantly reducing carbon footprint of WWTPs as well as positively impact the operational costs that are vital to two key factors to maintain sustainability in operations, and increase the utility plant and revenue generation toward maintenance. Therefore, achieving energy efficiency in WWTP operation on daily basis is one of the main objectives that need to consider all possible methods and means to achieve the goals. These goals can be achieved following four complementary strategies in WWTP operations such as (i) Real-time monitoring of energy consumption and control which enables the minimization of loss and optimization of new technologies to reduce energy consumption, (ii) Designing and development of innovative biological reactors which operate on optimized aeration system, (iii) Following innovative strategies to get optimized biogas production that can help in improving the production and storage of greener energy before dispatching for end-use, finally (iv) Installing high-performance photovoltaic panels to harness solar energy to generate electricity that can suffice the energy needs of WWTPs and surplus can be used for commercial purposes.

In all cases, digital footprint and adoption of AI, IoT, and ML tools can lead to a technological transformation of water treatment sector toward improved output. Revolutionary developments in materials science, nanotechnology, and device miniaturization are expected in every sector including water purification and treatment segment. The digital twin will further enable the performance of the WWTPs to be monitored and analyzed in real-time for addressing potential operation-related complexities and provide in-time solutions for anticipated problems. Further, modern tools in WWTPs are expected to assist in theoretical calculation, prediction, and biokinetic mathematical modelling, which helps in designing plants and enables self-calibration algorithms with ML and IoT tools that are optimized through the simulation studies on real-time behavior of biological and physicochemical processes and thus maximizing the plant performance and efficiency.

8.5 Emerging Desalination Technologies

8.5.1 Forward Osmosis

FO process is an osmotically driven process, which mainly depends on the simple natural principle of osmosis phenomenon through which a solvent (water) moves from a low concentration solution (seawater feed) to a high-concentration draw solution (DS). In FO process, a DS is a concentrated solution used at the permeate side to create exceptionally high osmotic gradient to draw water from the feed side across the semipermeable membrane. In this process, three important components play a vital role in achieving optimal freshwater production capacitates, namely (i) Effective semipermeable FO membrane, (ii) Type of draw solute/solution, and (iii) Recovery and reuse of draw solute/solution. Typically, depending on the FO membrane orientation, FO process can be operated in two modes (i) active layer membrane facing the feed solution (AL-FO mode) and (ii) active layer is facing the DS which is also known as pressure-retarded osmosis (AL-PRO mode). In both cases, water transportation from seawater, saline water, or wastewater to DS takes place due to the concentration gradient and does not involve any applied pressure. Therefore, FO operates based on naturally occurring phenomenon due to which process is energy efficient compared to pressure-driven RO processes. Also, as FO membrane experiences less applied pressure and stress during the operation, it experiences lower tendency toward organic and biofouling. However, FO process operates in two-step operation to produce freshwater in which first step draws the water from feed saline solution and second step is required to separate freshwater from the draw solute/solution. This simultaneous freshwater separation and recovery of draw solute/solution is essential to make overall process sustainable. Therefore, desalination or water separation from feed in FO process mainly relies not only on osmotic transportation of water molecules through an FO membrane using a highly osmotic gradient DS but also on subsequent recovery and reuse of the DS. Here, second step in which recovery of both products, water and draw solute/solution is energy-intensive step, which is frequently overlooked during the process development. However, recent studies have used easy-to-recover and reuse macromolecular draw solutes based on simple chemistry and efficient functional membrane materials that significantly contribute to reducing energy consumption and make FO process an economically competitive process for desalination as well as wastewater treatment [78].

8.5.2 Capacitive Deionization

CDI is an emerging water treatment technology which is also known as a membrane-less process widely used for desalination applications. The CDI technology uses inexpensive and abundantly found electrode materials like activated carbon as the active element in simple desalination operations. The CDI cell contains a high surface area and high conductive electrodes to leverage ionic species attraction from contaminated water sources to deliver sustainable desalinated water. The electrodes in the CDI cell can be tunable to control the movement of ionic species based on ion selectivity parameters. Interestingly, CDI water treatment process does not require any type of membrane to separate water from contaminants. The CDI cell can be prepared using simple conductive and high surface area materials such as activated carbon [79, 80].

A typical water treatment or desalination operation in CDI involves the use of a constant voltage or current between two electrodes in the presence of contaminated feed stream or saline water. The applied voltage to the CDI cell system stimulates the ions in the feed to electromigrate into the electrodes, where they are captured electrostatically in electric double layer (EDL) formation in the microporous structure of electrodes. Capacity of ion capture or storage mainly depends on the surface area of the electrodes and the conductivity of the electrodes influences the electromigration rate. Once electrodes are fully charged, the ionic species from electrodes are separated to regenerate the electrodes electrically either by reversing the applied current or short-circuiting, which leads to the discharge of ions from the EDL formation. Further, released highly concentrated ionic species are then separated from the cell as concentrated brine waste stream. Therefore, CDI technique can be a promising emerging water treatment process for the brackish water desalination, groundwater treatment, and ion-selective separations of low-to-medium contamination process streams. However, CDI has shown limited potential for highly saline or highly concentrated feed streams in which electrode recycling and regeneration are a major challenge in which repeated reversal of process shortens the lifetime of electrodes. In this process, a major process limit occurs as the electrode degradation is enhanced with Faradaic side reactions, thereby oxidizing the anode and reducing DO at the cathode side. Nevertheless, there have been several reports that make use of alternate electrode materials instead of only activated carbon that rely on intercalation processes. Also, performance of CDI cells can be enhanced by the use of novel materials that can also undergo redox conversion reactions in addition to EDL mechanism to achieve higher deionization capacity.

Further, recent advances in CDI have utilized several noncarbon-based electrode materials having greater ion adsorption capacities that have shown promising energy efficiency in CDI desalination operations. In particular, researchers have adopted a new concept of a "desalination battery," which makes use of high ionic capacity battery electrode materials as an alternative stock of electrode materials. A desalination battery working is similar to that of CDI concept as both use electrical energy input to extract salt ions from seawater. However, in desalination battery, not only the ions are physically held but also stored in the bulk of the high surface area electrode material through chemical bonding, whereas ion storage in CDI is a surface phenomenon (electrode/electrolyte interface). Characteristically, the use of battery-type materials or Faradaic-responsive materials that facilitate the ion-intercalation distinguishes the desalination battery systems from CDI. Therefore, desalination batteries are referred to as next-generation-CDI in which ions from feed are captured both on surface as well as in the bulk of the electrode materials.

The CDI technology discussed in the previous section has exhibited a high potential for lower salinity or groundwater desalination capabilities and has shown several limitations when applied to high salinity feed solutions. This is mainly due to (i) carbon-based electrodes currently used have limited low ion capacity (<30 mg/g), which drastically limits salt retention capabilities in CDI, and (ii) CDI with conventional electrodes is energy-intensive process for handling relatively high salinity feed and sensitive to variations in feed salinity. In CDI, salt retention on the electrode is also limited by the cell voltage. Use of lower voltage restricts the desalination performance. However, higher voltage causes undesirable Faradaic reactions damaging the overall electrode assemblies. This limits the maximum salt rejection in CDIs up to 5000 and 10,000 5000 mg/L with adsorption efficiencies of 55 and 28%, respectively. However, recent studies reveal that the energy efficiency of CDI process does not depend on salt rejection, which means CDI can be energetically preferred for desalinating feed streams to achieve low salt rejection (<25%) with an average water flux of ~10 L/m^2/h [81–83].

8.5.3 Membrane Distillation

Recent study reports suggest that advanced water treatment technology such as MD is gaining substantial attention in desalination and wastewater treatment sector. MD technique is a thermally driven process to generate water vapors from feed saline water or wastewater, which is then separated using a microporous hydrophobic membrane. On the other hand, the nonvolatile constituents are retained in the aqueous feed solution hydrophobic surface, whereas the volatile constituent (water vapor) is permeated across the porous structure of the membrane and then collected as a distillate in a pure form. Herein, the vapor pressure difference (ΔP) created by the temperature difference (ΔT) drives the vapor transfer water vapor across the membrane. This temperature gradient combined with a selective membrane barrier causes the separation of water from nonvolatile solutes at lower operating temperatures. As MD operates at low operating temperatures, the process can be easily developed by making use of surplus heat, exhaust waste heat in a relatively larger manufacturing unit, low-grade waste heat, and/or solar energy as an energy source, which is available as a sustainable alternative. Unlike pressure-driven membrane processes like NF and RO, MD process can be operated at atmospheric pressure and show remarkably high (99.9%) rejection for ion and nonvolatile compounds from feed solution. These advantages make MD process superior to other membrane-based separation processes for a wide range of applications including desalination.

Now MD technology has matured with several commercial-scale plants operating in various parts of the world with varying output capacities. Among many, Polytetrafluoroethylene (PTFE) is the most commonly used membrane material for various MD-based separation and purification applications. Today,

MD process has been widely used in applications like removal of volatile substances from aqueous solutions, textile industry, food industry, pharmaceutical industry, water treatment plants, radioactive wastes, and desalination. However, MD process also suffers from some limitations such as wetting and temperature polarization (TP) that are closely related to the lifespan and overall process efficiencies. Nevertheless, several researchers are working on developing an ideal MD membrane aiming to achieve optimal wetting resistance with low TP [84].

8.5.4 Hybrid Desalination Technologies

Several technologies have been developed for effective desalination applications including thermal processes and membrane-based separation processes. Even though many of these technologies have been effective in achieving optimal desalination performance, however, technologists have adopted a combination of processes to achieve optimal desalination performance. In this direction, thermal RO systems like multistage-flash distillation and reverse osmosis (MSF-RO) and multieffect distillation-RO (MED-RO) have been designed to produce high-quality and high-productive desalination. Similarly, FO and RP (FO-RO) and FO and nanofiltration (FO-NF) have shown numerous potentials in achieving optimal desalination. FO as a standalone does not produce freshwater directly. Instead it requires additional draw solute/solution regeneration step to achieve further desalinated water from the DS. Therefore, cost of desalination using FO depends mainly on the use of efficient draw solute/solution and its recovery. Here, recovery of draw solute/solution requires appropriate second step in hybrid system involving either NF or RO, which determines the overall cost of desalinating process. Even though RO can be directly used to regenerate high osmotic gradient DS. However, due to higher cost process involvement, the hybridization process (FO-RO) is not beneficial for seawater desalination.

However, research is advancing in integrating FO with low-operating pressure RO to treat wastewater with seawater as the DS. In this interesting approach, RO reject is also strategically used as DSs against relatively low-concentration wastewater feed to make a hybrid process FO-RO feasible. Although the capital cost of hybrid FO-RO is higher than standalone seawater reverse osmosis (SWRO), however, hybrid process operates at 56% lower OPEX costs. On the other hand, MD has shown greater potential with FO-MD combination than RO for resource recovery as a new combination of processes can be effectively used to treat highly saline solutions. For instance, FO-MD can be used for nutrient recovery and brine concentration. Here, highly concentrated brine can be used as DS for following FO process, making this overall process a sustainable and continuous process. Also, electrodialysis (ED) and RO have been used as hybrid systems targeting either increased overall water recovery of desalinated water in RO step or direct desalination process [85, 86].

8.5.5 Low-Carbon Footprint Desalination

Historically, modern desalination technologies have been widely used to meet the ever-increasing freshwater demands. Commercially successful MSF technologies and RO processes have played a key role in providing freshwater to millions of people across the globe. However, currently followed practices and sources of energy in most of the cases are often not sustainable. On the other hand, supplying millions of litters of freshwater produced through modern technologies using only sustainable sources may not be feasible. It is expected that freshwater demand is going to exponentially increase in coming decades due to population growth, which is expected to reach ~10 billion by 2050. On the other hand, majority of the desalination technologies discharge vast amounts of brine or concentrates and solid/liquid wastewater, which can cause serious concern to aquatic life. In worst cases, concentrate discharges into aquatic ecosystem may destroy life and neighborhood around water treatment or desalination plants may experience severe environmental conditions without the appropriate measures in places. On the other hand, generation of large amounts of energy, which are required to operate desalination plants and WWTPs from nonrenewable sources, is another major concern and adds severe carbon footprint to environment and may severely affect ecosystems in the long term. Therefore, it is a pressing point to adopt newer technologies that operate on use of renewable sources in carbon-free operations. It is also highly anticipated that hybrid technologies be adopted to generate sufficient power to sustain plant operations through renewable means. However, despite rigors efforts in making rapid advances toward energy-efficient wastewater treatment and desalination, presently used technologies remain energy-intensive and generate a considerable amount of greenhouse gas (GHG) emissions. According to several estimations, carbon footprint of SWRO operation poses as high as 6.7 kg carbon dioxide eq/m^3 and it is estimated even higher for thermal-based desalination processes. Therefore, it is essential to make appropriate policies and technology development toward achieving near-zero-carbon footprint targets for future water treatment technologies. Even though several efforts have already been made to create awareness among researchers, technologists, and innovators to adopt low-carbon-footprint technologies, however, CAPEX and OPEX considerations of technologies have always been a hurdle in adopting newer and cleaner sources of energy in water treatment sector [87, 88].

8.6 Renewable Energy in Desalination

It is inevitable that greener and cleaner energy sources will replace conventional power sources in every sector including water treatment and desalination. Increasing demand for adopting cleaner technologies is already forcing technologists to design water treatment processes integrated with renewable energy tools, which is an important step toward the decarbonization of water treatment and desalination processes. At the global level, policies are becoming stringent and increased awareness among nation states have formulated a series of regulations and norms on GHG emissions fueled by environmental concerns that have amplified importance of renewable energy in water treatment and desalination processes. In this direction, several attempts have already shown fruitful results in reducing carbon footprint with the use of renewable energy sources such as wind, solar, geothermal, tidal, and hybrid systems in the vicinity of desalination plants. Among these, solar-powered desalination systems have shown greater potentials and feasibilities due to the availability of matured and well-established solar energy technologies. On the other hand, solar energy is abundant and can be easily accessed from the water stress areas and a combination of advanced solar technology with abundance makes solar energy an attractive choice for desalination processes. Matured solar energy technology can also be adapted using various forms of advanced direct electrical energy from photovoltaics or through thermal energy using solar collectors depending on the site/location and the demand of the desalination plant. However, extensive use of solar power technologies is yet to be implemented in modern water treatment and desalination units. According to some estimations, at present, merely less than 1% of the desalination plants use solar energy in their operations and maintenance. This is expected to change drastically in the near future [89, 90].

On the other hand, conventional thermal technologies can still be used to obtain greener energy sources with innovative approaches and process modifications, which can substantially reduce GHG emissions. Electricity generated following thermal or mechanical energy conversion or generation technologies still contributes substantially to the energy needs. Therefore, energy from a modified conventional process can still supply sufficient cleaner power for desalination. However, a major concern with some renewable energy sources is their uncertainty in providing a stable or continuous power supply. Most of the renewable energy sources lack power grid connection. In such a case, existing desalination plants working utilizing grid-connected power supply and/or energy storage devices miss the opportunity to integrate renewable sources into their existing operations. Nonetheless, several scattered attempts have already been made to use different types of renewable energy sources for operating membrane-based separation processes like FO, ED, MD, and RO that are integrated with the photovoltaic technologies making them some of the earliest sustainable and greener plants. The most commonly found hybrid technology is photovoltaic power-assisted RO (PV-RO) process.

On the other hand, conventional power sources provide an uninterrupted supply of energy. However, renewable technologies lack reliability. For instance,

RO plants operated using solar and wind energy sources have often experienced interruption in their operations due to technical challenges associated with the integration of processes. This is known to cause severe damage to continuous plant operations leading to shortened membrane life. As a result of intermittent operation, membrane life drastically reduces leading to subsequent overall plant output. In this situation, demand for additional energy storage devices to supplement the operation also arises, which is an added cost and expensive storage and conversion devices add substantially to the CAPEX and OPEX of the plant. However, advanced solar technologies, which generate substantial amounts of power through their concentrating solar power (CSP), have shown reliable and largely uninterrupted supply of power that are sufficient to run relatively low energy-intensive processes like MD and FO plants. Similarly, photovoltaic systems have been successfully integrated to operate ED and RO processes and coupling of PV and geothermal energy has been adopted for MD process operations.

Nevertheless, operating WWTPs with medium to large capacity is a challenging task as the increased operational costs pose serious questions on the sustainability of overall approach and efficiencies of the technologies. This concern is also elevated when the issues are associated with additional waste management costs, costs involved in post-treatment, necessity to adopt appropriate route for activated sludge treatment, and so on. This situation turns further serious when stringent water treatment regulations and management norms apply on the effluent discharges concerning maintenance of permissible level contaminants in the treated products such as phosphorus, total nitrogen, and COD/BOD limits that require appropriate post-treatment operations to maintain discharge quality.

However, recent focus on recovery and reuse of value-added products from wastewater sources has led to a paradigm shift. Now, waste is being treated as new and valuable source materials terming WWTPs as "resource recovery and water recycling centers." There have been several innovative approaches and ideas which have been implemented to recover value-added products during wastewater treatment and post-treatment of wastewater. Highly concentrated sludge or wastewater concentrate has now been considered a rich source for a number of endangered substances such as Cr, proteins, and energy materials. Today, many research groups and private companies are extensively working on concepts of product recovery from waste sources and these efforts are gaining wider acceptance across the sectors. These efforts are expected to reduce the burden on OPEX as well as the investment in new research. Utilizing these concepts, researchers are developing a wide range of integrated technologies that could enable a high rate of water recovery, high-quality product concentration, less burden on concentrate waste management, and reduced energy demand, drastically lowering the capital costs, and assisting in supplementing operational costs.

Currently, waste management companies booming and the wastewater industry is continuously evolving and opportunities are staggering with the choice of technologies. At the same time, investment in the field of waste management and developing technologies in the sector is exponentially increasing. Water treatment industry is integrating advanced tools like AI, ML, and IoT with the help of a wide range of modelling and decision-making tools. Even though these efforts are in early stages, but efforts can give some sense of direction in which overall water treatment sector is heading. All these efforts are expected to result in improving the energy efficiency of water treatment systems and smoother integrations of renewable energy technologies, which create more opportunities in renewable energy-assisted water treatment. Nonetheless, there is a great opportunity to develop green water treatment sites by integrating conventional high-performing technologies that operate on renewable energy sources. At the same time, focusing on innovative technologies which can effectively and efficiently recover high-quality reusable water as well as value-added products from existing WWTPs can revolutionize the water sector beyond limits.

8.7 Commercial Potentials and Challenges of Advanced Water Filtration Systems

Currently, the most commonly used and commercially successful desalination techniques fall under two main categories: (i) thermal desalination and (ii) membrane-based separation and purification processes. As discussed in detail in previous chapters, thermal desalination processes are some of the simplest forms consisting of evaporating saline water followed by condensation of resulting steam to produce potable quality freshwater. Thermal desalination methods have been historically used for desalination of saline water for centuries to obtain potable water. Thermal desalination techniques have been recorded as early as the 4th century BCE. However, industrial revolutions in various parts of the world in the mid-18th and 19th centuries revolutionized distillation technology to produce potable water. Distillation technologies were widely used for the first time on ships during the long era of colonization. Further, the advent of the steam engine led to improvements in technologies associated with evaporators, which enabled the process to reach remote areas in quick time (1928 in the Netherlands). With time, distillation technology reached households for freshwater production as distillation apparatus achieved optimal commercial success (capacity of $60\,m^3/$day). Today, majority of middle-east region depended on large-scale thermal distillation plants for sustained freshwater supplies. As discussed in one of the previous chapters, MED or multistage flash (MSF) processes have played a major role in freshwater supplies to middle-east region.

On the other hand, RO desalination invented in mid-1960s struggled initially to establish as a commercially viable freshwater production process option due to the challenges in obtaining suitable semipermeable membranes. Lately, advent of composite membranes (CMs) and thin-film nanocomposite (TFNC) RO membranes have drastically improved freshwater production capabilities of RO plants. Innovations in CM and TFNC RO membrane materials and modules have revolutionized the seawater desalination process as a commercially viable and sustainable technique. Even though initially RO technology was criticized for its energy-intensive operations, however, due to the innovations in thin-film composite RO membranes in the early 2000s drastically reduced energy consumption even compared to traditional phase-change desalination processes. This led to technological breakthroughs in achieving seawater desalination and robust commercialization of RO membrane-based desalination technology [91, 92].

Today, the seawater desalination industry is successfully producing billions of liters of freshwater every year. According to some estimations, currently, the global installed desalination capacity stands at over 90 million m^3/day making it a major contributor (up >60%) to the world's installed seawater desalination capacity. With this, RO technology is the fastest-growing seawater desalination segment across water treatment market. RO technology has several advantages in comparison to thermal desalination techniques in which novel membranes consume low energy to separate freshwater from salty components in simple and facile operations.

Nonetheless, the demand for novel and efficient smaller desalination technologies is continuously growing, which is forcing researchers and technologists to explore alternative desalination technologies. Conventional RO technology efficiently operates at a large scale, which consumes relatively high energy inputs to operate plant economically. Even though larger operations have benefitted from RO technology, however, lack of energy-efficient small and mid-range pumps that operate low-energy inputs is still a challenge. Nevertheless, increased feed salinity requires efficient membranes and high operating pressure and energy inputs, which is still a challenging task. Another major issue associated with RO technology is the rejection of vast amounts of concentrates, which is also known as "hypersaline reject brine" from a desalination plant. The hypersaline rejection poses another set of challenges as the safe disposal creates large aquatic disasters and serious environmental concerns. Therefore, researchers need to pay attention to addressing serious issues of disposing and managing hypersaline rejection through complementary techniques. In this direction, development of waste minimization or management techniques in which complementary processes would make a significant impact on disposal and management of concentrated saline feed will play significant role in achieving sustainable goals; for instance, implementation of zero liquid discharge (ZLD) strategy through which brine management, secondary treatment of hypersaline water, extraction of value-added products from concentrates such as Epsom salt and other high-value components. In this direction, initial findings show that futuristic emerging desalination technologies possess remarkable promising features. These targets can be achieved through the development of processes that operate at low energy requirements. However, development and transition to efficient desalination technologies on full scale depends on future scientific advances and innovations in engineering to achieve threshold desalination performance and energy efficiency [93–95].

Alternatively, researchers have designed a membrane-based desalination separation process that exploits the huge osmotic energy stored in the saline feed systems to reduce the overall energy consumption in reverse osmosis processes. Use of naturally occurring high osmotic gradient due to the presence of high-concentration saline feed has been used to extract freshwater through the osmotically driven separation mechanism known as FO. Even though researchers realized the potential of FO process at the beginning of the 21st century, however, process is relatively tedious compared to straightforward RO process. Unlike RO technique, which uses a semipermeable membrane and high applied pressure to separate freshwater from seawater, FO process operates on concentration gradient of another freshwater withdrawing solution known as DS and the efficiency and the economics of FO process mainly depend on the freshwater drawing capacity and recovery and reusability of DS.

FO itself is a low-energy process, which in an average operation consumes roughly between 0.10 and 0.55 kWh/m^3 of energy. However, these numbers greatly depend on the type of draw solute/solution and method followed for the recovery of draw solute. On the other hand, energy consumption in FO desalination is estimated between 3 and 8 kWh/m^3. Direct reuse of diluted DS can save additional energy that goes into the recovery and regeneration process. For instance, fertilizer minerals have been used as draw solutes for dewatering contaminated water. In such a case, diluted draw solute/solution can be directly used as a liquid fertilizer making it a direct benefit product. Similarly, micronutrients can be used to induce osmotic gradient, thereby drawing a high rate of quality getting diluted and subsequent use in agricultural activities.

Recent advances in finding alternative and sustainable water treatment technologies have revived several conventional treatment technologies such as electrochemical desalination processes in which CDI techniques have been substantially improved. On the other hand, rapid developments in engineering unique properties in carbonaceous materials and non-carbonaceous conductive materials have substantially improved their adaptability in CDI technology. Simultaneously, due to growing demand for replacing fossil fuel-based energy with renewable energy, development of electro-desalination technologies integrated with renewable energy sources is expected to gain more momentum in the near future.

However, work has already been initiated in this direction in which various emerging desalination technologies have been modified to accommodate industrial and municipal wastewater treatment at medium- to large-scale operations. In this direction, trends in membrane-assisted processes earlier only used for desalination such as CDI, MD, FO, and PRO have been tested for their wastewater treatment potential and CAPEX and OPEX are being studied in detail. Desalination technology using RO process has matured in a large-scale operation, using SWRO as standard. Many attempts have been made to analyze the status of the same membrane-assisted electro-desalination technologies for their water treatment potentials. Similarly, studies are also being conducted to assess the present challenges and opportunities for adopting commercially successful processes such as SWRO used for only desalination purposes to wastewater treatment applications. These studies are also being extended to assess the environmental impact of emerging technologies such as ED/EDR, conventional adsorption-absorption, freeze desalination (FD), electrochemical desalination processes, FO, MD, and their hybrid combinations in comparison to SWRO.

Therefore, ever-growing demand for new and efficient water treatment and desalination technologies has forced researchers to develop novel filters such as nanomaterials, protein-based filters, carbon-based filters, polymer composite-based filters, advanced catalytic materials useful for multiple utility, novel mechanical filters, solar-assisted distillation units, humidity water filters, paper-based filters, biosand filters, and solar water disinfectors. Nonetheless, remote locations and arid regions will continue to depend on traditional technologies such as thermal distillation, chlorination, ceramic filters, chlorination membrane filters, coagulation, flocculation, sedimentation, and various disinfection methods. However, water treatment systems will witness revolutionary changes in materials development and system design and many of these efforts will focus on wider implementation of these technologies as standalone or in hybrid combinations [96, 97].

8.8 Water Treatment Prospective in 2030 and 2050

According to the UN reports, global freshwater demand is expected to increase by 40% by 2030. The world water crisis is imminent and expected to worsen due to increased consumption. Before a water emergency hits the world, nations are devising crisis management mechanisms as most of them are highly dependent on dwindling water resources. In addition to this, overexploitation of water supplies, overuse of freshwater, pollution, changing weather patterns, and the climate crisis threaten water-dependent economic activities. Among them, sectors like manufacturing units, leisure industries, utility industries that include water and power generation sector, farming and fishery, municipal consumption, activities related to conversion of arable lands, and so on may experience increased stress. On the other hand, a major hurdle in implementing new and advanced technologies is the longevity of currently used infrastructure and assets. Majority of the presently used WWTPs are decades old. However, most of them still deliver a satisfactory water treatment output. In such a situation, so replacing large infrastructure or upgrading overall unit operation is a complex and challenging task. Even though some of the plants have adopted minor and incremental changes, however, medium- to large-scale plants remain unchanged. Therefore, it is anticipated that even in 2030, the use of conventional primary, secondary, and tertiary wastewater treatment will remain with minor changes.

Nevertheless, in the coming years, drastic development and implementation of newer and innovative technologies involving resource recovery steps and engineering systems will intensify achieving higher water recovery, effluent quality, and attending NET ZERO targets with a minimum carbon footprint. Some of these novel technologies are predicted to realize in the near future the combination of preliminary and primary treatment involving specific resource recovery; for instance, innovative technology like Cellvation process in which high-value cellulose recovery is achieved from wastewater sources, thereby drastically decreasing organic load subsequent secondary treatment step. This approach can lead to generating additional resources through recovered cellulose, which goes into the manufacturing of a wide range of composite materials. Similarly, secondary treatment technique will be integrated with the inclusion of biogas production through anerobic digestion process in WWTPs, which is another innovative approach toward implementing novel technologies and adding more value to the plant operations. Similar approaches will be made to address phosphorus and total nitrogen removal to achieve high effluent quality using sequencing batch reactors, which can avoid the additional steps like secondary clarifiers, which rely on the additional use of granular biomass that is essential to treat high loading rates. Further, it is expected that the use of aerobic granular sludge, which is a rich source of organic substances, can be used for the extraction of polymers that are useful for a range of applications such as agriculture activities in particular seed coating through which to achieve higher germination success.

Similarly, the majority of upcoming water treatment systems may integrate additional steps to ensure resource recovery such as various nutrients useful for agriculture activities such as phosphorus, potassium, magnesium, sulfur, and nitrogen. Particularly, wastewaters are characterized by their high concentrations of phosphorus, which is known as a "disappearing nutrient." Nitrogen and phosphorus are widely found in sewage sludge ash and biosolids that can be easily concentrated and extracted from various wastewater sources. Majority of these nutrients (K, P, Mg, and S) are essential for increasing fertility of soil and extraction from wastewater sources might prove a valuable proposition for rejuvenating fertile land and waste mitigation. In presence of suitable technology, concentration and recovery like K, P, Mg, S, and N can be extremely valuable as natural reserves for these nutrients are extremely rare. Nevertheless, innovation in suitable nutrient-enriching technologies such as crystallization of struvite and recovery of phosphorus from biochar and/or ash produced from sludge after thermal treatment is still a challenging task. In this direction, several technologies have already shown greater potential and innovation in hybrid technologies which are advancing at a faster pace to achieve enhanced and efficient enrichment of nutrients and their recovery from the wastewater sources. These efforts will make overall water treatment process economically feasible making them a sustainable alternative for providing a series of valuable products to supply chains and markets.

Similarly, adaptation of wider ML and IoT tools in WWTPs operation and extending wider digitalization infrastructure will enable reliability in the resource recovery process of WWTPs. At present, digitization processes in desalination technologies have widely adopted in particular potable water production units in the form of analytical tools, early warning systems/devices, operation and supply chain network monitoring, variable operation parameter operations such as pressure control, leakage detection, and so on. Coming years will witness similar digitization trends in medium- to large-scale operations of WWTPs. Another significant leap in wastewater treatment technologies will occur with the acceptance of meaningful big data analytical tools and devices, and integration of artificial intelligence technologies with wider automation of WWTPs. Even though these efforts require huge investments to expand or replace the present complex infrastructure with advanced tools, however, increasing economic prosperities and prioritization in the waste management efforts may attract sufficient input toward making advanced infrastructure a reality. Similarly, efforts will be made to design the next generation of technologies that are carbon-neutral or with substantially reduced carbon emissions/footprints. These measures are necessary to satisfy the increased restriction to follow stringent regulations. In all these cases, major efforts will be made to create a parallel market for reuse of recovered water. All these efforts driven by circular economic goals will be a reality by 2050.

8.9 Value Addition in Future Water Treatment Plants

Hydrogen, ammonia, methane, and carbon dioxide are now being perceived as new sources of fuel and raw materials for industrial chemical production, respectively. Due to increased demand for green fuel, hydrogen and ammonia have been considered alternative fuels for future and carbon dioxide as raw material for industrial chemicals. Wastewater is a rich source of ammonia, methane, hydrogen, and carbon dioxide. To extract these renewable fuels from wastewater sources, it is estimated that transformative wastewater and resource recovery technologies will be energy-efficient by 2050. These transformative wastewater treatment and resource recovery technologies will focus on extraction of industrially important gases from WWTPs. Hydrogen, ammonia, methane, and carbon dioxide can be extracted during wastewater treatment, making it a rich and safer source, which can be further processed to obtain large-scale renewable energy or raw chemicals. The organic carbon in wastewater can be processed and extracted as biogas through the anerobic membrane reactors (anMBRs) having ~80% methane content simultaneously recovering high-quality water for reuse purposes. In this direction, anMBR technologies are expected to mature to extract safe and clean dissolved methane from a sealed reactor leading to arrest of GHG emissions. The final treated effluents using anMBR process have already been characterized for their rich nutrient content such as ammonia that can be recovered as a gas following additional stripping or concentrated step in liquid streams.

On the other hand, green hydrogen can be produced from wastewater sources following biological fermentation and electrolysis techniques. Similarly, methane and carbon dioxide can be extracted using additional extraction devices which can be further used in the production of a range of high-value chemicals. In this direction, efforts have already been underway to adopt novel and state-of-the-art biological processes in which microbial diversity will be explored to harness valuable gaseous products. Efforts in biotechnology sector will accelerate in producing microorganisms that are capable of producing highly specific and active functional materials and will continue to evolve to produce raw chemicals. Highly valuable biominerals such as struvite can be extracted from wastewater using specific bacterial activities through which concentrations of ionic species are achieved leading to the precipitation of valuable salts. In this infinite opportunity, microbial extraction of carbonate biominerals can lead to carbon dioxide sequestration, which further suppresses GHG emission and the other recovered concentrated products can be directly used as biological cement for sustainable construction activities through suitable composite formation and nanomaterial recovery (bionanomaterials, abundant oxides, endangered metals, and so on). Overall, today's actions and efforts in developing advanced and smart technologies aiming to solve issues associated with desalination and wastewater treatment are expected to yield fruitful end results in the next 2–3 decades. The development of the next-generation desalination and WWTPs encompassing all modern and smart tools will release pressures on the water purification and wastewater treatment industries. Many small steps taken today will change our future toward achieving cleaner and greener environment and sustaining safe and healthier water resources to pass on future to the next generation [98, 99].

8.10 Conclusions

Future of water purification technologies still has to fill several gaps, which requires serious disruptions from both researchers and innovators. Generally, many of the modern wastewaters are very complex, hazardous, and highly stable systems. Materials and methods used for such wastewater treatment require highly reactive, reliable, thermally stable, efficient, inexpensive, reusable, eco-friendly, and easy-to-apply prospects. Materials used for water treatment should also possess unique transformative abilities under recycling and degradation conditions that are useful for real-world applications. Several studies have shown that multifunctional nanomaterials in water purification and ecological remediation have the potential to provide customized solutions. While nanotechnology-assisted solutions are expected to undoubtedly deliver efficient water purification solutions, at the same time, several concerns associated with nanotoxicity and relatively unknown secondary pollutant issues must be properly resolved before nanotechnology can be considered harmless and accessible. In addition to this, use of traditional treatment technologies will witness transformative changes and/or alterations with the adaptation of smart and intelligent devices/technologies. Other major concerns with conventional technologies existing today are the size and location of treatment plants that are generally large and stationary in nature. In the future, water treatment technologies are expected to be flexible and should be offered as "point-of-use" technologies to make them more affordable and feasible. Nonetheless, energy-efficient and low-cost desalination and wastewater treatment are critical to achieving sustainable development goals set by the United Nations. To achieve these goals in attaining a high level of technological maturity, cutting-edge research and innovative approaches are key. In this direction, national desalination and wastewater treatment regulations, and research programs initiated in several countries intend to push technology-related developmental activities from proof-of-concept and lab-scale operations to commercialization stage. Even though these efforts in developing sustainable water treatment technologies targeting full-scale commercialization gained momentum, however, efforts still require breakthroughs in overcoming operational challenges like upgradation, replacement, adaptation, modernization, digitization, and use of smart technologies in future desalination and wastewater treatment sectors.

References

1 Silva, J.A. (2023). Water supply and wastewater treatment and reuse in future cities: a systematic literature review. *Water 15*: 3064.

2 Liao, Z., Chen, Z., Xu, A.O. et al. (2021). Wastewater treatment and reuse situations and influential factors in major Asian countries. *J. Environ. Manag. 282*: 111976.

3 Kharraz, J.A., Khanzada, N.K., Farid, M.U. et al. (2022). Membrane distillation bioreactor (MDBR) for wastewater treatment, water reuse, and resource recovery: a review. *J. Water Process Eng. 47*: 102687.

4 Sharma, M., Yadav, A., Mandal, M.K. et al. (2022). Wastewater treatment and sludge management strategies for environmental sustainability. In: *Circular Economy and Sustainability*, vol. 2 (ed. A. Stefanakis and I. Nikolaou), 97–112. Amsterdam, The Netherlands: Elsevier.

5 Lyu, S., Chen, W., Zhang, W. et al. (2016). Wastewater reclamation, and reuse in China: opportunities and challenges. *J. Environ. Sci. 39*: 86–96.

6 Awasthi, M. and Moten, T. (2022). Recent advancement and efficiency hindering factors in the wastewater treatment plant: a review. In: *Recent Trends in Wastewater Treatment* (ed. S. Madhav, P. Singh, V. Mishra, et al.). Cham: Springer.

7 Eyitayo, S.I., Watson, M.C., and Kolawole, O. (2023). Produced water management and utilization: challenges and future directions. *SPE Prod. Oper. 10*: 1–16.

8 Jodar-Abellan, A., López-Ortiz, M.I., and Melgarejo-Moreno, J. (2019). Wastewater treatment and water reuse in Spain. Current Situation and Perspectives. *Water* 11: 1551.

9 Luthy, R.G., Wolfand, J.M., and Bradshaw, J.L. (2020). Urban water revolution: sustainable water futures for California cities. *J. Environ. Eng.* 146: 4020065.

10 Radcliffe, J.C. and Page, D. (2020). Water reuse and recycling in Australia–history, current situation and future perspectives. *Water Cycle* 1: 19–40.

11 Bauer, S. and Wagner, M. (2022). Possibilities and challenges of wastewater reuse—planning aspects and realized examples. *Water 14*: 1619.

12 Meese, A.F., Kim, D.J., Wu, X. et al. (2022). Opportunities and challenges for industrial water treatment and reuse. *ACS ES&T Eng.* 2 (3): 465–488.

13 Beveridge, R., Moss, T., and Naumann, M. (2017). Sociospatial understanding of water politics: tracing the multidimensionality of water reuse. *Water Altern.* 10 (1).

14 Gupta, G.K. and Shukla, P. (2020). Insights into the resources generation from pulp and paper industry wastes: challenges, perspectives and innovations. *Bioresour. Technol.* 297: 297.

15 Navarro, D., Cantero, R., Valls, E., and Puig, R. (2020). Circular economy: The case of a shared wastewater treatment plant and its adaptation to changes of the industrial zone over time. *J. Clean. Prod.* 261: 121242–121250.

16 Singh, R.L. and Singh, R.P. (2019). *Advances in Biological Treatment of Industrial Waste Water and their Recycling for a Sustainable Future*. Singapore: Springer.

17 Thakur, R.R., McMillan, H.L., and Jones, D.S. (2014). Solvent induced phase inversion-based in situ forming controlled release drug delivery implants. *J. Control. Release* 28 (176): 8–23.

18 Mazinani, S., Darvishmanesh, S., Ehsanzadeh, A., and van der Bruggen, B. (2017). Phase separation analysis of extem/solvent/non-solvent systems and relation with membrane morphology. *J. Membr. Sci.* 526: 301–314.

19 Wang, D.-M. and Lai, J.-Y. (2013). Recent advances in preparation and morphology control of polymeric membranes formed by nonsolvent induced phase separation. *Curr. Opin. Chem. Eng.* 2: 229–237.

20 Baker, R.W. (2012). *Membrane Technology and Applications*, 3e. Wiley.

21 Kahrs, C. and Schwellenbach, J. (2020). Membrane formation via non-solvent induced phase separation using sustainable solvents: a comparative study. *Polymer* 186: 122071.

22 Wan, C.T., Jacquemond, R.R., Chiang, Y.M. et al. (2021). Non-solvent induced phase separation enables designer redox flow battery electrodes. *Adv. Mater.* 33: 2006716.

23 Aivazidou, E., Banias, G., Lampridi, M. et al. (2021). Smart technologies for sustainable water management: an urban analysis. *Sustainability (Basel)* 13 (24): 13940.

24 Kumar, A., Srivastav, A.L., Dubey, A.K. et al. *Innovations in Machine Learning and IoT for Water Management*,. https://doi.org/10.4018/979-8-3693-1194-3.ch001. Pennsylvania, USA: IGI Global.

25 Ahansal, Y., Bouziani, M., Yaagoubi, R. et al. (2022). Towards smart irrigation: a literature review on the use of geospatial technologies and machine learning in the management of water resources in arboriculture. *Agronomy (Basel)* 12 (2): 297.

26 Bhardwaj, A., Dagar, V., Khan, M.O. et al. (2022). Smart IoT and machine learning-based framework for water quality assessment and device component monitoring. *Environ. Sci. Pollut. Res. Int.* 29 (30): 46018–46036. Advance online publication. 10.1007/s11356-022-19014-335165843.

27 Chiu, M.-C., Yan, W.-M., Bhat, S.A., and Huang, N.-F. (2022). Development of smart aquaculture farm management system using IoT and AI-based surrogate models. *J. Agric. Food Res.* 9: 100357. https://doi.org/10.1016/j.jafr.2022.100357.

28 Elbeltagi, A., Srivastava, A., Deng, J. et al. (2023). Forecasting vapor pressure deficit for agricultural water management using machine learning in semi-arid environments. *Agric. Water Manag.* 283: 108302. https://doi.org/10.1016/j.agwat.2023.108302.

29 Ighalo, J.O., Adeniyi, A.G., and Marques, G. (2021). Internet of Things for water quality monitoring and assessment: a comprehensive review. In: *Artificial Intelligence for Sustainable Development: Theory, Practice and Future Applications*, Studies in Computational Intelligence, vol. 912 (ed. A. Hassanien, R. Bhatnagar, and A. Darwish). Cham: Springer.

30 Kanade, P. and Prasad, J.P. (2021). Arduino-based machine learning and IoT smart irrigation system. *Int. J. Soft Comput. Eng.* 10 (4): 1–5.

31 Lowe, M., Qin, R., and Mao, X. (2022). A review on machine learning, artificial intelligence, and smart technology in water treatment and monitoring. *Water (Basel)* 14 (9): 1384.

32 da Silva, F.R.M., Fonsêca, D.A.M., da Silva, W.L.A. et al. (2019). System of sensors and actuators for the production of water used in the manufacture of medicines. *Sensors (Basel)* 19 (20): 4488.

33 Calderón Godoy, A. and González Pérez, I. (2018). Integration of sensor and actuator networks and the scada system to promote the migration of the legacy flexible manufacturing system towards the industry 4.0 concept. *J. Sens. Actuator Netw.* 7: 23.

34 Gilchrist, A. (2016). *Industry 4.0: The Industrial Internet of Things*. New York, NY, USA: Apress.

35 Kadajji, V.G. and Betageri, G.V. (2011). Water soluble polymers for pharmaceutical applications. *Polymers* 3: 1972–2009.

36 Zainurin, S.N., Wan Ismail, W.Z., Mahamud, S.N.I. et al. (2022). Advancements in monitoring water quality based on various sensing methods: a systematic review. *Int. J. Environ. Res. Public Health 19*: 14080.

37 Burbery, L., Abraham, P., Wood, D., and de Lima, S. (2021). Applications of a UV optical nitrate sensor in a surface water/groundwater quality field study. *Environ. Monit. Assess.* 193: 303.

38 Li, Y., Wang, X.-K., Zhang, H.-Y. et al. (2021). An integrated regional water quality assessment method considering interrelationships among monitoring indicators. *Environ. Monit. Assess.* 193: 223.

39 Horvat, M., Horvat, Z., and Pastor, K. (2021). Multivariate analysis of water quality parameters in Lake Palic, Serbia. *Environ. Monit. Assess.* 193: 410.

40 Hasan, M.N., Rahman, K., and Tajmunnaher, B.M.R. (2020). Assessment of ground water quality in the vicinity of Sylhet City, Bangladesh: a multivariate analysis. *Sustainable Water Resour. Manage.* 6: 88.

41 Khatri, N., Tyagi, S., Rawtani, D., and Tharmavaram, M. (2020). Assessment of river water quality through application of indices: a case study River Sabarmati, Gujarat, India. *Sustainable Water Resour. Manage.* 6: 101.

42 Gayathri, S., Krishnan, K.A., Krishnakumar, A. et al. (2021). Monitoring of heavy metal contamination in Netravati river basin: overview of pollution indices and risk assessment. *Sustainable Water Resour. Manage.* 7: 20.

43 Hettige, N.D., Rohasliney, H., Ashaari, Z.H.B. et al. (2021). Application of GIS for water quality monitoring in the aquaculture impacted Rawang sub-basin of the Selangor river, Malaysia. *IOP Conf. Ser.: Earth Environ. Sci.* 711: 012002.

44 Wan Ismail, W.Z., Hurot, C., and Dawes, J.M. (2020). Properties of random lasers in a hollow core photonic crystal fiber. *Laser Phys.* 30: 035002.

45 Wan Ismail, W.Z., Kamil, W.M.W.A., and Dawes, J.M. (2019). Enhancement of random laser properties on solid polymer films by increasing scattering effect. *J. Russ. Laser Res.* 40: 364–369.

46 Jiang, X. and Meng, Q. Design of optical fiber SPR sensing system for water quality monitoring. *Proceedings of the 2015 International Conference on*

Computational Science and Engineering, Qingdao, China (20–21 July 2015); Volume 17, pp. 123–127.

47 Murphy, K., Heery, B., Sullivan, T. et al. (2014). A low-cost autonomous optical sensor for water quality monitoring. *Talanta* 132: 520–527.

48 Wan Ismail, W.Z., Goldys, E.M., and Dawes, J.M. (2016). Extended emission wavelength of random dye lasers by exploiting radiative and non-radiative energy transfer. *Appl. Phys. B Lasers Opt.* 122: 40.

49 Fan, Y., Xu, Z., Huang, Y. et al. (2020). Long-term continuous and real-time in situ monitoring of Pb (II) toxic contaminants in wastewater using solid-state ion selective membrane (S-ISM) Pb and pH auto-correction assembly. *J. Hazard. Mater.* 400: 123299.

50 Haimi, H., Mulas, M., Corona, F., and Vahala, R. (2013). Data-derived soft-sensors for biological wastewater treatment plants: an overview. *Environ. Model Softw.* 47: 88–107.

51 Therrien, J.D., Nicolaï, N., and Vanrolleghem, P.A. (2020). A critical review of the data pipeline: how wastewater system operation flows from data to intelligence. *Water Sci. Technol.* 82 (12): 2613–2634.

52 Wang, T., Xu, Z., Huang, Y. et al. (2020). Real-time in situ auto-correction of K+ interference for continuous and long-term NH^{4+} monitoring in wastewater using solid-state ion selective membrane (S-ISM) sensor assembly. *Environ. Res.* 189: 109891.

53 Water Environment Federation (2011). *Information Technology in Water and Wastewater Utilities: WEF MoP No. 33*, 1e. New York: McGraw-Hill.

54 Zhang, S., Jin, Y., Chen, W. et al. (2023). Artificial intelligence in wastewater treatment: a data-driven analysis of status and trends. *Chemosphere* 336: 139163.

55 Vakili, M., Mojiri, A., Kindaichi, T. et al. (2019). Cross-linked chitosan/zeolite as a fixed-bed column for organic micropollutants removal from aqueous solution, optimization with RSM and artificial neural network. *J. Environ. Manag.* 250: 109434.

56 Lotfi, K., Bonakdari, H., Ebtehaj, I. et al. (2019). Predicting wastewater treatment plant quality parameters using a novel hybrid linear-nonlinear methodology. *J. Environ. Manag.* 240: 463–474.

57 Basile, A., Alfredo Cassano, M.R., and Rahimpour, M.A.M. (2023). *Advanced Technologies in Wastewater Treatment*. Elsevier.

58 Kanafin, Y.N. and Collinson, S.R. (2021). Advanced technologies in water treatment. In: *Clean Water and Sanitation. Encyclopedia of the UN Sustainable Development Goals* (ed. W. Leal Filho, A.M. Azul, L. Brandli, et al.). Springer Nature.

59 Krishnan, R.Y., Manikandan, S., Subbaiya, R. et al. (2023). Recent approaches and advanced wastewater treatment technologies for mitigating emerging microplastics contamination – a critical review. *Sci. Total Environ.* 858 (Part 1): 159681.

60 Lai, C.Y., Groth, A., Gray, S., and Duke, M. (2014). Enhanced abrasion resistant PVDF/nanoclay hollow fibre composite membranes for water treatment. *J. Membr. Sci.* 449: 146–157.

61 Mahto, A., Halakarni, M.A., Maraddi, A. et al. (2022). Upcycling cellulose acetate from discarded cigarette butts: conversion of contaminated microfibers into loose-nanofiltration membranes for selective separation. *Desalination* 535: 115807.

62 Halakarni, M.A., Samage, A., Mahto, A. et al. (2023). Forward osmosis process for energy materials recovery from industrial wastewater with simultaneous recovery of reusable water: a sustainable approach. *Mater. Today Sustainability* 22: 100361.

63 Mahto, A., Aruchamy, K., Meena, R. et al. (2021). Forward osmosis for industrial effluents treatment – sustainability considerations. *Sep. Purif. Technol.* 254: 117568.

64 Santhosh, K.N., Samage, A., Mahadevaprasad, K.N. et al. (2024). Harnessing deep eutectic solvents for upcycling waste membranes into high-performance adsorbents and energy storage materials. *Chem. Eng. J.* 484: 149747.

65 Mahadevaprasad, K.N., Aditya, D.S., Santhosh, K.N. et al. (2023). Design and preparation of zirconium-induced bio-foam for the removal of fluoride, chromate, and other emerging pollutants: a scale-up approach. *Chem. Eng. J.* 75: 146112.

66 Maraddi, A.S., Mruthunjayappa, M.H., Kamath, S.V. et al. (2023). *A sustainable waste plastic valorisation: conversion of discarded polyurethane into an active micro-cleaner using a DES system*, vol. 25, 10538–10548. Green Chem.

67 Mruthunjayappa, M.H., Sharma, V.T., Dharmalingam, K. et al. (2020). Engineering a biopolymer-based ultrafast permeable aerogel membrane decorated with task-specific Fe–Al nanocomposites for robust water purification. *ACS Appl. Bio Mater.* 3 (8): 5233–5243.

68 Sharma, V.T., Manohara, H.M., Kamath, S.V. et al. (2020). Fe–Al based nanocomposite reinforced hydrothermal carbon: efficient and robust absorbent for anionic dyes. *Chemosphere* 259: 127421.

69 Wen, H., Cheng, D., Chen, Y. et al. (2024). *Review on Ultrasonic Technology Enhanced Biological Treatment of Wastewater*, 171260. Science of The Total Environment.

70 Rani, N., Sangwan, P., Joshi, M. et al. (2019). Chapter 5 – microbes: a key player in industrial wastewater treatment. In: *Microbial Wastewater Treatment* (ed. M.P. Shah and S. Rodriguez-Couto), 83–102. Elsevier.

71 Kumar, V., Sehgal, R., and Gupta, R. (2023). Chapter 13 – Microbes and wastewater treatment. In: *Development in Wastewater Treatment Research and Processes* (ed. M.P. Shah, S. Rodriguez-Couto, A.K. Nadda, and A. Daverey), 239–255. Elsevier.

72 Yemane, G., Asfaha, A.K.T., and Zewge, F. (2021). Hybrid process of electrocoagulation and electrooxidation system for wastewater treatment: a review. *Cleaner Eng. Technol.* 4: 100261.

73 Zahmatkesh, S., Kassian, T.T., Amesho, M.S., and Wang, C. (2022). Integration of renewable energy in wastewater treatment during COVID-19 pandemic: challenges, opportunities, and progressive research trends. *Cleaner Chem. Eng.* 3: 100036.

74 Ali, S.M.H., Lenzen, M., Sack, F., and Yousefzadeh, M. (2020). Electricity generation and demand flexibility in wastewater treatment plants: benefits for 100% renewable electricity grids. *Appl. Energy* 268: 114960.

75 Bhandari, B., Lee, K.-T., Lee, C.S. et al. (2014). A novel off-grid hybrid power system comprised of solar photovoltaic, wind, and hydro energy sources. *Appl. Energy* 133: 236–242.

76 Kollmann, R., Neugebauer, G., Kretschmer, F. et al. (2017). Renewable energy from wastewater-practical aspects of integrating a wastewater treatment plant into local energy supply concepts. *J. Clean. Prod.* 155: 119–129.

77 Marangon, B.B., Silva, T.A., Calijuri, M.L. et al. (2020). Reuse of treated municipal wastewater in productive activities in Brazil's semi-arid regions. *J. Water Process Eng.* 37: 101483.

78 Wang, J. and Liu, X. (2021). Forward osmosis technology for water treatment: recent advances and future perspectives. *J. Clean. Prod.* 280 (Part 1): 124354.

79 Porada, S., Zhao, R., van der Wal, A. et al. (2013). Review on the science and technology of water desalination by capacitive deionization. *Prog. Mater. Sci.* 58 (8): 1388–1442.

80 Zhang, X., Zuo, K., Zhang, X. et al. (2020). Selective ion separation by capacitive deionization (CDI) based technologies: a state-of-the-art review. *Environ. Sci. Water Res. Technol.* 6: 243–257.

81 Biesheuvel, P.M. and van der Wal, A. (2009). Membrane capacitive deionization. *J. Membr. Sci.* 346: 256–262.

82 Evans, S., Accomazzo, M.A., and Accomazzo, J.E. (1969). Electrochemically controlled ion exchange. *J. Electrochem. Soc.* 116: 307–309.

83 Datar, S.D., Mane, R., and Jha, N. (2022). Recent progress in materials and architectures for capacitive deionization: a comprehensive review. *Water Environ. Res.* 94 (3): e10696.

84 López-Porfiri, P., Ramos-Paredes, S., Núñez, P. et al. (2023). Towards the technological maturity of membrane distillation: the MD module performance curve. *npj Clean Water* 6: 18.

85 Obotey Ezugbe, E. and Rathilal, S. (2020). Membrane technologies in wastewater treatment: a review. *Membranes (Basel)* 10 (5): 89.

86 Davoodbeygi, Y., Askari, M., Salehi, E., and Kheirieh, S. (2023). A review on hybrid membrane-adsorption systems for intensified water and wastewater treatment: process configurations, separation targets, and materials applied. *J. Environ. Manag.* 335: 117577.

87 Biswas, W.K. and Yek, P. (2016). Improving the carbon footprint of water treatment with renewable energy: a Western Australian case study. *Renewables* 3: 14.

88 Ziping, W., Duan, H., Li, K., and Ye, L. (2022). A comprehensive carbon footprint analysis of different wastewater treatment plant configurations. *Environ. Res.* 214 (Part 2).

89 Soteris, A. (2005). Kalogirou, seawater desalination using renewable energy sources. *Prog. Energy Combust. Sci.* 31 (3): 242–281.

90 Sayed, E.T., Olabi, A.G., Elsaid, K. et al. (2023). Recent progress in renewable energy based-desalination in the Middle East and North Africa MENA region. *J. Adv. Res.* 48: 125–156.

91 Feijoo, S., Kamali, M., and Dewil, R. (2023). A review of wastewater treatment technologies for the degradation of pharmaceutically active compounds: carbamazepine as a case study. *Chem. Eng. J.* 455: 140589.

92 Mahon, J.M. and Gill, L.W. (2018). Sustainability of novel water treatment technologies in developing countries: lessons learned from research trials on a pilot continuous flow solar water disinfection system in rural Kenya. *Dev. Eng.* 3: 47–59.

93 Date, M., Patyal, V., Jaspal, D. et al. (2022). Zero liquid discharge technology for recovery, reuse, and reclamation of wastewater: a critical review. *J. Water Process Eng.* 49: 103129.

94 Pundir, A., Thakur, M.S., Radha, G.B. et al. (2024). Innovations in textile wastewater management: a review of zero liquid discharge technology. *Environ. Sci. Pollut. Res. Int.* 31 (9): 12597–12616.

95 Sun, L., Dong, H., Lu, Y. et al. (2023). A hydrate-based zero liquid discharge method for high-concentration organic wastewater: resource recovery and water reclamation. *npj Clean Water* 6: 49.

96 Wu, J., Cao, M., Tong, D. et al. (2021). A critical review of point-of-use drinking water treatment in the United States. *npj Clean Water* 4: 40.

97 Selihin, N.M. and Tay, M.G. (2022). A review on future wastewater treatment technologies: micro-nanobubbles, hybrid electro-Fenton processes, photocatalytic fuel cells, and microbial fuel cells. *Water Sci. Technol.* 85 (1): 319–341.

98 Kamath, S.V., Attokkaran, J.R., Maraddi, A.S. et al. (2024). Cello-MOF filters for multifaceted emerging pollutant abatement and their value addition. *Chem. Eng. J.* 479: 147805.

99 D'Souza, G.B., Maraddi, A., Juno Rose, K.N. et al. (2023). Green carbon induced super hydrophilic bioFoam filters for hazardous emerging pollutants removal and their sustainable value addition. *Sustain. Mater. Technol.* 37: e00676.

Index

Materials and Methods for Industrial Wastewater and Groundwater Treatment, First Edition. S. K. Nataraj.

C

d

f

g

n

o

p

u

v

w

Printed in the USA
CPSIA information can be obtained
at www.ICGtesting.com
CBHW051209160924
13844CB00002B/3

9 781119 901501